Construction Management of Healthcare Projects

About the International Code Council

The International Code Council (ICC) is a member-focused association. It is dedicated to developing model codes and standards used in the design, build, and compliance process to construct safe, sustainable, affordable, and resilient structures. Most U.S. communities and many global markets choose the International Codes. ICC Evaluation Service (ICC-ES) is the industry leader in performing technical evaluations for code compliance fostering safe and sustainable design and construction.

Headquarters: 500 New Jersey Avenue NW, 6th Floor, Washington, DC 20001-2070

District Offices: Birmingham, AL; Chicago, IL; Los Angeles, CA

1-888-422-7233; www.iccsafe.org

About the American Society for Healthcare Engineering

The American Society for Healthcare Engineering (ASHE) is a personal membership group of the American Hospital Association that represents more than 11,000 professionals involved with planning, design, construction, operations, and maintenance of healthcare facilities. ASHE offers education courses, annual conferences, and publications to keep members up-to-date on the rapidly changing healthcare field and engages in the development of codes and standards used to regulate healthcare construction. ASHE also offers certification preparation courses and certificate programs to help members, including those involved in healthcare construction, stand out from the field in a crowded marketplace. For more information, visit www.ashe.org.

Construction Management of Healthcare Projects

Sanjiv Gokhale, Ph.D., P.E., F.ASCE
Thomas C. Gormley, LEED AP, CHC

New York Chicago San Francisco
Athens London Madrid
Mexico City Milan New Delhi
Singapore Sydney Toronto

McGraw-Hill Education books are available at special quantity discounts to use as premiums and sales promotions or for use in corporate training programs. To contact a representative, please visit the Contact Us page at www.mhprofessional.com.

Construction Management of Healthcare Projects

1 2 3 4 5 6 7 8 9 0 DOC/DOC 1 9 8 7 6 5 4 3

ISBN 978-0-07-178191-6
MHID 0-07-178191-9

This book is printed on acid-free paper.

Sponsoring Editor
Michael McCabe

Editing Supervisor
Stephen M. Smith

Production Supervisor
Lynn M. Messina

Acquisitions Coordinator
Bridget L. Thoreson

Project Manager
Vipra Fauzdar, MPS Limited

Copy Editor
Surrendra Shivam, MPS Limited

Proofreader
David Stern

Indexer
Diana Witt

Art Director, Cover
Jeff Weeks

Composition
MPS Limited

About the Authors

Sanjiv Gokhale, Ph.D., P.E., F.ASCE, Professor of Civil Engineering and Director of the Graduate Program in Construction Management, Department of Civil and Environmental Engineering, Vanderbilt University (engineering.vanderbilt.edu/cee/faculty-staff/sanjiv-gokhale.php), has taught construction project management courses to undergraduate and graduate students since 2001. He is the recipient of the 2009 Distinguished Professor Award from the Construction Industry Institute (CII) and the Teaching Excellence Award from the School of Engineering, Vanderbilt University, in 2005. Dr. Gokhale is coauthor of *Trenchless Technology: Pipeline and Utility Design, Construction, and Renewal* (McGraw-Hill Professional, 2005).

Thomas C. Gormley, LEED AP, CHC, Associate Professor and Director of the Commercial Construction Management Program, Middle Tennessee State University (www.mtsu.edu/commconstruction/index.php), recently started teaching after nearly 40 years in hospital construction. He has worked as a contractor, an owner, and a program manager in the United States, Europe, and the Middle East. He is working on his dissertation for a Ph.D. in environmental engineering from Vanderbilt University. For over 10 years, Mr. Gormley has served on the committee that writes the Guidelines for Design and Construction of Health Care Facilities—the nationally recognized standard for hospitals. He is a board member of the American Hospital Association's Certification Center and served on the committee that wrote the test for Certified Healthcare Constructor, a nationally recognized certification for hospital builders.

Contents

Foreword

Whether it is a home, a school, or a healthcare facility, constructing safe, resilient, sustainable, and affordable structures is paramount for creating safe communities and providing public safety.

According to McGraw-Hill Construction, the construction value of institutional facilities hovers around $100 billion annually, which is approximately 22 percent of the nation's total building construction value. As such, the implications of the new Affordable Care Act for the healthcare construction industry could be profound.

Accordingly, development of healthcare facilities—from the initial concept to planning, design, construction, completion, commissioning, and maintenance—has gained critical importance in the national economy. While healthcare projects have much in common with other building construction types, they also have many unique features. Elements such as medical gas systems, specialized medical equipment, safety of patients incapable of self-preservation or self-evacuation, accessibility for patients in hospital beds and wheelchairs, and the requirement that healthcare facilities remain functioning and operational during severe natural events and disasters are only some of the unique issues of healthcare projects.

In this context, the importance of state-of-the-art codes and standards becomes evident. The International Code Council (ICC), publisher of the family of International Codes, and the American Society for Healthcare Engineering (ASHE) of the American Hospital Association have been working closely to address needed advances in code requirements and have decided to partner with McGraw-Hill Professional in support of this book.

The great value of *Construction Management of Healthcare Projects* is its skillful coverage of several important issues dealing with healthcare projects. Authors Andrew Collignon, Sanjiv Gokhale, Tom Gormley, Thomas Koulouris, Terry Miller, Christopher Payne, Rusty Ross, and Rick Wood, each an expert in a specific area, have joined together to produce an invaluable source for all building construction professionals. Codes and standards provisions of the *International Building Code*, NFPA, and ASHRAE related to healthcare projects are featured, along with many other relevant topics, including planning and predesign, budgeting, business planning, financing, traditional construction delivery methods, new delivery methods and alternate approaches, challenges of additions and renovations, specialized mechanical and electrical systems, medical technology and information systems, safety and infection control during construction, commissioning, occupying the project, and the future of healthcare construction.

I strongly recommend this publication to all those involved in planning, financing, budgeting, designing, building, commissioning, and maintaining new and renovated healthcare projects. State-of-the-art technology, efficient project management, and the effective conclusion of projects will undoubtedly result in safer buildings and in saving millions of dollars.

Dominic Sims, CBO
Chief Executive Officer
International Code Council

Foreword

In *Construction Management of Healthcare Projects*, Sanjiv Gokhale and Thomas Gormley have created a comprehensive body of knowledge on healthcare construction. By covering every stage of the unique healthcare construction process, the authors have created a solid foundation for those new to the field and an invaluable desk reference for experienced operational professionals.

This book speaks to everyone who has a role in the construction process. For design and construction professionals seeking to learn about the unique environment, business, and culture of healthcare, the authors provide an accurate first-hand account of the complexities and challenges they are likely to face. For seasoned healthcare facility professionals charged with delivering projects on time and on budget, every chapter offers a candid and insightful perspective of each player, each process, and each issue to be addressed during construction and occupancy.

The text includes a wealth of practical applications and examples of real-world projects, challenges, and solutions. The authors cover complex financing options, provide fundamental tools such as a utility shutdown form, and give important guidance on the move-in and start-up process. Given the wealth of practical knowledge and expertise captured in this single volume, I suspect this will become a fundamental text for university programs and for aspiring construction professionals, as well as a reference for those preparing for the Certified Healthcare Facility Manager and the Certified Healthcare Constructor credentialing exams.

Healthcare construction is difficult and complex. But understanding the process is easier when learning from experts whose entire careers have been in healthcare. Applying the knowledge in this book will help professionals create healing environments that will serve patients, their families, and our communities for years to come.

Dale Woodin, F.ASHE, CHFM
Senior Executive Director
American Society for Healthcare Engineering of the American Hospital Association

Foreword

I am pleased to present the first comprehensive book focused directly on the unique challenges of healthcare construction. Sanjiv and Tom have organized this publication to address key issues, such as codes, safety and infection control, new delivery methods, and the difficulties of working around healthcare activities on addition/renovation projects. To address the complex issues of healthcare construction, they have assembled a comprehensive team of experts, many of whom I have known for years, to explore the specific complex topics of medical equipment, financing, commissioning, project planning, and budgeting.

This book provides basic and practical information on the construction process from the perspective of the person responsible for managing a healthcare construction project. In my opinion, it will serve to help teach entry-level managers as well as provide a guide for healthcare personnel not familiar with construction. I have worked with Tom on the Health Guidelines Revision Committee that writes the Guidelines for Design and Construction of Health Care Facilities for over 10 years, and I have presented at the Vanderbilt Healthcare Workshop that Sanjiv organizes annually. Readers of this book will benefit from the information and experience of the experts they have assembled to develop the content.

Congratulations on a wonderful accomplishment.

Douglas S. Erickson, F.ASHE, CHFM, HFDP, CHC
Chairman of the Guidelines for Design and
Construction of Health Care Facilities
CEO, Facility Guidelines Institute

Acknowledgments

This book has been a "labor of love." The idea was born out of a series of highly successful workshops and symposia that have been hosted by the Graduate Program in Construction Management, Vanderbilt University, Nashville, Tennessee, since 2004. These annual events bring together the best and brightest of the design and construction industry focused on delivery of healthcare projects. Many of these individuals are contributors to this book. They bring their specific viewpoints and their vast experience to bear in the various chapters of this project. This book is intended as a text for advanced undergraduates and graduate students, and a reference for professionals continuing their education in management of healthcare construction projects.

We are grateful for the support of numerous individuals who contributed to this effort. We would like to thank Michael McCabe, Senior Editor—Engineering, McGraw-Hill Professional, for his guidance and patience. We appreciate the help from Shannon Stegall with editing and word processing. We would also like to thank our families; without their love and support this book would not have been possible.

Sanjiv Gokhale, Ph.D., P.E., F.ASCE
Thomas C. Gormley, LEED AP, CHC

Contributors

Andrew Collignon, J.D., AIA, GGP, President, Starting Point Health Facility Planning, LLC

Thomas Koulouris, MBA, CGC (www.tomkoulouris.com)

Terry Miller, BSEE, Executive Vice President, GBA (www.gbainc.com)

Christopher Payne, Managing Director, Ponder & Co. (www.ponderco.com)

Rusty Ross, P.E., Director of Commissioning Services, Smith Seckman Reid, Inc. (www.ssr-inc.com)

Rick Wood, P.E., Senior Engineer, Smith Seckman Reid, Inc. (www.ssr-inc.com)

Construction Management of Healthcare Projects

CHAPTER 1

What Is Different About Hospitals?

Hospitals are similar in many respects to other types of buildings or structures. Common elements can be quickly identified by the lay person or the construction expert. All buildings have some type of foundation, structural elements, and envelopes or exterior skins. Other common basic building elements include plumbing, electrical, and HVAC systems. Many have conveying systems, such as elevators or escalators, and life safety systems, such as fire alarm or fire sprinkler systems. Hospitals have many of these common elements; however, they are also unique in many ways. These differences include both building components, such as a medical gas system, and functions, such as providing emergency medical services. As a result of these many unique features, the design and construction of hospitals requires special considerations, skills, and processes.

The purpose of this book is to explain these unique construction requirements, to educate the reader on best practices and techniques, and to explore the latest trends and innovations in the hospital construction industry. The information here will help owners, builders, design professionals, and students understand how to successfully deliver healthcare projects by meeting and exceeding the near universal standards of

- Safety
- Quality
- Good relationships
- Budget
- Schedule

The challenge is to consistently deliver all of these items on the different types of healthcare projects across various locations in the United States and the world. Unlike manufacturing with typically fixed plants, a relatively consistent workforce and similar products, each construction project is unique. This is due to the fact that projects are in different geographic locations with a team of designers, builders, and subcontractors that are in the vast majority of cases assembled for that one particular project. Many times the contractors and subcontractors are selected on the basis of lowest cost, which often does not support the goals mentioned above. So in addition to the typical construction

This chapter was written by Tom Gormley and Sanjiv Gokhale.

opportunities and risks, the unique aspects of healthcare facilities functions and systems further add to the complexity. While there are many building components fabricated off-site and some prefabrication of rooms and systems, such as corridor walls that will be explored later, the finished healthcare construction project is generally a "stick-built," custom-made product. It is similar but, at the same time, unlike any other, due to the differences in the design, the variation in the codes by location, and the amount of assembly required by construction workers on site at each project. This challenge is further compounded when the project involves adding to or renovating an operational healthcare facility, which generally occurs more often than new projects. In almost all cases, the constructor will be required to maintain access and functionality for the hospital throughout the construction process. In most healthcare construction careers, individuals will work on many more additions and renovations than new, greenfield facilities. Health Facilities Management (HFM) and American Society For Healthcare Engineering's (ASHE) Annual Construction and Design Survey published in February of 2013 showed only 19% of construction projects either currently under construction or planned within the next 3 years to be new facilities. Modern Healthcare's Annual Construction and Design Survey published in March 2013 showed 161 "new acute care hospitals" completed in 2012 while there were 1596 "acute care expansions and renovations completed that year." With 80 to 90 percent of the projects requiring work around an existing facility, this is an important topic that will be addressed in Chap. 9 on additions and renovations.

Hospitals and healthcare facilities are unique from several perspectives that are explored below.

Incapable of Self-Preservation

Many remember the mandatory fire drills from K-12 years. While these were often perceived as opportunities to escape the "boring" classroom or to engage in some mischief, they were an important function that saved lives. Emergency evacuations for a hospital are much more complicated and those who have participated know they are not as much fun as the K-12 experience. While the regulatory authorities, such as State Fire Marshals, Licensing Agencies, and Accrediting organizations, require detailed plans and routine evacuation drills, the practical aspects of lining up hospital staff and patients and relocating to the parking lot are quite challenging.

In the neonatal intensive care unit (NICU), there are premature infants often weighing as little as 1 pound that are dependent on complex incubators that provide warmth, light, and a stable environment. Most require oxygen, medicine, and fluids. Few have a chance for survival outside this highly staffed and technical hospital space. In the operating suite, there are patients with open sites for surgical procedures that may require literally hours of intense efforts by teams of skilled caregivers. Thankfully, the emergency systems in hospitals work the vast majority of the time so surgeries and other life-saving work can be safely completed. There are rare occasions when the normal power fails, emergency generators fail, and the surgeons must complete their work by flashlight. Surgeons tell of the pressure and stress experienced when the room goes dark and continues to stay that way. The seconds required to locate flashlights seem like an eternity, but the patients are generally fine due to the quick response by the highly trained staff.

The NICU and operating rooms are just two examples of many that occur in facilities every day across the country when the patient is "incapable of self-preservation." This is the term used throughout the healthcare regulations including the Facility

Guidelines Institute (FGI) guidelines for the design and construction of healthcare facilities. Having many patients incapable of self-preservation is very much the norm for hospitals, so the design and construction of hospitals must be different from other building types. These unique building components include backup power, additional medical gas outlets, extensive fire and smoke wall systems, fire sprinklers, and fire alarm systems.

The approach recommended for hospitals is to "defend in place," so when the fire alarms go off the first step for critical patients and staff is *not* to line up and exit the hospital. The first step is to quickly communicate with the emergency response team to develop a safe course of action.

This reality of occupants that are incapable of self-preservation and cannot evacuate in an emergency is one of the unique aspects of a healthcare facility construction. This requires additional construction expertise during installation, tight quality control, and even more care during renovations with patients often adjacent to the work activities.

24/7 Operations

Hospitals are one of the few building types that do not close. These facilities are open 365 days per year, all day and all night. Healthcare does not take a holiday. The continuous operation of hospitals during construction on an addition or renovation leads to special requirements and techniques. These include Infection Control and Risk Assessments (ICRA), Interim Life Safety Measures (ISLM), shutdown procedures, and emergency response plans. One of the worst scenarios on a project is an unplanned disruption of services. One of the book contributors has seen electricity, medical gas, and water abruptly shut off to a hospital. Fortunately, no one was injured during these incidents, but this is not always the case. Several lessons were learned. On the power outage, closer supervision of an aggressive backhoe driver was needed. The medical gas line break led to not trusting old as-built plans and hand-digging around critical services. The water line break taught a different lesson regarding better communication as the key to minimizing impact. The problem with as-built plans for the water lines was identified as a risk factor during an early planning meeting, which led to better preparation for the risk of a disruption and less political outfall. Learning to deal with both the technical and political aspects of working around and in a fully operational healthcare facility is critical for any successful construction team.

Healthcare Facilities Have Unique Systems

Most healthcare facilities and especially hospitals have special systems that are required for the diagnosis, treatment, and safety of the patients and staff. These systems include medical imaging that may range from a simple mobile C-Arm requiring only an electrical outlet to MRI (magnetic resonance imaging) requiring construction of an elaborate steel enclosed box with special electromagnetic shielding. Another clinical system is the physiological monitoring typically seen in the intensive care unit (ICU), which also requires special coordination with construction. One of the most extensive systems in hospitals that require special expertise and coordination during construction is medical gas. This typically includes oxygen, nitrous oxide, vacuum, and air. The complex web of piping, valves, and outlets extends throughout most hospitals and is essential to the effective clinical care of patients. Emergency power systems are also essential components of a life safety support required in a hospital.

People Come to the Hospital in Times of Disasters or Trauma

With most buildings, the occupants are evacuated and sent home or to shelters in the event of a disaster, such as hurricane, earthquake, or terrorist attack. With hospitals, there is often a large surge of people prior to and after disastrous events. In preparation for hurricanes, additional staff is called in to be ready for the potential increase in patients. In many cases with the numerous hurricanes that hit Florida in 2004, the hospital staff also brought family members and even pets. One facility had to set up an area in the maintenance department with cages for birds, cats, and dogs belonging to staff members who were needed but did not want to leave their pets home alone. Another hospital in west Florida set up sleeping quarters for a National Guard unit that was providing emergency services in an area that was devastated by one of the hurricanes.

These additional people need space, food, water, and other basic services so the hospital may need to be constructed to support these requirements. The building components and systems must be properly installed and tested to withstand these additional pressures which are often beyond their design capacities. In some cases, hospitals in hurricane-prone areas are designed with "hardening" features to help withstand these dangerous storms. These items may include shutters for openings, tie-downs for rooftop equipment and substantial exterior skins, such as precast concrete to withstand the wind loads and flying objects. Other unique features at some facilities are water connections or shower heads on the exterior or in canopies for mass decontamination in the event of a biological terrorist attack. A large-scale attack could lead to a massive influx of people to the area hospitals. Emergency response plans call for decontamination outside the building to avoid contaminating the emergency department (ED) as this contamination could lead to the inability to care for any of the patients.

Hurricane Katrina Case Study (Urban Land Institute, *After Katrina*, July 2006)

Hurricane Katrina presented New Orleans and its hospitals with the effects of two related but distinctive events. The first was the hurricane itself, which arrived on Monday morning, August 29, 2005, with heavy rain and sustained winds of 120 to 130 mph, with gusts up to 160 mph. Electrical and communications services were disrupted by the destruction of landlines and the toppling of cell phone and radio repeater towers, but hospitals and other large buildings suffered only superficial damage.

For hospitals, the problems created by the storm would have been minor were it not for the second event—the failure Monday night of the levees protecting New Orleans from Lake Pontchartrain and the Mississippi River. By Tuesday morning, large sections of the city were under as much as 15 to 20 feet of water, far exceeding the capacity of the city's pumping system (which was designed to pump water into the very canals whose walls had been breached). Evacuation became essential in the flooded areas. The situation was particularly urgent for the hospitals that lost power, communications, and water/sewerage service, and that couldn't resupply such essentials as drugs, blood, linens, and food. According to figures assembled by the Louisiana

Hospital Association (LHA) during the storm, 1,749 patients occupied the 11 hospitals surrounded by floodwaters.

Many of these beleaguered hospitals received much publicity during the crisis—Charity Hospital, University Hospital, Tulane University Hospital, Veterans' Affairs Medical Center, Lindy Boggs Medical Center, and Memorial Medical Center. The LHA's compilation also showed that these 11 hospitals housed more than 7,600 people in addition to their patients. Some were staff members, but hospitals, like the Superdome and convention center, became refuges for patients' families and for thousands of others who left their homes. Hospitals also housed pets. Personnel at Lindy Boggs Medical Center dealt with 45 dogs, 15 cats, and a pair of guinea pigs brought in by staff and patients to ride out the storm. Conventional modes of transportation were used to evacuate a dozen or so hospitals that were not isolated by water. But evacuation from the 11 flood-bound hospitals posed the most difficult problems, requiring the use of boats or helicopters.

Special Codes and Regulations

In addition to complying with the typical building codes, such as the International Building Code (IBC) and National Fire Protection Agency (NFPA), hospitals are required to comply with several specialized codes. The Centers for Medicare and Medicaid Services (CMS) requires facilities to comply with NFPA 101: Life Safety Code as a condition for participation in their program. Given that over 50 million people are enrolled in Medicare or Medicaid in the U.S., compliance with CMS rules is of utmost importance to healthcare facilities. In addition to CMS requirements, each state typically has its own agency rules to follow to obtain a license to provide hospital services. These laws and regulations vary state to state and may be forms of a national healthcare guideline or several states have their own customized hospital regulations.

California has its own hospital code and agency, Office of Statewide Health Planning and Development (OSHPD), set up to regulate the design and construction of hospitals. The earthquake legislation there, commonly known as SB 1953, spawned a multibillion dollar hospital building program as the hospitals and systems in the state replaced old facilities that could not meet the new requirements. This state law was the result of significant damage and loss of use of entire hospitals after the Northridge earthquake in 1994. This situation helped point out the vulnerability of hospitals and the need for their essential services in the time after a disaster.

Another populous state, Florida, also has a dedicated state agency to regulate the design and construction of hospitals, diagnostic and treatment centers, and nursing homes. The Agency for Healthcare Administration (AHCA) has a unique and customized set of regulations for healthcare facilities. It is supplemented by HB 911, a state law, which requires special design and construction measures for hurricane resistance. This law was passed by the legislature after hurricane Andrew destroyed one hospital and damaged several others.

While California and Florida are considered by most in the industry to be the most highly regulated, other states have a patchwork of regulations. Many adopt all or portions of the Guidelines for the Design and Construction of Health Care Facilities, produced by the Facilities Guidelines Institute (FGI). New York recently adopted the 2010 Guidelines in their entirety, while Arkansas unofficially uses a version of the 2001 Guidelines. The variations continue across the states so it is necessary to check both state and local requirements before building in an area. In addition to the state laws and regulations, the NFPA publishes NFPA 99: Healthcare Facilities Code. Other NFPA regulations also have specific requirements related to hospital design and construction. Examples of this are the fire alarm sections NFPA 72: National Fire Alarm and Signaling Code and NFPA 241: Standard for Safeguarding Construction, Alteration, and Demolition Operations for renovations of existing facilities. IBC's Chap. 4 has special detailed requirements for Group I-2 occupancies (includes hospitals and nursing homes). NFPA 70: National Electrical Code (NEC) and the American Society of Heating, Refrigeration and Air Conditioning Engineers (ASHRAE) also have specific sections addressing healthcare facilities.

Other organizations that have a significant impact on the design, construction, and operation of healthcare facilities are the Joint Commission (TJC), Det Norske Veritas (DNV), and Healthcare Facilities Accreditation Program (HFAP). Each accrediting organization is voluntary for the hospital, but to be accredited by any one of the organizations compliance with their standards must be addressed and in particular the requirements for working in a functioning hospital.

Life Saving and Emergency Medical Care

Healthcare facilities and especially medical/surgical hospitals provide a very unique service for the community. It is an essential service and unlike many others, it is one that most everyone will use at some point in their lives. This service, ranging from scheduled inpatient surgery to an unplanned trip to the emergency room can literally be a matter of life or death for an individual. Hospitals are places where people experience a wide range of emotions from incredible joy with the birth of new baby to incredible sadness with the loss of a loved one. The hospital construction team must share in the responsibility to provide quality healthcare as the facility strives to provide these services. This is even more critical with additions and renovations so patients have access to services and staff has the resources needed to provide medical care. The quality of construction must be at a high level, so that it is reliable and functional to enable the caregivers to deliver their best to the community.

Range of Services

Hospitals provide a wide range of services for the varying needs of the staff, patient, and families. These different services require many types of spaces from a full-service kitchen and dining facilities to a morgue and autopsy area. Between these extremes are overnight sleeping accommodations for patients, staff, and often family members, pharmacies, laboratories, and loading docks. Pharmacies typically have unique requirements for special hoods to safely mix IV solutions, while the labs may require acid-resistant piping, and the docks may require levelers for different height vehicles. There is often a wide range of clinical spaces as well, from simple exam rooms to complex hybrid operating rooms that also include imaging equipment. This wide range of requirements is not often found in one facility, except in healthcare construction. It requires the builder to have knowledge and expertise in many different types of building components.

Conclusion

These differentiators with a hospital and some other types of healthcare projects drive the need for specialized training and experience to successfully complete the construction phase of a capital development project. As in many industries, capital investment in facilities is essential to provide the latest technology, to upgrade aging facilities, to adjust to changing market conditions, and to develop the capacity for the continuing demand for healthcare. As the "baby boomers" continue to age, the need for healthcare services and facilities will continue as there is strong correlation between age and hospital utilization. The statistics are staggering:

- The estimated 79 million "baby boomers" started turning 65 in 2011 and 10,000 per day will turn 65 until 2030 when they all reach that age. (From the PEW Research website http://www.pewresearch.org/daily-number/baby-boomers-retire/)
- Between 2010 and 2030 the population over 65 will grow from 40.2 million to 72.1 million or 79 percent.
- Between 2010 and 2030 the population over 85 will grow from 6.1 million to 9.6 million or 57 percent.
- Between 2010 and 2030 the overall U.S. population is projected to grow from 310.2 million to only 373.5 million or 19 percent. (From the Department of Health and Human Services Administration on Aging website http://www.aoa.gov/Aging_Statistics/Census_Population/Index.aspx)

Many experts predict this significant growth of older people along with the overall population growth will continue to fuel hospital construction. While the types of facilities and projects may change due to many factors, the overall need will most likely continue to grow. There are many factors driving the changes such as: technological advances allowing more outpatient procedures, increased use of stents and interventional cardiology reducing open heart surgeries, telemedicine to decrease inpatient stays, new medicines, the potential for gene therapy, and many more. Also, government intervention and reimbursement are huge factors that affect the delivery of healthcare and consequently the type of healthcare construction projects in the marketplace in the future.

In this book, an understanding of the complexities and unique aspects of healthcare construction projects will be provided to enable the reader to better prepare for these challenges. While the focus will be on construction, it is always important for construction builders and managers to learn about and understand the delivery of healthcare. Most will never be healthcare providers, but will need to communicate with physicians, nurses, staff, administrators, department heads, and others in the healthcare industry. A better understanding of the vernacular and medical operations will improve communication and enable the construction manager to deliver the quality facilities needed to deliver quality healthcare. In addition to learning about the business of healthcare, it is important to educate the healthcare community about the construction process. Through patient explanations of construction terms and processes, construction professionals help the "hospital people" better understand why certain activities must occur and the steps required to deliver a new facility or a renovation on time. The medical community may not fully comprehend why the HVAC (heating, ventilation, and air-conditioning) system has to be shut down in the operating room, or why the emergency entrance has to be rerouted; however, the process can be much smoother if the hospital staff understands and can work with construction professionals. In many cases, the staff can make recommendations and even help develop solutions that can ease the "pain" of the construction process. Contractors and healthcare professionals collaborating during new construction and renovations produce optimal results.

Meeting and communicating with the end users to discuss options are a critical part of the construction process. Mock-ups of key spaces, such as operating rooms or patient rooms, are an important tool on many projects to communicate the intended design and to test the functionality of the space. An important lesson to learn is that an end user, doctor, or administrator may be nodding their heads as they view plans, but it does not necessarily mean they really understand what the space will look like or if it will meet their needs. This is where the mock-ups and computer models can be valuable. The ultimate goal is to safely deliver a facility that is functional and meets the needs of caregivers allowing them to provide the best healthcare possible while staying in budget and on schedule. The most successful approach is to work in a collaborative manner with the end users, the designers, the contractors, the subcontractors, the suppliers, and the many other players on the healthcare project team. True collaboration is not unique to healthcare projects, but it typically leads to the best outcomes for all projects when it can be achieved.

CHAPTER 2

Regulations and Codes Impacting Hospitals

Introduction

Today the design and construction of almost all facilities are regulated by a myriad of codes and regulations. These range from local zoning and building codes for houses to complex regulations and multiple agencies for nuclear power plants. Permit approval times for these extremes on the continuum can vary from one day to several years. Generally, the codes and regulations are successful in achieving the basic goal of maintaining public safety and providing a minimum standard for all. There are always notable exceptions that draw the attention from the press and politicians leading to calls for improvements. The Fukushima Daiichi nuclear plant in Japan that was damaged by a tsunami in 2011 provides a recent example of a catastrophic failure that resulted in significant changes to the codes and standards both in Japan and United States. Smaller, but no less tragic for the victims, are the many house and apartment fires that occur throughout the country on daily basis.

Many of today's building codes are a result of tragic events that led to greater scrutiny by public officials. These responses generally involve changing existing codes, developing new regulations, creating new enforcement agencies and prosecuting negligent or guilty parties. There were several catastrophic events that were milestones leading to significant changes and development of the current codes. The "Great Chicago Fire" in 1871 and the 3-day fire in San Francisco after the earthquake in 1906 led to code changes involving building separations and fire walls to avoid multiple buildings becoming involved in a fire leading to neighborhood or citywide events. The Iroquois Theatre Fire in 1903 was another tragedy that led to changes in the fire codes. This event cost over 600 people their lives as a spark from a light set the flammable stage curtain on fire. Many of these victims were trapped at the inadequate and poorly marked exits, where bodies were piled 5 feet deep or were asphyxiated in their seats due to the poisonous gasses from the flammable materials. The skylights that were to open and exhaust the smoke were not operating properly. This event led to requirements for minimum corridor widths, easily visible exit signs, and nonflammable stage scenery.

Other fires led to code changes involving the interior elements of buildings that continue to apply to hospital projects. One such event was the Triangle Shirtwaist

---

This chapter was written by Tom Gormley.

Factory fire in New York City in 1907. In this tragic event over 140 people died. The majority were young immigrant girls working as seamstresses. In just 10 terrifying minutes, the fire spread through the 8th, 9th, and 10th floors where 1200 workers were crowded in the workspace with flammable fabric scraps covering the floor and locked fire exits. As a result of this tragedy, significant building code changes were made in New York City which in turn were replicated in many other cities across the country. These regulations were a precursor to the National Fire Protection Codes or NFPA (National Fire Protection Association). These changes included requirements for sprinklers and exit doors that swing out to avoid trapping people as they surge to escape.

There were other significant fires over the years that also impacted life safety requirements, such as the Coconut Grove Night Club fire in Boston in 1942 in which 491 people were killed. While the official occupancy allowed in the club was 600, it was estimated that over 1000 people were actually in attendance. This disaster led to stricter enforcement of the occupancy limits and more fire inspectors. The Winecoff Hotel fire in Atlanta in 1946 resulted in 119 deaths. This tragedy led to the requirements for multiple exits and self-closing fire doors. Another milestone that led to significant improvements in the codes related to the number of occupants and flame spread of building materials was the Beverly Hills Supper Club fire. This occurred in a nightclub in Covington, KY, in 1977 and left 165 dead. The club was overcrowded with too few fire exits and some of which were locked. The highly flammable interior finishes also led to the rapid spread of the fire and loss of life due to smoke inhalation. The MGM fire in 1980 cost the lives of 85 people and was largely a result of the mechanical systems moving smoke into other areas of the building as the fire was on the first floor and most of the fatalities were on upper floors. Most died as a result of carbon monoxide poisoning. The lack of a fire sprinkler system also contributed to this tragedy, as well as the fact that no fire alarm was sounded in the facility during the fire.

Many of these tragic events led to changes that indirectly still have an impact on hospital construction today. The common elements in many of these fires were

- The inability of the occupants to safely egress the building due to poor marking of exits, locked doors, doors swinging to the inside and inadequate number of exits
- Lack of a fire alarm or fire sprinkler systems
- Number of occupants exceeding the safe limits
- Presence of flammable materials that spread the fire quickly and produced toxic fumes

While man-made tragedies have led to many code improvements, two natural disasters led to some of the most significant state laws that directly affected the design and construction of hospitals. One was the Northridge earthquake that occurred on January 17, 1994, in Reseda, a neighborhood in Los Angeles, California, lasting for about 10 to 20 seconds. This 6.7 magnitude quake led to significant damage to the Olive View Hospital and the Northridge Hospital. These and six other hospitals had to be evacuated and hundreds of patients relocated. Ironically, Olive View had already been upgraded structurally to meet an earlier seismic law after a 1971 quake. While it continued to stand and did not fail structurally, Olive View and the other hospitals were rendered useless due to infrastructure damage, such as lack of water and power. Again, the reaction by the community and the public officials led to code changes, in this case in the form of Senate Bill 1953 (aka SB 1953). This state law required hospitals in California to be evaluated and upgraded so they would continue to be operational after an earthquake. This process was to have been completed by 2013. The cost to upgrade the hospitals across the state has been estimated from $24 billion by the San Francisco Department of Public Health to $41 billion by the California Healthcare Foundation. Due to this staggering cost, the lack of any funding in the bill, and the political response to the potential for having many hospitals closed due to noncompliance, this requirement was postponed several times. The law did lead to the construction of several new hospitals and the upgrade work on existing hospitals continues.

Another cataclysmic natural disaster in a populous state led to a law called Senate Bill 911 to improve the design and construction of hospitals in Florida. In August 1992, hurricane Andrew made landfall as a Category 5 storm just south of Miami. Numerous hospitals were damaged and one in Deering, Florida, was nearly destroyed, leading to a full evacuation. Hundreds of patients in the area had to be relocated due to the closures related to damage from the storm. Deering Hospital was so badly damaged that a new facility had to be constructed. Senate Bill 911 requires all new hospitals and new additions to meet the most recent standards. It did not address upgrading existing facilities. These minimum standards include

- Additional attachments for roof-mounted equipment that blew off during the storm and led to flooding and the loss of use of fans and HVAC units
- On-site potable water storage for 3 days for patients and staff
- The ability for the exterior skin of the hospital to withstand the impact of an 8-foot-long 2 × 4 shot from an air cannon at 30 mph

Regulating Healthcare Facilities

As evidenced by Fig. 2.1 provided by the American Society for Healthcare Engineers (ASHE), there are many organizations regulating hospitals in the United States today. Most regulate the operations of the healthcare organizations, but there are still many that control the design and construction of healthcare facilities. While the design is generally the legal responsibility of the architects and engineers, most owners will appreciate and, hopefully, reward a hospital builder or construction manager who understands code requirements and can help avoid problems. It is often also their contractual responsibility to build the facility in compliance with the applicable codes and regulations, so a good understanding is critical to a successful project. This chapter explores and explains important requirements of many different codes, laws, and regulations for hospitals.

Certificate of Need

Certificate of need (CON) laws are intended to control healthcare costs and allow for statewide coordinated planning of new services and construction. New York was actually the first state in 1964 to enact a law requiring approval for construction of a new hospital or nursing home. Many of these laws were put into effect across the nation as part of the federal National Health Planning Resources Development Act of 1974 (Public Law 93-64). In spite of many changes since then,36 states retain some type of

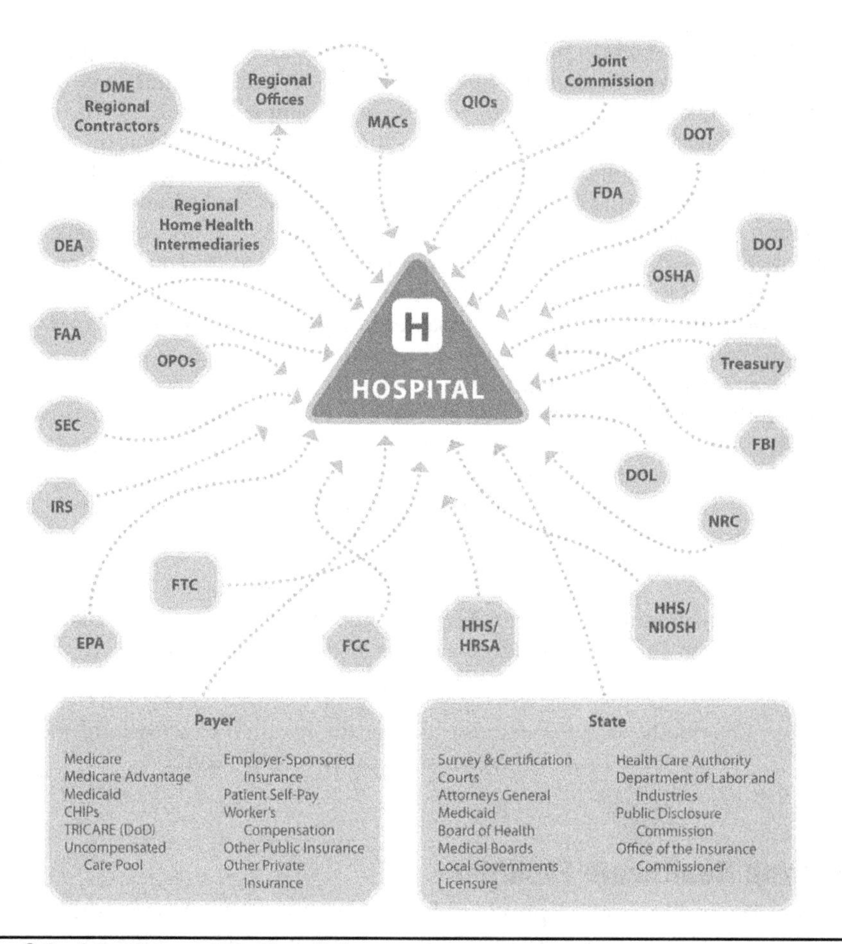

FIGURE 2.1 Agencies regulating hospitals.

CON program, law, or agency as of December 2011. The CON laws are more prevalent in the eastern half of the United States than in the west.

In this 1974 Act, three distinct existing programs were consolidated:

- Hill-Burton
- Regional Medical Program
- Comprehensive Health Planning Act

Congress recognized that the provision of federal funds for the construction of new healthcare facilities was contributing to increasing healthcare costs by generating duplication of facilities. The intent of Congress in passing this act was to create throughout the United States a strengthened and improved federal-, state-, and area-wide system of health planning and resources development that would help provide solutions to several identified problems. The perceived problems the act was intended to address were as follows:

- Lack of equal access to quality healthcare at reasonable cost
- Infusion of federal funds into the healthcare system, contributing to inflationary healthcare costs while failing to produce an adequate supply or distribution of health resources
- Lack of a comprehensive, rational approach by the public and private sectors to create uniformly effective methods of delivering healthcare
- Misdistribution of healthcare facilities and manpower
- Increasing and uncontrolled inflation of healthcare costs, particularly the costs associated with hospital stays
- Inadequate incentives for the use of alternative levels of healthcare and the substitution of ambulatory and intermediate care for inpatient hospital care
- Lack of basic knowledge regarding proper personal healthcare and methods for effective use of available health services in large segments of the population

The 1974 Federal law required all states to have a process for the submission of proposed projects and approval from a state-wide health planning organization prior to starting major capital projects, such as new hospitals, additions/renovations, or installing medical equipment. The limits on the cost vary significantly by state. Many implemented the programs because of the incentive of receiving federal funding although that was stopped in 1987. The CON laws seem to be in decline in many states, but are still strong in some and can have a significant impact on approvals, timing, and market competition. (Information from the National Conference of State Legislatures website, http://www.ncsl.org/issues-research/health/con-certificate-of-need-state-laws.aspx.)

The CON laws will typically have minimal impact on the actual construction process; however, they can affect the owner's decision on what services to provide, the capital cost that can be expended on the project, and the approval time to begin construction. There are cost limits in CON laws in some states. The construction manager needs to be aware of this process as they are sometimes asked to provide input on schedule and cost as the owner prepares the package to submit to the state for approval. The formal submission may include preliminary plans, budgets, cost breakdowns, demographic studies, and analysis of hospital utilization rates.

Summary of the CON Process

The following narrative provides a brief summary of the CON process for the state of North Carolina. This should help the reader understand the typical steps, but it should be noted that these laws vary by state. Applicants and other interested people should always refer to the applicable statute (G.S. 131E, Article 9, 175–190) and administrative rules (10A NCAC Subchapter 14C) for more complete information.

1. *Allocation of beds and services.* At the beginning of each calendar year, a new State Medical Facilities Plan is published by the Medical Facilities Planning Branch of DHSR, which sets forth the maximum number of health service facility beds by category, operating rooms, medical equipment, home health offices, and other services that may be approved by the CON Section.
2. *Review schedule.* In order for competitive applications to be reviewed at the same time, the CON Section has adopted a system to review applications

according to a batched review schedule. Under this system, applications for similar services in the same geographic area are reviewed at the same time. Review schedules are found in the State Medical Facilities Plan.

3. *Pre-application procedure.* Each applicant submitting an application must submit a letter of intent (LOI) to the CON Section no later than the date the application is due. Most applicants submit an LOI as the first step in the application process. In response to an LOI submitted before the beginning of the review period, the CON Section forwards a letter to the applicant that indicates whether a CON review is required. If so, the CON Section explains the category, in which the application will be reviewed, provides the beginning review dates for that category, and provides the necessary application forms. An applicant may meet with representatives of the CON Section for a pre-application conference to discuss any questions relative to the CON process.
4. *Application submittal.* Applications must be received by 5:30 p.m. on the 15th day of the month prior to the beginning of the applicable review period. When the 15th of the month falls on a weekend or holiday, the filing deadline is 5:30 p.m. on the next business day. The filing deadline is absolute. After an application is submitted, it cannot be amended; however, the CON Section may request that an applicant submit clarifying information. The CON Section reviews each application to determine if it is complete. An application is deemed complete if the correct fee is paid and the original signature page is provided. If an application is deemed incomplete, within 5 days the CON Section will notify the applicant of the items needed to make the application complete.
5. *Public comment period.* During the first 30 days of the review period, any person may file written comments or letters of support concerning the proposals under review.
6. *Public hearing.* A public hearing is no longer required to be conducted for each proposal under review. However, under certain circumstances as set forth in G.S. 131E-185(a1)(2), a public hearing is required to be conducted by the CON Section in the service area affected by the application no more than 20 days from the conclusion of the written comment period.
7. *Application review.* The CON Section has from 90 to 150 days to review an application for a certificate of need. Each application is reviewed against G.S. 131E-183 Review criteria and any applicable criteria and standards in the administrative rules. All written comments and presentations at the public hearing are also taken into consideration by the CON Section during the review of an application. An application must be conforming or conditionally conforming with all applicable criteria and standards in order to be approved.
8. *Appeals of decision.* Within 30 days after the date of a decision, any affected person may file a petition for a contested case hearing with the Office of Administrative Hearings (OAH). The administrative law judge must make his decision within 270 days after the petition is filed. This decision may be appealed to the N.C. Court of Appeals.
9. *Monitoring.* After the certificate is issued, the CON Section will monitor the development of the project through review of progress reports submitted by

the applicant. In accordance with G.S. 131E-189, the CON Section may withdraw a certificate if the holder of the certificate fails to develop and operate the service consistent with the representations made in the application or with any conditions the CON Section placed on the certificate of need.

Health Service Facilities Regulated by the CON Law [G.S.131E-175 (Article 9)]

1. Acute Care Hospitals [G.S. 131E-176(13)]
2. Inpatient Psychiatric Hospitals [G.S. 131E-176(21)]
3. Inpatient Rehabilitation Hospitals [G.S. 131E-176(22)]
4. Nursing Homes [G.S. 131E-176(17b)]
5. Kidney Disease Treatment Centers (i.e., Certified End-Stage Renal Disease Facilities) [G.S. 131E-176(14e)]
6. Intermediate Care Facilities for the Mentally Retarded [G.S. 131E-176(14a)]
7. Certified Home Health Agency Offices [G.S. 131E-176(12)]
8. Chemical Dependency Treatment Facilities (inpatient & residential) [G.S. 131E-176(5a)]
9. Diagnostic Centers [G.S. 131E-176(7a)]
10. Hospice Programs [G.S. 131E-176(13a)]
11. Hospice Inpatient Facilities [G.S. 131E-176(13b)]
12. Hospice Residential Care Facilities [G.S. 131E-176(13c)]
13. Ambulatory Surgical Facilities [G.S. 131E-176(1b)]
14. Adult Care Homes [G.S. 131E-176(1)]
15. Long-Term Care Hospitals [G.S. 131E-176(14k)]

Guidelines for Design and Construction of Health Care Facilities

Another governmental step led to significant improvements and a continuing impact on hospital requirements. In 1946, the U.S. Congress passed the Hill-Burton Act to encourage the construction of new hospitals in underserved areas. This act provided funding and resulted in the construction of numerous hospitals in the 1950s and 1960s. As part of the Hill-Burton Act, standards were developed to guide the design and construction of these hospitals. These standards were originally called Public Health Service Minimum Requirements. These Guidelines were eventually adopted by many states in full or in part and were promulgated into law for new hospitals as well as additions and renovations and other types of healthcare facilities.

The Facility Guidelines Institute (FGI) was formed in 1998 (www.fgiguidelines.org) and assumed responsibility for maintaining and updating, what is now called, the "Guidelines for Design and Construction of Health Care Facilities." This document provides minimum standards for several different types of healthcare facilities, such as medical/surgical hospitals, psychiatric hospitals, ambulatory care facilities, residential healthcare facilities, and most recently "critical access" hospitals. The Guidelines are a consensus document developed by a committee of experts in healthcare facility operations, design, and construction called the Health Guidelines Revision Committee

(HGRC). The Guidelines are updated every four years by the committee. During the updates, the public, as well as industry professionals, have the opportunity to propose changes and improvements to the document. During the 2006 and 2010 cycle, the HGRC Steering Committee added a subcommittee to evaluate the potential costs and benefits of the proposed changes in order to increase the fiscal responsibility in the standards development process.

The FGI Guidelines have been adopted by over 30 U.S. states and are also used for many international projects. They are written as minimum standards and some local or state codes may require more stringent requirements than the Guidelines. They address a wide range of topics, such as size of rooms, number of medical gas outlets, and a critical process for all renovations called ICRA (Infection Control Risk Assessment). The Guidelines have partnered with ASHRAE (American Society of Heating, Refrigeration, and Air Conditioning Engineers) and the American Society for Healthcare Engineering to include the ANSI/ASHRAE/ASHE Standard 170-2008 for the Ventilation of Healthcare Facilities. This addresses HVAC issues, such as required number of air changes, pressure relationships, and equipment start-up. A copy of the applicable Guidelines is an essential requirement for anyone responsible for or involved in managing a hospital construction project. Familiarity with these requirements and a proactive approach to compliance will avoid many problems. Verification that the design and construction actually meets these standards, or whatever code is applicable, is essential to delivering a quality facility that will receive timely approval for occupancy by the authority having jurisdiction, often referred to as AHJ.

NFPA 101: Life Safety Code

In addition to the Guidelines, the National Fire Protection Association (NFPA) (www.nfpa.org) has a number of publications with requirements related to hospitals. One of the primary documents that focuses on public protection is NFPA 101 Life Safety Code. This comprehensive publication addresses many aspects of a facility's layout and building components for a wide variety of structure types. There are separate chapters for new healthcare facilities and for existing healthcare facilities.

The Life Safety Code contains a key provision called the "Total Concept" that addresses the underlying principle of life safety in hospitals. This is the idea that due to medical reasons the patient may not be safer by being evacuated for a fire emergency. This is one of the fundamental differences from other types of buildings as evacuation of the public is typically the first priority. If a patient is evacuated without the necessary life-saving medical equipment or during a medical procedure, they may be in more danger or at greater risk than with a fire emergency. Thus, the stated objective is "limiting the development and spread of a fire emergency to the room of origin and reducing the need for occupant evacuation, except from the room of fire origin." The approach then is to protect the "occupants" or patients through the "appropriate arrangement of facilities, adequate staffing and development of operating and maintenance procedures." The arrangement of the facilities involves the design, construction, and the concept of compartmentalization. This involves developing distinct fire and smoke zones or compartments to isolate the fire emergency and provide the patients with protected zones. Another requirement is for automatic "detection, alarm, and extinguishment" with fire alarm and fire sprinkler systems. In addition, the code requires fire prevention measures, training and drills on isolation of fires, transfer of patients to safe refuge, or in the last case—evacuation.

Hospital builders need to understand this concept as it impacts many design decisions, construction materials, building techniques, and testing or commissioning at the completion of the work. It is especially important for construction and hospital processes used for renovations and additions. The steps required for renovations are extensive due to patients being potentially exposed to the additional risks of adjacent construction activities. Chapter 9 of this book specifically covers additions and renovations of operational hospitals.

The NFPA 101 chapter on new healthcare occupancies also addresses means of egress, corridor walls and doors, wall finishes and flame resistant furnishings through reference to NFPA 701. The fire sprinkler system requirements are covered and this is the code that requires "quick response" heads in "patient sleeping rooms" (Fig. 2.2). These special types of sprinkler heads must be used by the hospital builder. Hazardous areas and required protection are defined as well as the requirements for fire/life safety plans and fire drills.

The Life Safety Code chapter on Existing Healthcare Occupancies requires the means of egress during "construction, repair, or improvements" to be "inspected daily," which is typically the responsibility of the contractor or construction manager. This is a very important assignment with high risk that should be well documented and supervised diligently. The egress must include not only clear exit ways of the appropriate size, but also proper lighting, stairs, ramps, or covered walkways so the occupants can safely move away from the building. Hospitals that are not fully sprinkled must meet additional requirements. Typically additions are separated by 2-hour fire walls for fire safety and to minimize additional work in the existing facility. Some states, such as Florida, encourage additions, renovations, and improvements by minimizing the scope of upgrading the existing building to current codes.

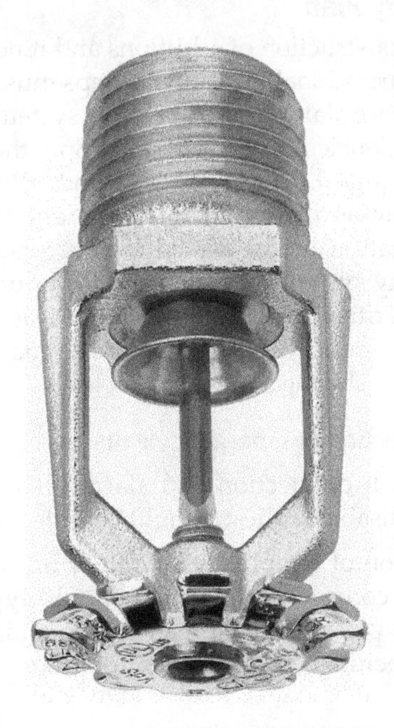

Figure 2.2 Quick response sprinkler head. (*Courtesy of Tyco Fire Protection Products. All rights reserved.*)

NFPA 101 Life Safety Code explains the defend-in-place strategy of creating smoke compartments for safe areas to allow people to remain in the location. Moving people out of the building is only used as a last resort, and the facility must include features that help prevent, detect, and suppress fires. The defend-in-place approach is a fundamental concept that all healthcare builders must understand. In addition to grasping this strategic approach, the builder must be able to recognize and apply the tactical steps needed to protect patients rather than moving them in the event of a fire. These measures are seen throughout hospital facilities and span the work of many different building trades. The building components that comprise this can range from fire walls installed by the drywall subcontractor, to fire dampers installed by the mechanical subcontractor, to rated doors often installed by the general contractor. Two very important elements are (1) the fire sprinkler system, which is usually in the mechanical specifications, and (2) the fire alarm system, which is typically installed by the electrical subcontractor.

The hospital builder must comprehend how these varied systems are woven together by the architects and engineers to create the safety net for patients and avoid the necessity to evacuate them. If a patient is seriously ill, it may be more dangerous to move them than to remain in the hospital protected in a fire/smoke compartment. The movement of seriously ill patients, at times, may be the only option during natural disasters, such as a hurricane.

During the construction process, the builder must understand how to install and test these *features*. On renovations, it is even more critical that the builder implement steps to make sure the life safety systems are always operational. In the event these systems must be shut down to complete the work, backup plans must be defined, approved, and implemented.

Interim Life Safety Plan

At times, during construction of additions and renovations, some of the life safety systems may not be operational. Additional steps must be taken to protect the patient during these periods. Fire alarm or fire sprinkler systems may be shut down to make tie-ins or to upgrade deficiencies. If out of service more than 4 hours in a 24-hour period, then a "fire watch" is required. This is described in NFPA 101. At times, exits may be blocked while building new additions. In this case, alternative exits must have appropriate emergency illumination and be posted with egress signage.

When life safety systems are compromised during construction, the hospital is required to have a written Interim Life Safety Measures plan. The builder should work with the hospital to identify the potential deficiencies and develop measures to address the additional risks. The measures may include

- Daily inspection of emergency exits.
- Temporary fire detection and alarm systems in the areas where the primary system is disabled. These should be tested and documented monthly.
- Construction of temporary smoke or fire walls to protect compartments that have been compromised. These should typically be built of metal studs and drywall to prevent the spread of fire and smoke. Careful attention must be paid to openings in the walls for doors or ductwork creating continuous protection.
- Increased monitoring of construction areas and the overall facility is needed to identify potential life safety risks. This includes storage, demolished material,

and trash. Special attention should be given to combustible and hazardous materials, such as paint, chemicals, wood, and paper.

- Workers and hospital staff must have additional training in the use of fire-fighting equipment. The builder should provide specific training not only to his own forces but also to subcontractors and vendors so that they can understand the importance of fire and life safety measures while working in the hospital. This training should be done by certified professionals with proper documentation that can be provided via paper or electronically, if required, to the regulatory authorities.
- Additional fire drills may be required due to the extent of the renovation or addition and the need to reeducate the hospital staff on life safety measures and means of egress that have been impacted by construction. The Joint Commission requires at a minimum one additional fire drill per shift, per quarter.

With complicated, multiphased renovations, these measures may be constantly changing as the construction work is done in sections at various points in the hospital. Mechanical and electrical systems as well as fire/smoke walls and emergency exits may sometimes have to be modified on a daily or weekly basis to accommodate the construction schedule and to keep the hospital operational. The development of a well-defined but flexible phasing plan is essential to successfully completing the construction work in a safe, efficient, and cost-effective manner. This plan must be developed in conjunction with the hospital staff, the architects, engineers, and the subcontractors so that it is workable and practical while supporting both activities—health care and construction. The phasing plan must be effectively communicated to the construction team as well as to the hospital staff. Close coordination, training, and education are essential to a successful project. This subject will be explored further in Chap. 9.

Compartmentalization

NFPA defines compartmentalization as the concept of using various building components (e.g., fire-rated walls and doors, smoke dampers, fire-rated floor slabs) to prevent the spread of fire and products of combustion so as to provide a safe area of refuge. This is the key element of the defend-in-place strategy along with fire sprinklers and fire alarm systems. The successful hospital builder must understand how to apply this concept by using the correct building products, such as fire dampers, doors, drywall, and the correct installation processes, such as using angles on fire dampers and having priority 2-hour walls extending through 1-hour intersecting walls. Many of the "rated" products and installation requirements are defined by the Underwriters Lab (UL) rating system. These building components are also generally called assemblies and must be used or assembled in strict accordance with the way they were tested. This ensures they will function properly in a fire situation. More explanation of UL assemblies and rating systems are provided later in this chapter.

Hospitals are designed to minimize the potential risk from fire and smoke. Design elements include fire- and smoke-rated walls that must extend from slab to slab and from exterior wall to exterior wall. Intersections of rated walls and exterior walls or slabs are an important area that requires special attention by the builder. These could possibly allow fire or smoke penetration and are often closely inspected by the authority having jurisdiction (AHJ). Figures 2.3 and 2.4 are sketches showing examples of these critical intersections.

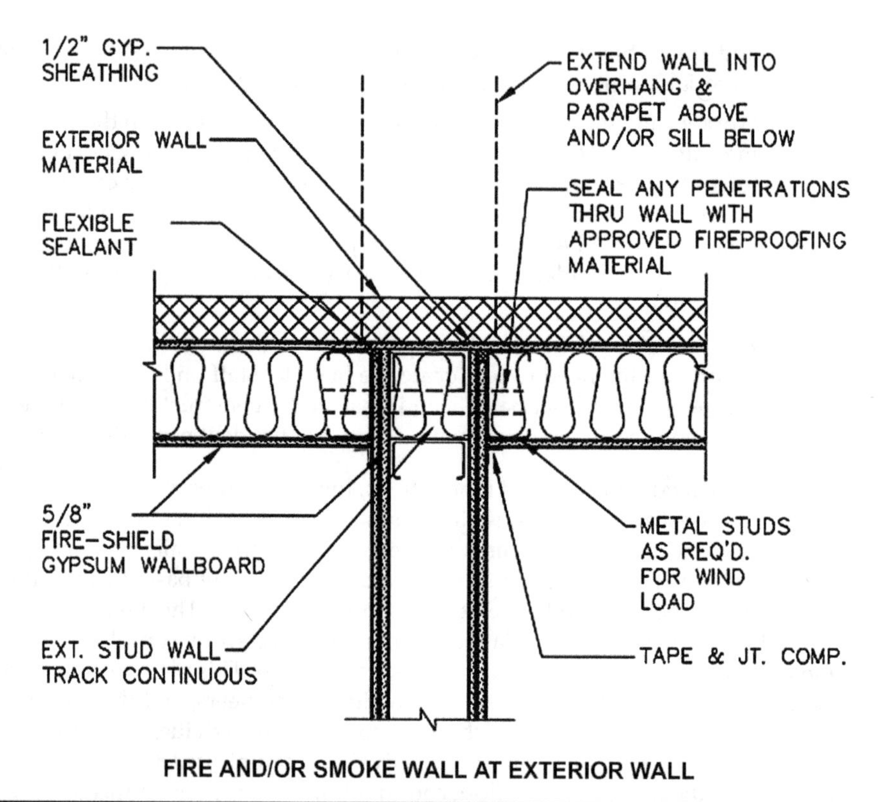

Figure 2.3 Fire wall intersections at exterior wall—fire wall must extend to exterior sheathing of the wall to avoid passage of fire or smoke around it. (*Courtesy of Brian Lee, USG Nashville, TN.*)

Any openings in "rated" walls or slabs must be protected from the passage of fire and/or smoke. These include doors, ductwork, and the space or cracks around conduits, cables, pneumatic tubes, etc. Doors or ducts that penetrate 2-hour walls or slabs must have 1½-hour-rated doors and hardware or fire/smoke dampers. The doors are required to have self-closing hardware and positive latching devices to make sure the openings remain closed in the event of a fire. The cracks or spaces around other penetrations must be sealed using an approved fire-rated material, which is usually a "fire caulk" product.

The hospital constructor must understand these requirements, know which products provide the necessary ratings and manage the proper installation of these complex systems. This work must be carefully planned and inspected during installation to verify it complies with NFPA, the specified UL assemblies, and the manufacturer's installation directives. This is very important to protect the lives of the patients, staff, and visitors, but also must be done correctly and documented properly for the AHJ to approve the facility for occupancy meeting the builder's obligation to the owner. It is good practice to keep binders or electronic files indicating all products that have been installed and to take photographs of the work as it is put in place in order to document the way critical elements or intersections were built properly. Preconstruction conferences to help plan the work and mock-ups to show proper installation are also preventative methods to avoid problems later.

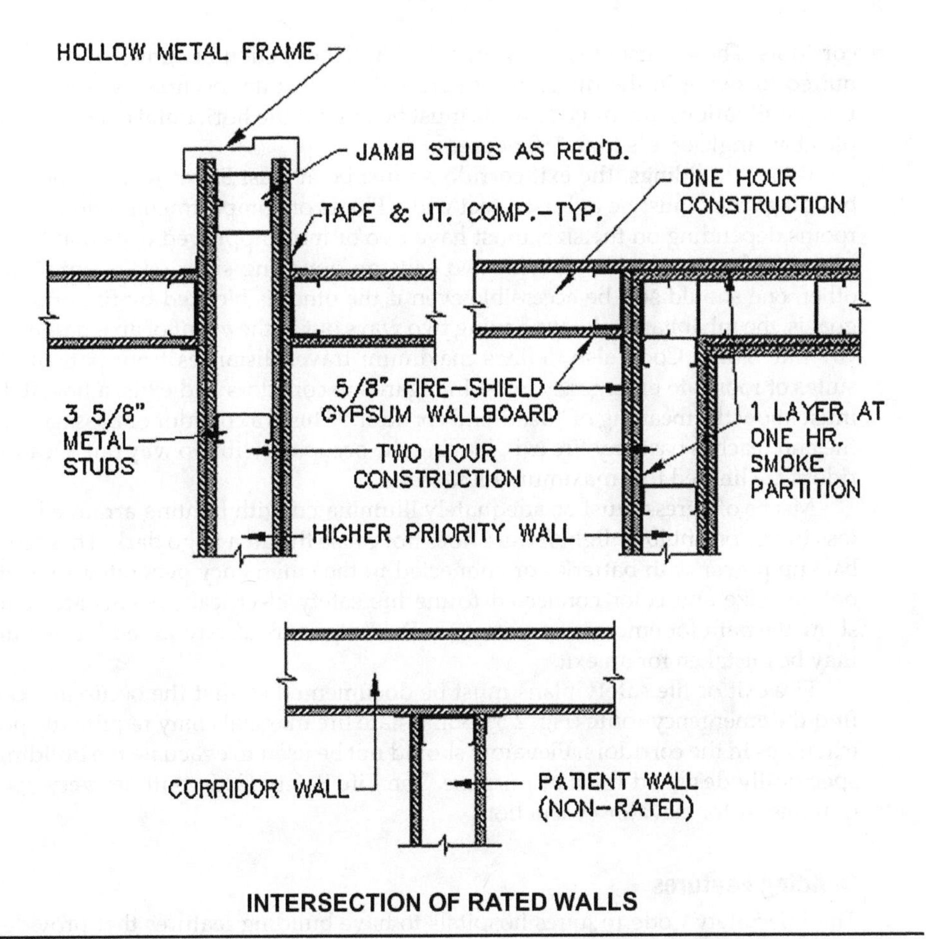

Figure 2.4 Details of priority wall intersections—higher priority walls, such as 2-hour or corridor walls must have priority and extend continuously through the intersection with other walls. (*Courtesy of Brian Lee, USG Nashville, TN.*)

Means of Egress

The means of egress is another important element in fire safety for hospitals. The architect designs the space with an adequate number of exits and places them to strategically protect people from fire and smoke.

The means of egress is designed to evacuate people to the exterior of the building so they are clear of any danger. The means of egress is composed of three distinct parts: the exit access, the exit, and the exit discharge. On upper floors, occupants need to be able to get to a protected pathway to the exterior (the exit access), which is generally a protected corridor. As they obviously cannot exit directly to the outside, protected stairwells (the exit) are provided with walls rated for 2 or more hours, with the stairs exiting directly to the exterior. A clear path to an area outside and away from the building needs to lead to the public way (exit discharge). In some cases, the stairs or even a first floor space cannot be exited directly to the outside so a 2-hour protected passageway or corridor must be provided. This is referred to as an "exit passageway." Fire-rated doors are required as the means of egress in stairwells, horizontal exits, and non-rated corridors. These must always be unlocked in the direction of egress. These are also required to swing in the direction of egress if the hospital occupancy is 50 or more. In unique situations, the outside walls must be rated if the horizontal exits discharge people at an angle of less than 180 degrees.

In new buildings, the exit corridors must be at least 8 feet wide while in existing buildings they must be at least 4 feet wide. Floors or compartments, and in some cases rooms depending on the size, must have two or more approved exits that "are located remotely from each other." With two exits on opposing sides or "remote" from each other, one should still be accessible even if the other is blocked by fire or smoke. The goal is, the inhabitants always having two ways out in the event of an emergency. NFPA 101 Life Safety Code also defines maximum travel distances from patient rooms or suites of rooms to emergency exits. In regards to corridors and exits, a hospital builder must know the meaning of "dead-end corridor." This is a corridor or passageway where the end reaches past any fire exit, leaving the occupant with no way out. Dead-end corridors are limited to a maximum of 30 feet.

Means of egress must be adequately illuminated with lighting arranged so that the loss in any one bulb or light fixture does not cause the area to go dark. These must have backup power with batteries or connected to the emergency generator. Exit signs of a defined size and color connected to the life safety electrical systems are required to show the path for emergency exits. "No Exit" signs are also required for openings that may be mistaken for an exit.

Fire exit or life safety plans must be documented so that the occupants can easily find the emergency route (Fig. 2.5). Some state fire marshals may require the posting of exit maps in the corridors. Elevators should not be used to evacuate the building unless specifically designed for that purpose. The Life Safety Code outlines very specific requirements for elevator evacuation.

Building Features

The Life Safety Code requires hospitals to have building features that provide protection for patients. In addition to the compartmentalization approach, openings in fire/smoke walls must be protected. These openings may include doors, ductwork, elevator shafts, and other vertical or horizontal openings. Stairs must be protected by 2-hour walls. With other vertical openings, 1-hour walls are required if they connect three or fewer floors, while more floors than this requires 2-hour walls.

Areas that are deemed hazardous must be protected by fire-rated walls and doors as well. The ratings vary with the type and size of spaces. Some of the spaces defined as hazardous areas are

- Boiler/fuel-fired heater rooms
- Central laundries
- Flammable liquid storage rooms
- Laboratories
- Repair and paint shops
- Soiled linen rooms
- Storage and trash collection rooms
- Gift shops

FIGURE 2.5 Emergency exit or life safety plan. (*Courtesy of Thomas, Miller & Partners, Brentwood, TN.*)

Interior finishes, such as vinyl wall covering, fabric materials, decorations, wood or plastic wall coverings, or other material, must be rated to limit smoke development or contributing to flame spread. Existing finishes must be Class A or B, while new must be Class A and have a Class 1 radiant flux rating. The hospital builder should get the flame spread rating documentation from the subcontractor or the owner, if it is not part of the construction contract, and have these in a binder and in an electronic format for the final and periodic inspection. The AHJ may want to review the information prior to approval of occupancy. In the event the owner is planning to install some material that is not properly rated, the builder must work with the owner and/or designer to resolve the issues prior to final inspections as this could impact the builder's ability to meet their contractual schedule obligations.

Corridor doors and doors in smoke barriers are another part of the life safety system that provides for protection in place for patients or protected egress. The doors need the appropriate fire rating and hardware as well as being fitted properly. The door undercuts or space between the bottom of the door and floor must be no more than 1 inch on corridor doors and ¾ inch on smoke doors. The NFPA documents use a phrase—"positive latching hardware." This means a door handle and fastening assembly mechanism that fits solidly into the strike plate in the frame to prevent the door from opening due to pressure or heat from fire and smoke. Roller latches or closers only are not permitted. Corridor doors are also required to have automatic closers, unless the wall opening is a part of a smoke barrier, exit enclosure, hazardous area enclosure, or vertical opening that keep the doors in place for protection. When a closer is used, it is permitted to have the doors held open with a magnetic device tied to the fire alarm system that releases when the alarm is activated.

NFPA 101 defines the need for at least two smoke compartments for every floor with patients sleeping or treatment areas, although the requirements vary for some new and existing buildings. The size of the compartments is defined as well as the smoke barriers to protect these compartments. One of the key points for the healthcare builder is the need for the smoke walls to be continuous from floor slab to floor or roof slab and from inside the exterior wall on one side of the hospital to inside the exterior wall on the other side. While this sounds straightforward, it can become complicated in the field with ceilings, canopies, structural framing, interstitial spaces, mechanical ductwork or piping, and other building elements. Careful attention must be paid to wall construction and protection of the many penetrations required in the smoke barriers.

NFPA 72: National Fire Alarm and Signaling Code

The fire alarm system is a very important component in the overall hospital life safety system. NFPA 72 National Fire Alarm and Signaling Code addresses the fire alarm system. The master fire alarm panel must be located in a 1-hour fire protected space that is monitored continuously. This is typically the telephone operator room or emergency room check-in as these areas are functioning 24 × 7. A remote ancillary panel is also required in case the master panel is inaccessible due to a fire or other emergency. This is often located in the facility engineering department or the powerhouse.

The fire alarm system monitors for smoke and fire in occupied space as well as some unoccupied, such as the elevator shafts. This system is also connected to the fire sprinkler system and goes into alarm if the water starts flowing, indicating a sprinkler head has been activated. The system must automatically notify the local fire department of the alarm.

The standards address other building elements or equipment to provide fire life safety. These include the requirement for the placement of a fire extinguisher within 75 feet or less at any point in the hospital. Special Class K extinguishers are required near cooking locations which produce grease. These cooking areas also require exhaust hoods with fire protection systems, connections to the fire alarm and automatic shut-offs for the fuel sources. Linen and waste chutes, not common in today's modern facilities, have specific additional requirements. These may be encountered with renovations of existing buildings, and may require upgrading if within the scope of the project.

NFPA 99: Health Care Facilities Code

NFPA 99 Healthcare Facilities Code is dedicated entirely to the unique aspects of healthcare facilities from a life safety and systems standpoint. This document provides a comprehensive minimum standard for design and construction and for some aspects of operations.

The chapter on electrical systems sets forth some of the unique requirements for hospitals. There the different types of electrical systems and the functions are defined in detail. The two systems required are "normal" power and the "essential electrical system." The normal power system is the portion connected to the local utility that functions during routine operations and conditions. The essential system is *divided into* three separate branches, the Life Safety branch, Critical branch, and the Equipment Branch. The Life Safety branch supplies power for important components, such as illumination of the means of egress, exit lights, fire alarm, emergency communication systems, and more. The Critical branch feeds important components related to patient care such as task lighting, nurse call systems, blood and tissue banks and selected receptacles (red plugs), and more. The equipment branch serves emergency generator accessories, smoke control systems and stair pressurization, heating equipment for patient care areas, HVAC equipment for isolation rooms, selected elevators, and more.

The essential electrical systems must have two independent sources of power: normal and an alternate, which is usually a diesel- or natural-gas-fired generator that must be located on the premises. NFPA 99 determines which type and class of power source is required and NFPA 110 provides the specifications for those power sources. This is where the critical requirement for the generator to "pick up the load" within 10 seconds is found. This means that all electrical devices and circuits required to automatically be on emergency power must be operational through power from the generator within this timeframe. It is one of the very important items that must be tested by the builder and designers during the commissioning of any new facility or a renovation involving the generator. The electrical systems are further regulated by NFPA 70 National Electrical Code, which addresses materials, installation, and testing. The emergency generator system and automatic transfer switches are further regulated by NFPA 110 Standard for Emergency and Standby Power Systems.

The chapter on gas and vacuum systems is another portion of NFPA 99 regulating a system that is important to patient care and somewhat unique to hospitals. The medical gas system is a complex and extensive network of outlets, piping, valves, pumps, and tanks that must be installed and maintained properly for the safety of the patients and staff. It recognizes early in the chapter the potential for "fire and explosion hazards associated with medical gas central piping systems and medical-surgical vacuum systems." The other hazards noted are

- Unplanned shutdowns that are a concern with renovations
- Inability to perform in emergency situations due to secondary system failures, such as power
- Potential for cross-connections

This comprehensive chapter addresses the complete medical gas system that typically includes medical air, carbon dioxide, nitrogen, nitrous oxide, oxygen, waste anesthesia gas disposal, and vacuum. The requirements address a wide range of items such as bulk storage, alarms, valves, material, installation, labeling, and testing. While

NFPA 99 covers the installation and performance of all of these items, the actual number of outlets required in each type of patient care room is defined in the Guidelines.

There are several definitions used in the regulations that are important to understand for those involved in the construction of hospitals. Examples include

- Authority having jurisdiction (AHJ)—the organization or individual responsible for approving the design, equipment, or the work and, perhaps more importantly, for interpreting or granting exceptions to the requirements.
- Flame resistant—required for many materials in a hospital and further defined in NFPA 701 Standard Methods of Fire Tests for Flame Retardant Textiles and Fabrics. Typically, the documentation of the flame resistance for everything from trash cans to curtains will be required for the final inspections.
- Patient care areas—(a) general care areas such as patient rooms and exam rooms where the patient will only be in contact with ordinary appliances and (b) critical care areas such as operating rooms and intensive care units (ICUs) where the patients will undergo invasive procedures and/or be connected to patient care appliances.
- Limited care facility—a building used on a 24-hour basis where four or more persons who are incapable of self-preservation are housed.

In addition to the NFPA documents mentioned above, there are several other documents that have referenced sections that impact or regulate hospital design and construction. These include: NFPA 20: Standard for the Installation of Stationary Pumps for Fire Protection, NFPA 90A: Standard for the Installation of Air-Conditioning and Ventilating Systems, NFPA 111: Standard on Stored Electrical Energy Emergency and Standby Power Systems, NFPA 13: Standard for the Installation of Sprinkler Systems, NFPA 72: National Fire Alarm and Signaling Code, NFPA 241: Standard for Safeguarding Construction, Alteration, and Demolition Operations.

Construction professionals responsible for hospital projects should have a good understanding of and access to the NFPA documents to help verify compliance during the bidding, submittal, installation, and commissioning phases. One should also verify the appropriate edition is being used for the project as code officials may have not adopted the latest version in some localities.

Accrediting Organizations

The Joint Commission

The Joint Commission (www.jointcommission.org) is the most widely used of several organizations that accredit hospitals for reimbursement for federal healthcare programs, such as Medicare and Medicaid. Joint Commission accreditation is an alternate method for facilities to show conformance to the center for Medicare/Medicaid conditions of participation as well as show to consumers that they have excelled in areas of management and leadership. The accreditation is validated by CMS on a periodic basis. This comprehensive survey and accreditation process is mostly focused on the operations, governance, and quality of patient care; however, there are also aspects which relate to the facilities and construction operations especially on hospital renovations. The Environment of Care, Emergency Management, and Life Safety sections contain most of these requirements.

Environment of Care (EC)

This section requires hospitals to have written plans for managing the following:

- Environmental safety of patients and everyone else who enters the hospital facilities
- Security of everyone who enters the hospital's facilities
- Hazardous materials and waste
- Fire safety
- Medical equipment
- Utility systems

The environment of care addresses three basic elements:

1. The building
2. Equipment
3. People

Obviously, hospital construction, whether new or additions/renovations, can impact these required plans. Section EC.02 addresses fire drills and testing of mechanical and electrical systems, which also may be affected by construction activities. In many cases, EC.02 references other codes, such as the Guidelines for air pressure relationships, air exchange rates, and filtration, as well as numerous NFPA documents. This section provides the requirements for a pre-construction risk assessment, which includes ICRA or Infection Control Risk Assessment, to be discussed later, as well as other risks or hazards that could affect "care, treatment, and services."

Emergency Management (EM)

This section requires hospitals to address four areas: mitigation, preparedness, response, and recovery. The assessment is called hazard vulnerability analysis (HVA) and major construction projects on-site or a renovation could trigger this requirement and necessitate collaboration between the constructor and the hospital staff. Just a few examples of potential construction emergencies that have happened include accidentally cutting off power or medical gases, cranes falling on the facility, introduction of toxic vapors in the HVAC system, fire, and many more. The potential for these risks needs to be assessed and the builder in cooperation with the hospital must develop mitigation measures, preparedness plans, and protocols for shutdowns of utilities, as well as an emergency response plan and recovery methods.

Hazard Vulnerability Analysis

Evaluate every potential event in each of the three categories: probability, risk, and preparedness. Add additional events as necessary.

Issues to consider for probability include, but are not limited to

- Known risk
- Historical data
- Manufacturer/vendor statistics

Issues to consider for risk include, but are not limited to

- Threat to life and/or health
- Disruption of services
- Damage/failure possibilities
- Loss of community trust
- Financial impact
- Legal issues

Issues to consider for preparedness include, but are not limited to

- Status of current plans
- Training status
- Insurance
- Availability of backup systems
- Community resources

Life Safety (LS)

The Joint Commission recognizes that accidental damage or other situations during construction can at times prevent the requirements of the Life Safety Code from being met. This section requires the hospital to correct the deficiency immediately or manage it through one of these options:

- Management process to resolve the situation in 45 days
- Submitting an "equivalency" to "demonstrate that alternative building features exist that complies with the intent"
- Submitting a plan for improvement (PFI)

The Joint Commission suggests using a building maintenance program (BMP) to track and manage deficiencies. The hospital builder needs to be aware of steps the hospital may have implemented to be sure that their construction activities do not negatively impact the hospital's ability to manage these building deficiencies or meet the obligations of plans submitted by the hospital.

The Joint Commission Standards address the following key elements of a hospital facility:

- General life safety design and building construction
- Means of egress, including design of space, travel distances, egress illumination, and signage
- Protection provided by door features, fire windows, stairs and other vertical openings, corridors, smoke barriers, and interior finishes
- Fire alarm notification including audible and coded alarms
- Suppression of fires, including sprinkler systems
- Building services including elevator and chutes
- Decorations, furnishings, and portable heaters

A successful hospital constructor should understand all of these terms and concepts and their implications related to fire and life safety. These will be further explained in the chapter as the different sections are explored.

The Life Safety section also provides an important clarification regarding the application of the Life Safety Code. There is no question on inpatient buildings, which include hospitals, nursing homes, and limited care facilities. While the Life Safety Code definition of an ambulatory healthcare occupancy makes the distinction that anesthesia or outpatient services must be provided to four or more patients, at the same time, making them incapable of saving themselves in the event of an emergency, CMS has adopted a more stringent rule. All ambulatory surgical centers seeking accreditation for Medicare/Medicaid certification purposes regardless of the number of patients incapable of saving themselves in the event of an emergency shall comply with the Ambulatory Healthcare Section of the Life Safety Code.

Routine Surveys

The accreditation standards are a critical part of most hospitals' routine operating procedures. The accrediting organization typically surveys a facility every few years, but can also hold unplanned inspections or respond to significant negative issues. The Joint Commission calls these "sentinel" events and may be due to an accident, an unusual death, some catastrophe or a complaint. It is important for a hospital builder to understand the requirements of the accrediting organization to help make sure the project is constructed properly. Also, the builder may be on-site for a renovation project during a survey. This may require them to help the hospital prepare and make sure there are no construction-related issues. The hospitals have to prepare a documentation such as the "statement of conditions" in which the facility and life safety systems are evaluated. The builder may be asked to help in the process by providing documents or correcting deficiencies. Additionally, the state health agency responsible for licensing the hospital may do routine surveys to verify compliance. These surveys are equally as important and they validate that the facilities meet the conditions of participation so that the hospital can receive Medicare/Medicaid reimbursement.

International Building Code

In 2000, the first edition of the International Building Code (IBC) (www.iccsafe.org) was promulgated by the International Code Council to provide a comprehensive and unified code. Prior to this, there were several different codes utilized regionally by different regulatory agencies. These included the Standard Building Code (SBC), published by Southern Building Code Congress, the National Building Code (NBC), published by Building Officials & Code Administration (BOCA), and the Uniform Building Code (UBC) published by International Conference of Building Officials. The preface to the IBC states that "the intent was to draft a comprehensive set of regulations for building systems consistent with and inclusive of the scope of the existing model codes." The principles defined are

- Protect public safety, health, and welfare
- Do not unnecessarily increase construction cost

- Do not restrict the use of new products
- Do not give preference to particular materials, products, or methods

The IBC addresses "structural strength, means of egress, sanitation, adequate lighting and ventilation, accessibility, energy conservation and life safety in regards to new and existing buildings." These minimum standards are updated every 3 years.

The IBC utilizes a "Use and Occupancy Classification" system to define the different types of buildings and the applicable minimum standards. Hospitals fall under the Institutional Group "I" classification. As with the NFPA regulations, once again the concept of people "who are not capable of self-preservation without physical assistance" is introduced as the basis for determining building requirements. Institutional Group "I" occupancies are divided into 4 divisions of I-1, I-2. I-3, and I-4. I-3 includes correctional facilities "in which the liberty of the occupants is restricted," thus making them incapable of providing for their own safety. Medical facilities in jails or prisons pose unique challenges given the dual restrictions from both a medical and security perspective. Additions and renovations to these facilities while in operation require extensive planning and careful execution.

Hospitals fall under the Institutional Group I-2 which includes buildings providing medical care "on a 24-hour basis for more than five persons who are incapable of self-preservation." This varies from the NFPA standard definition. The minimum standards are then defined by Occupancy Group for all of the different building elements addressed by the IBC as listed in the opening paragraph of this section, such as structure strength, means of egress, etc. For example, the lighting required for emergency exits for hospitals as well as other building types would be found in Chap. 10 of the IBC, "Means of Egress," Sec. 1006 about Illumination. Exceptions for specific groups are noted in each section if there are any. An example of an exception can also be found in Chap. 10, Sec. 1003.5 with the standard for elevation changes in means of egress. For most Occupancy Groups, steps are allowed; however, there is a note for Group I-2 occupancy, which is for hospitals, that states "any change in elevation in position of the means of egress that serve non-ambulatory persons shall be by means of a ramp." So hospital patients that cannot walk and must be evacuated in wheel chairs or hospital beds will have ramps as they cannot practically be moved via steps. Review of Sec. 1010 regarding ramps would determine the minimum standards for a ramp which limits the slope to 2 percent and requires a slip-resistant floor material.

Special Requirements

Chapter 4 of the IBC provides special detailed requirements for unique use and occupancies. These include malls, atriums, parking garages, motion picture projection rooms, and others. If a hospital contained an open mall or atrium, it would have to comply with this section. Section 407 addresses Group I-2 Occupancies in hospitals. This provides special requirements and exceptions for hospitals. For example, it requires corridors in Group I-2 to be "continuous to the exits" and protected by smoke walls. This is to protect the patients and staff during evacuations as moving sick or incapacitated people may take longer than the healthier, ambulatory public. It also provides exceptions to meet the practical operational needs of hospitals. For example, Sec. 407.2.2 allows for the nurses' station to be open to the corridor for better access by physicians and staff, better visibility of patients, and faster emergency response as long as the corridors have smoke walls. Otherwise, the nurses' station would have to be a separate room behind a smoke wall, which would be difficult for the reasons mentioned above.

The chapter under Sec. 407.5.1 also requires a "refuge area" in each smoke compartment to accommodate the patients and occupants from adjacent smoke compartments. There are specific requirements on the amount of space mandated for the "refuge area." This concept is utilized to provide a safe place in the hospital to avoid having to evacuate patients as this may harm them more and is part of the "defend in place" strategy. These areas of refuge can also be used to protect a patient until they can be safely evacuated. Chapter 4 under Sec. 413, Combustible Storage, and Sec. 414, Hazardous Materials, may also have applications in hospitals. There are areas of the hospital where the medical gases are stored and areas where radioactive materials are located. Also, facility maintenance areas may have paint or chemicals accumulated. These areas would need to comply with these sections. Section 424, Children's Play Structures, may also apply in the case of a play area at a children's hospital. So the hospital builder needs to understand how the IBC would apply depending on what elements are contained in a particular hospital facility.

Building Characteristics

Chapter 5 of the IBC, Building Heights and Areas, defines the permitted size of a facility. It is controlled in two ways: (1) the number of stories or floors and (2) the size or square footage on each floor. This chapter works in conjunction with Chap. 6 of the IBC, "Types of Construction," which defines five types based on the fire-resistance ratings of key building elements, such as structural frame, load bearing walls, and floor and roof construction. In Type 1 construction, which is most common for hospitals, there are two fire resistance levels: (1) 3-hour rating for structural frame and load bearing walls and (2) 2-hour rating for the same building components. Using these construction types IA and IB, one can use Table 503 on page 96 of the 2012 IBC to determine the allowable height and floor area. Using Type IA, which is the most stringent rating, the table shows UL-unlimited for number of stories and size of the floors. For Type IB, the hospital would be limited to four stories but still have unlimited floor areas. To further clarify why Type IA is typically used for hospitals, the table shows for Type IIB, the hospital could only be one-story with a maximum floor area of 11,000 SF and with Type IIIB, it is NP (not permitted) at all for hospitals.

This chapter of the IBC has further restrictions and clarifications that sometimes come into play in a hospital setting. One of these is with mixed use, such as an academic medical center that may contain physician offices or clinic spaces. Typically, the entire building must meet the most stringent requirements for the "main occupancy," which could be the hospital. For this reason, physician offices are typically housed in an attached medical office space that can be built to less stringent and therefore less costly building codes. The only requirements are for the two occupancies—Hospital Group I-2 Institutional and Office Building Group B Business—to be separated by fire barriers that are defined in Chap. 7 of the IBC, "Fire and Smoke Protection Features."

Chapter 7 serves to "govern the materials, systems, and assemblies" used to protect structures and the public from the spread of fire and smoke. This is another important chapter for hospitals, as fire and life safety protection is critical for people that are "incapable of self-preservation" and could be harmed more by emergency evacuation given their medical condition. This chapter focuses on protecting the structure from failure or collapse as well as preventing the spread of fire from one building to another to avoid the catastrophic fires that engulfed entire cities, such as Chicago and San Francisco.

The requirements are defined for fire walls and barriers, smoke barriers, and partitions, as well as the opening or penetrations in these for doors, ductwork, pipes, etc. While these minimum standards are not unique to hospitals, there are some critical requirements that must be properly executed. "Continuity" is the concept that fire barriers must extend from the top of the rated floor to the underside of the rated assembly above (floor or roof). They must go through suspended ceilings or other concealed space. This same concept applies to hospital fire/smoke barriers as they must extend from exterior sheathing to exterior sheathing and not stop at the face of an exterior wall. These intersections through the building should be carefully planned and inspected by the hospital builder as they will often be receiving special attention by the AHJ during final inspection. Taking pictures of these during construction is a good practice to have documentation for the AHJ during inspections.

Vertical openings in building, such as elevators, stairs, and atriums that penetrate through the rated concrete floor assemblies, also deserve special attention as they can allow fire and smoke to easily spread to multiple floor levels. Sections 712 and 713 address these unique situations and risks that are common in hospitals. Typically, 2-hour-rated drywall or concrete block walls are used with fire rated or "labeled" doors. Pipes, conduits, and ductwork not serving these spaces typically are not allowed to pass through them.

Protection of Penetrations

Protection of penetrations is covered in Sec. 714. In hospitals, there are typically many fire-rated assemblies, such as fire and smoke walls or rated floors and roofs, dampers, doors, etc. An assembly is a combination of different building materials used or assembled into one system and tested as a whole. For example, this might be a drywall partition used as a firewall that is made up of metal track and studs, gypsum wall board, metal screws, tape, and drywall mud. These assemblies are tested as a whole in actual fire conditions to verify they can meet the required hourly rating to stay in place providing protection. The subject of assemblies will be discussed more in this chapter under the UL Assembly section.

The penetration section defines how to protect openings in the rated assemblies for conduit, pipes, ductwork, structural members, and a myriad of other building components. In hospital construction, there will be many, many penetrations that need to be addressed to meet the IBC and the NFPA requirements. In addition to resisting the spread of fire, the protection of the penetrations must be capable of resisting the pressure created by a fire and hot gases. This pressure is defined as a "minimum positive pressure differential of 0.01 inch (2.49 Pa) of water at the location of the penetration." The code specifically defines the size of openings and the material that can be used to fill the "annular" space. This is the space between the pipes, conduit, or other building component penetrating the assembly. It is best to have the subcontractors cut the material in the assembly, typically drywall or block, as close as possible to the penetrating object to keep the annular space small. Several different materials or fire stop systems are available in the market. A red intumescent fire caulk is the most common in the industry. Intumescent material swells when exposed to heat, thus filling the space so that smoke cannot penetrate the assembly. This material expands when exposed to heat to fill the annular space and is a poor conductor of heat so it retards the spread of fire.

Doors in fire/smoke-rated walls must also be rated and the size of glass openings in fire doors is limited by this section of the IBC. The frames, doors, and hardware are all part of the assembly and must be rated as well.

Section 717 ducts and air transfer openings address supply, return, and exhaust ducts that cross through fire and/or smoke walls. At these points, a fire or fire/smoke damper must be installed to protect the hole in the fire or smoke wall. During fires, the heat and pressure may cause the duct to collapse and pull away from the wall. The damper is held in place in the wall with steel angles. It has multiple blades that close automatically triggered by the fire alarm system or a fusible link that burns through. With these blades closed, the damper seals the opening in the wall or floor so fire and/or smoke cannot escape from one level or compartment to another. Access doors are required at the dampers for maintenance. The fire/smoke dampers are also rated assemblies that have been tested by an independent lab, such as Underwriters Lab (UL) or American Society for Testing Materials (ASTM). See Figs. 2.6 and 2.7 for fire/smoke damper construction and installation.

Interior Finishes

Interior finishes, such as wall coverings, decorations, and trim, can contribute to the spread of fire, add to smoke development, and in some cases produce toxic gases when burned. For this reason, the code limits the materials that can be used, and provides a rating system based on flame spread and smoke developed indexes. There are three classes for interior wall and ceiling finishes:

FIGURE 2.6 Fire damper in open position. (*Courtesy of Ruskin Air and Sound Control. All rights reserved.*)

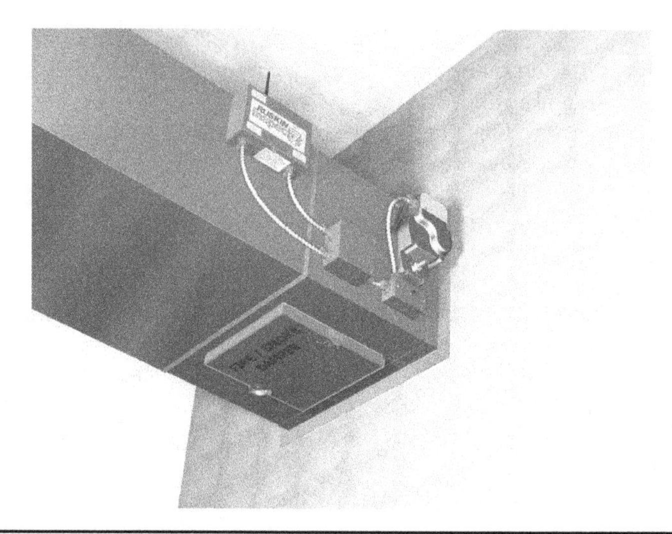

Figure 2.7 Fire damper installed with steel angle attached against the wall to hold it in place if the duct falls away and access door for maintenance. (*Courtesy of Ruskin Air and Sound Control. All rights reserved.*)

Class A—Flame spread index 0–25; smoke developed index 0–450

Class B—Flame spread index 26–75; smoke developed index 0–450

Class C—Flame spread index 76–200; smoke developed index 0–450

These are tested per NFPA 286.

Table 803.9 provides the required class of interior wall and ceiling finishes based on the Occupancy Group and whether it is protected by a fire sprinkler system. As learned earlier, a hospital is Group I-2. In a fully sprinklered hospital, Class B finishes are required in all rooms. There are two exceptions where Class C material may be used: administrative spaces or rooms that have capacity of four persons or less.

Fire Sprinkler and Fire Alarm Systems

The IBC requires hospital occupancies to have a fire sprinkler system throughout the building for new construction. Due to the age of many facilities, there are some that still have areas that may not have fire sprinklers. Typically, as upgrades and renovations are done the code officials require hospitals to provide fully sprinklered buildings. The IBC requires the sprinkler system to be installed per NFPA 13.

Portable fire extinguishers are also required by the IBC. Again, the NFPA is used as a reference by the IBC and it is NFPA 10 that governs the type of fire suppression equipment required.

Fire alarm and detection systems are an important element in the life safety systems in a hospital. The IBC in Section 907.2.6 requires fire alarm systems and references NFPA 72. The requirements for the fire alarm systems are complex including both the detection of fire and/or smoke and the notification of the building occupants. There are specific standards for location of the detectors in the hospital spaces as well as in mechanical equipment, elevator hoist ways, and concealed spaces. The fire alarm system also must detect the flow of water in the fire sprinkler system and trigger alarms. Manual pull stations are mounted near fire exits for occupants to activate the system during emergency evacuations. In addition to the detection, the fire alarm system must interface with other building controls systems to shut down air handling units, activate smoke evacuation systems, and put elevators into emergency response mode. Most of the deaths in the MGM Grand fire on November 21, 1980, were due to smoke that was moved to other areas by the heating and air-conditioning system. Another automatic function is to notify the fire department so they can respond. In addition to the items above, the fire alarm system provides emergency notification to the occupants through audible and visual alarms, as well as a public address system.

Further explanation of fire sprinklers and fire alarm systems is provided in Chap. 10 of this book on mechanical/electrical systems.

Egress in Special Patient Units

One part of the section on means of egress is specifically applicable to psychiatric hospitals or acute care hospitals that have dementia or psychiatric units. In this application, it is often necessary to lock the doors to prevent patients from leaving or getting lost as with dementia patients. These conflict with the general life safety concept that all doors out of the building (means of egress) should be visible, easy to use and always available for emergency evacuation. In case of these types of patient units, the code allows the doors in means of egress to be locked if they can be unlocked manually and "by at least one of the following measures":

- Activation of an automatic sprinkler
- Activation of a manual pull station
- "Signal from a constantly attended location"

This arrangement is commonly achieved by having an emergency push button at the constantly staffed nurses' station that will release the door. Connecting the door release to the fire alarm system which would achieve #1 and #2 above is typically not practical as the patients quickly learn they can set off the fire alarm and leave the premises.

The balance of the means of egress chapter may apply to hospital facilities; however, it is not unique to hospitals. Constructors must meet the other requirements in this section as they would on any building. Means of egress is an important part of providing for the life safety of the occupants for any facility. The hospital builder should be familiar with all of the other requirements that include occupant loads, sizing and lighting of the egress path, areas of refuge, signage, doors, stairways, travel distances, and exit requirements.

General Chapters

Many of the chapters in IBC address common building elements or requirements that apply to a wide range of facilities including hospitals. These include structural elements, roofs, exterior walls, soils, and foundations. There are also chapters addressing specific common building materials and systems, such as steel, wood, masonry, drywall, plumbing, elevators, and safeguards during construction. Chapter 11 of the IBC addresses the scoping provisions of the access for disabled and references the ICC A117.1 Accessible and Usable Buildings and Facilities. Chapter 17 of the IBC addresses special inspections and structural observations. These are specialized inspections and observations programs that govern the quality, workmanship, and various construction materials.

Most Widely Accepted and Trusted

ICC-ES Evaluation Report **ESR-4999**

Issued September 1, 2013

This report is subject to renewal September 1, 2014.

www.icc-es.org | (800) 423-6587 | (562) 699-0543 *A Subsidiary of the International Code Council®*

DIVISION: 03 00 00—CONCRETE
Section: 03 15 00—CONCRETE ACCESSORIES

REPORT HOLDER:

XYZ INTERNATIONAL, INC.
1234 SAMPLE AVENUE
VANCOUVER, BRITISH COLUMBIA V8C 2Q3
CANADA
(605) 987-6543
www.XYZINT.com
info@XYZINT.com

EVALUATION SUBJECT:

XYZ INTERNAL MEMBRANE (XIM)

1.0 EVALUATION SCOPE

Compliance with the following codes:

- 2009 *International Building Code®* (2009 IBC)
- 2006 *International Building Code®* (2006 IBC)
- 1997 *Uniform Building Code™* (UBC)

Property evaluated:

As set forth in Section 3.2 and 3.3 of the ICC-ES Acceptance Criteria for Chemical Admixtures Used in Concrete (AC198).

2.0 USES

XYZ Internal Membrane (XIM) is a chemical admixture used for the treatment of reinforced and plain concrete containing portland cement. The admixture reduces the water demand for a given slump and retards the initial and final set times of concrete.

3.0 DESCRIPTION

The XIM admixture complies as a Type D (water-reducing and set-retarding) admixture in accordance with ASTM C494 / C494M 05a[01], with an extended set time. The dry cementitious admixture has an unlimited shelf life when stored in dry conditions in the original, unopened containers, unless otherwise printed on the packaging. Compatibility of XIM with other admixtures is beyond the scope of this report.

4.0 INSTALLATION

XIM admixture must be added to reinforced or plain concrete in accordance with XYZ's published instructions that are packaged with the product. The admixture must be proportioned into the concrete mix at a rate of 2 percent by dry weight of cementitious materials, to a maximum of 13.5 pounds per cubic yard (8 kg/m^3). Concrete mixtures must be proportioned in accordance with the applicable code. The expected water content must be reduced 5 to 10 percent, depending on slump requirement. The water-cement ratio must be maintained between 0.39 and 0.45. Water must be clean and free of deleterious amounts of acids, alkalies, or organic materials. The mixing time must be a minimum of 10 minutes. Concrete quality, mixing, and placing must be in accordance with Section 1905 of the IBC and Section 1905 of the UBC, except as noted in this report. The minimum initial and final set times of XIM-treated concrete are expected to be 8 hours and 9 hours, respectively, prior to concrete finishing procedures and removal of framework. The exact set times will vary depending on the concrete mixture and environmental conditions. The initial and final set times of XIM-treated concrete must be at least one hour longer than that for untreated concrete. The project scheduling for labor and equipment allocation, finishing, curing, and form removal must be adjusted based on extended set times in accordance with the XIM Best Practices Guide, which specifies ACI 308R-01, *Guide for Curing Concrete*, for concrete curing and ACI 306R-88, *Cold Weathering Concreting*, for cold weather concreting.

5.0 CONDITIONS OF USE

The XYZ Internal Membrane described in this report complies with, or is a suitable alternative to what is specified in, those codes listed in Section 1.0 of this report, subject to the following conditions:

5.1 Structural design of the concrete complies with the IBC or UBC, as applicable.

5.2 Use of the admixture in concrete under the UBC is subject to approval of the code official.

5.3 The admixture used in concrete under the IBC is subject to prior approval by the registered design professional.

6.0 EVIDENCE SUBMITTED

Data in accordance with the ICC-ES Acceptance Criteria for Chemical Admixtures Used in Concrete (AC198), dated January 2008 (editorially revised April 2010).

7.0 IDENTIFICATION

XYZ Internal Membrane (XIM) is packaged in 11- and 55-pound (5 and 25 kg) sealed pails and in 22- and 33-pound (10 and 15 kg) bags. The pails and bags of admixture are labeled with the XYZ International, Inc., name, address and contact information, the product name, the batch number, and the evaluation report number (ESR-4999).

Page 1 of 1

FIGURE 2.8 ICC-ES report.

The IBC makes reference to other codes for regulation of trades, such as the National Electrical Code (NEC) for electrical, the International Plumbing Code (IPC) for plumbing, the International Mechanical Code (IMC) for mechanical, the International Fire Code (IFC) for fire prevention, the International Energy Conservation Code (IECC) for energy efficiency, the International Fuel Gas Code (IFGC) for fuels, and the International Existing Building Code for existing buildings that undergo alternation, repair, or change of occupancy. Hospitals and other institutional facilities can also meet green and sustainable practices by complying with the International Green Construction Code (IgCC).

As previously mentioned, the IBC does not intend to prevent the installation of other materials and other methods of construction as long as the minimum safety levels intended are met. Section 104.11 addresses "Alternative materials, design, and methods of construction and equipment." Building officials and other AHJ rely heavily on evaluation reports issued by ICC Evaluation Service (ICC-ES) to make decisions for code compliance of alternative designs, materials, and methods of construction (www.icc-es.org). ICC-ES reports are available for free access online. A sample ICC-ES report is shown in Fig. 2.8.[1]

Americans with Disabilities Act

The Americans with Disabilities Act (ADA) was enacted by the U.S. Congress in 1990. It provides "comprehensive civil rights protection for individuals with disabilities." A disability is defined broadly as a "physical or mental impairment that substantially limits one or more major life activities." These activities include "caring for oneself, performing manual tasks, walking, seeing, hearing, speaking, breathing, learning, and working." The Title III Section of the ADA addresses public accommodations and commercial facilities. Hospitals and doctors' offices are specifically listed as a "place of public accommodation," so this is broadly interpreted to include all healthcare related facilities or those services that are often needed by people with disabilities.

The requirements state that those providing public accommodation must "design and construct new facilities and when undertaking alterations, alter existing facilities in accordance with the ADA Accessibility Guidelines." New hospitals or healthcare facilities must comply with the requirements of the ADA and this is typically enforced by the local or state regulatory agency through plan reviews and inspections. In some states, such as Texas, the inspections can come after final occupancy and lead to costly rework if the work is not in compliance. There are also access rights groups that will pursue civil litigation against facility owners if they perceive a disabled person has faced discrimination through lack of access. For these reasons, it is important for the builder to have a strong quality control plan in place to see that the work related to ADA is done properly to avoid the potential liability for repairs that could be required months or years later.

The ADA requires that "physical barriers to entering and using existing facilities must be removed when readily achievable." "Readily achievable" is defined as "easily accomplished and able to be carried out without much difficulty or expense." This is obviously very subjective and is "determined on a case-by-case basis in light of the resources available." Thus for a large hospital or system the cost to tear out

[1]For comprehensive coverage and a discussion of the IBC key provisions, readers may refer to the McGraw-Hill–ICC publication *2012 IBC Handbook*.

existing facilities and rebuild accessible toilets, for example, may cost tens of thousands of dollars or more. This may still be considered "readily achievable" given the millions or billions of dollars in revenues earned by the owners. General examples of barrier removal measures are given in the Guidelines, such as: installing ramps, adding curb cuts at sidewalks, widening doors, providing grab bars in toilets, or adding Braille letters to signage. The directive is to give people with disabilities "access to areas providing goods and services." If the modifications are not "readily achievable" to strictly meet the ADA Accessibility Guidelines, then other limited, safe measures must be done, such as installing a "slightly narrower door than required by the Guidelines."

The ADA Guidelines address several broad areas that can impact access to medical services or public spaces in a hospital environment. These include the following:

- A minimum of 50 percent of all public entrances must be accessible. This can involve accessible parking, sidewalks, ramps, doors, signage, and sometimes other building elements.
- Accessible public or common use toilets must be provided. This could indicate having a dedicated accessible toilet for each gender or having an accessible stall or stalls (if six or more stalls in the toilet) as well as grab bars, proper height sinks or counters, and required turning spaces for wheelchairs.
- Each floor in a building without a supervised fire sprinkler system is required to have an "area of rescue assistance." This is a space where people who cannot use stairs may wait safely during emergency evacuations.
- Five percent of fitting and dressing rooms are required to be accessible. There can never be less than one. This is common in hospitals in the imaging area, outpatient surgery, GI labs, or other spaces used for outpatient procedures. This involves properly sized rooms with adequate doors and signage.
- Ten percent of patient bedrooms and toilets must be accessible. This goes to 100 percent in "special facilities that treat conditions that affect mobility" and 50 percent in "long-term care facilities and nursing homes." This involves properly sized rooms and toilets with appropriate doors and hardware along with proper counter heights, grab bars, seats, and other toilet accessories. One of the complications that have continually challenged the industry is providing accessible tubs or showers without having water run all over the room. Options for this will be discussed later in this section.

The key requirements and issues that hospital builders must address are explained below:

Site

"Accessible routes" need to be defined from parking to the accessible entrance or, in the case of urban facilities, to the nearest public transportation. One of the common errors listed in the 1994 Workbook Supplement is "construction sites infringing on accessible routes with no provision for temporary access." This issue needs to be clearly addressed in planning site logistics for additions and renovations.

Another of the common site errors related to construction is the failure to provide the proper slopes for the parking spaces and ramps. The slopes required are very strict and often difficult to achieve in the field given the tolerances associated with placing asphalt pavement or concrete. Accessible parking spaces and adjacent "aisles" or loading/unloading areas must be sloped less than 2 percent. Ramps in sidewalks cannot exceed a slope of 1:16. Vertical rises along the accessible route cannot exceed ¼ inch; but if they must due to existing grades, they must be beveled as a 1:2 maximum slope up to a ½-inch vertical rise. Careful quality control measures combined with accurate design drawings are required for the hospital builder to meet these specifications.

While the design is technically and legally the responsibility of the architects and engineers, most owners will appreciate and, hopefully, reward a hospital builder that understands code requirements and can help avoid problems. Site-related ADA issues typically include

- Path of travel from accessible parking to entrance that goes behind parked cars
- Lack of detectable warning strips at curbs and water features
- Lack of proper signage at accessible parking
- Lack of proper landing areas at ramps

Public Toilets

There are three common types of toilets found in hospitals or healthcare facilities:

1. *Multiuser toilets.* These have multiple fixtures including commodes and urinals with toilet partitions and usually a counter top for the sinks. Typical locations are in the lobbies or waiting areas.
2. *Single-user toilets.* These may be men/women or unisex spaces and are often found within one of the hospital departments or near dressing rooms. This type may be staff or public toilets and needs to be accessible.
3. *Single-user family toilets.* These may be in public areas or in a department. They may be used by disabled people and have room for their spouse or helper to assist them without the impact of providing assistance in a multiuser space.

Accessible Toilets

Three key elements need to be addressed for a functional accessible toilet:

1. *Entrance.* This involves having the right size door, proper signage, and an accessible route to the toilet. The ADA also requires space for people in wheelchairs or on walkers to maneuver when opening the door. This is done by requiring 18 inches of clear space on the latch side of the door so they can open it easily out of the line of the swing. The door closer, if one is used, must require no more than 5 pounds of pressure to open and the handle must be easy to grasp.
2. *Accessible route.* Once in the toilet, there must be a clear, unobstructed 3-foot-wide path to access the fixtures. Also, a 5-foot-wide turning radius must be provided to facilitate movement of a wheelchair. An alternate is a T-shaped turning area. The accessible stall must be large enough to allow a 90-degree turn inside it.

3. *Accessible fixtures.* While the actual fixtures used are typically just commercially available sinks, urinals, and commodes, the mounting heights and access around the fixtures is critical. Accessible commodes must have a seat between 17 and 19 inches above the floor and be mounted 18 inches from the adjacent wall. Urinals must be 16 inches to the rim and 42 inches to the flush handle. Sinks must be 33 inches to the top of the rim and 29 inches to the bottom to allow the wheelchair and legs under the sink. The pipes and any sharp edges under the sink must be wrapped to protect a person's legs as they may not have feeling.

Accessible bath tubs and showers must have seats and enough space beside them to allow for a "transfer" from a wheelchair. Most of the seats fold down so the shower may be used by others when accessibility is not required. One of the challenges with the "roll-in" shower is that the lip or curb cannot exceed ½ inch to provide easy access. This small lip or edge often does not provide an adequate barrier to the water during a shower and it often runs all over the toilet floor creating a safety hazard. Many solutions have been tried, from sloping the floor to shower curtains with weighted bottoms. The most effective seems to be installing a "trench" or pool drain at the shower entrance so it can catch the water rolling out. Another choice that works well is a removable plastic curb that can quickly be put in place to keep the water in after the patient enters. The hospital builder should work closely with the design team and the hospital in developing a solution for the roll-in shower. Many have been done poorly in the industry; and it is expensive and disruptive to fix later.

There are many accessories included in the typical toilet in a hospital. These include soap and towel dispensers, toilet paper holders, robe hooks, mirrors, grab bars, and others. There are specific mounting heights and access areas defined in the ADA Rules and Regulations. Many of them are minimums and maximums so there are tolerances, but some are fixed dimensions. The hospital builder must identify the locations during the wall framing phase so that adequate blocking or supports may be installed prior to hanging drywall. In addition to specific locations, there are requirements on the weight that some of these accessories must hold. The regulations also include strict requirements concerning protruding objects, such as shelves or towel dispensers that could be a hazard for visually impaired people.

Drinking Fountains

As with other building elements that are utilized by the public, drinking fountains must also be accessible. The requirements are very specific and special accessible fixtures are commercially available. Typically, a "dual" unit is specified by the designer which has both an accessible fountain and a standard height fountain built together as one (Fig. 2.9).

These address the needs of all members of the public and are typically more cost effective to buy and operate. The accessible fountain must have the spout no higher than 36 inches above the floor. The spout must be in the front of the fountain and direct the water "in a trajectory that is parallel or nearly parallel to the front of the unit" (ADA). The water must flow at least 4 inches high to allow a cup or glass to be filled. The accessible units must also be sloped at the bottom to allow a wheelchair to pull under the fountain. The required dimensions are very specific and the builder should check the fountain dimensions against the ADA to make sure they comply. The law also requires a clear access area with a minimum of 48 inches in width and 30 inches in depth.

Figure 2.9 Combination accessible and normal height drinking fountain.

Elevators

The ADA has many specific requirements for the elevator equipment and operations. Freight elevators typically do not have to comply unless this is the only elevator on the accessible route. The requirements for access are numerous and include mounting heights, signage, and audio/visual displays. The elevator cars must be self-leveling and have a tolerance of ½ inch at the landings. The lobby call buttons must be mounted at 42 inches to the centerline above the finished floor. Nothing can be mounted under the button that protrudes more than 4 inches. Indicator lights at each floor are called "hall buttons" in the ADA and must be mounted at 72 inches to the centerline. The visual or lighted portion must be a least 2½ inches in the smallest direction and indicate up or down. Audible signals are also required for the visually impaired with sounds (once for up and twice for down).

There are specific time requirements for the door and signal timing as well as door opening. The doors must also have a self-opening device that will "stop and reopen" the doors if they hit an obstruction. The sensor for them must be at 5 inches and 29 inches above the floor.

Inside the elevator car, the button to access a floor must be no higher than 54 inches above the floor. An indicator light must show the floor level of the car and it must have an audible signal as it passes each floor so sight-impaired individuals can count the floors passed. The car has minimum inside dimensions depending on where the doors are located, but all doors (side or center) must be a minimum of 36 inches wide. An emergency communication system is also required.

The control buttons in the lobbies or corridors as well as in the elevator car must be raised and utilize Braille letters for easy identification by the visually impaired. At each floor level, there must also be a sign in the entrance door with the number in raised Braille lettering to identify the floor. This must be mounted at 60 inches and the letters must be 2 inches high.

The installation of the elevators is almost always subcontracted to a firm representing one of the major elevator manufacturers. It is essential for the hospital builder to understand these requirements and check for compliance on the shop drawings and during installation. Any firm can make a mistake. It is costly and could impact completion schedules if the elevator subcontractor has to make corrections near project completion. For this reason, the contractor or construction manager should work with the elevator subcontractor during the subcontracting phase or "buy-out", the submittal process, and installation to be clear which elevator(s) is/are to be accessible and that the ADA requirements are met. In a facility with multiple elevators, all do not have to meet ADA.

Service Counters

In hospitals and other healthcare facilities, there are often many locations that fall under the definition of "service counters." These would be places where the public has interaction at a table or counter with staff. These may be business transactions, such as with the accounting department, or medical, such as with a doctor on a patient floor. The typical locations are at the information desk in the lobby, registration positions in the emergency room or admissions, nurses' stations on the patient floors, consult rooms near surgery, and possibly more. These counters must be a minimum of 36 inches in length and no higher than 36 inches above the floor allowing a person in a wheelchair easy access to the counter. Typically, a separate section of the nurses' station, admitting desk, etc. is built at the lower, accessible height and the balance is at the standard height for convenience to staff and other visitors or patients (Fig. 2.10).

Alarms

Audible and visual alarms are both required to provide protection to people with hearing and visual disabilities. The requirement typically applies to the emergency notification from the fire alarm system, but there may be isolated cases of other alarms or notifications that apply such as emergency room waiting systems. Specific decibel levels and duration are required for audible alarms as well as color, flash ratio, and intensity for visual alarms. The minimum spacing requirements are typically 50 feet.

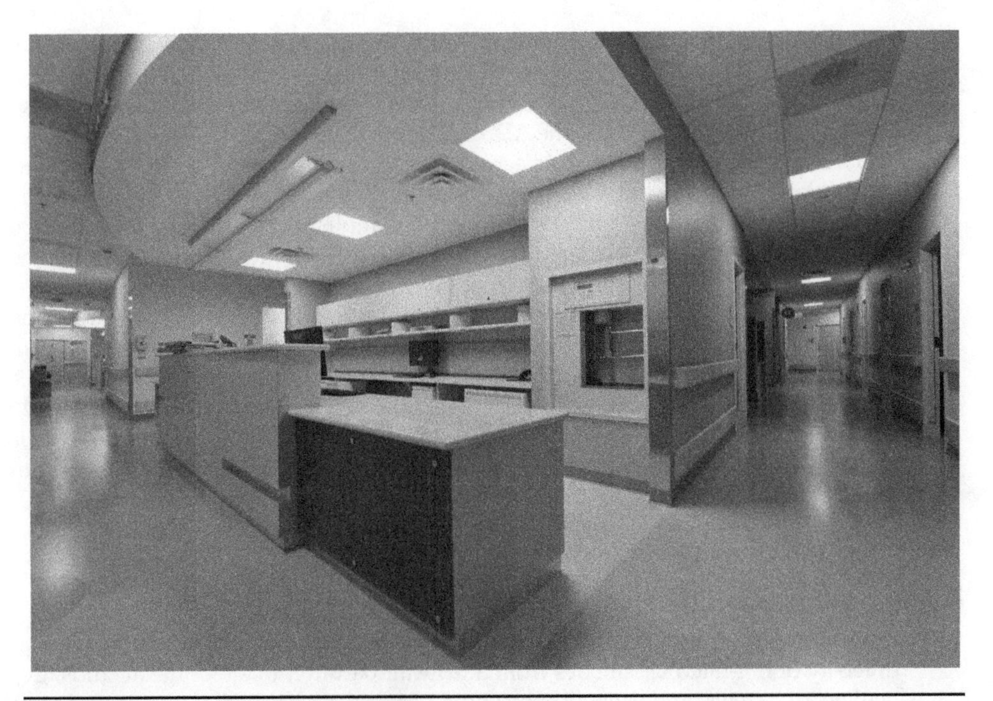

FIGURE 2.10 Accessible nurses' station.

Patient Rooms

The ADA has a specific section that addresses healthcare facilities in addition to the requirements that apply to most public buildings, as previously discussed. The specific medical care facility language applies to "hospitals where persons may need assistance in responding to an emergency and where the period of stay may exceed 24 hours." Offices for doctors and dentists are not included; however, ambulatory surgical and diagnostic centers are required to comply. General purpose hospitals and psychiatric facilities are required to have at least 10 percent of the patient rooms and toilets to be accessible as well as all "public use" and "common use" areas. In long-term care facilities and nursing homes, a minimum of 50 percent of patient rooms and toilets must be accessible. At least one accessible facility entrance must have weather protection with a canopy or roof overhang and include an ADA passenger loading zone. The patient bed rooms must have an accessible door, but can be exempted from the latch side maneuvering requirements if the door is a minimum of 44 inches wide. Each patient room must include a 60-inch diameter or 60 × 60 inch T-shaped space for maneuvering a wheelchair. These accessible rooms must also have a minimum of 36 inches clear space on each side of the bed along with an accessible route to the same on each side. The patient toilets must be accessible and on an accessible route, which means they must have clear space unobstructed by cabinets or furniture to allow wheelchair access. The toilet must meet all of the other requirements previously discussed for ADA compliance.

By the very nature of the services provided and the condition of people using a hospital, proper implementation of these ADA requirements is very important for quality and timely services. In addition, there has been significant litigation by access rights groups and individuals related to healthcare facilities. Several national class action lawsuits and settlements have been executed in the past 10 to 15 years. Failure to comply with the

ADA Law leaves the facility owner and, indirectly, the hospital builder open to future litigation and repairs. Both of these can be expensive and time consuming, so careful attention to details along with a robust quality control plan is essential. In spite of the significant cost and work required by the ADA litigation, there have been positive benefits. Most importantly, many of the hospitals across the country are more accessible to people with disabilities. During the litigation some of the hospital companies were successful in filing motions to allow for construction tolerance in compliance with the ADA. Several of the requirements in the ADA are very definitive and do not allow for the variations encountered in the fabrication and installation of building materials and components. For example, the ADA requires the commode in an accessible toilet to be 18 inches to the centerline from the adjacent wall. The adjacent wall, typically made of drywall and metal studs in a hospital, is not perfectly straight. All commodes are also not perfectly straight. As a result of these common and reasonable variations, it is nearly impossible to have every toilet exactly 18 inches from the adjacent wall. In one particular case, the federal court allowed a range of 17 to 18 inches to allow tolerance for these common variations. The hospital builder should identify these court-approved tolerances and use them to ease and speed up the construction while still delivering accessible, ADA-compliant facilities.

Rated Assemblies

An assembly is a system typically made up of multiple building materials. There is a wide diversity of fire-rated assemblies from a drywall partition made of metal studs, gypsum wall board, screws, and drywall compound to a roof assembly that may include steel bar joists, sprayed-on fireproofing, metal deck, insulation board, and a membrane roof. These assemblies are tested by independent labs in actual fire conditions to verify they will perform as designed to protect the occupants of a building. An assembly's fire resistance is defined as the period of time that it will effectively serve as a barrier to the spread of fire and the time period it can remain structurally sound after being exposed to a fire of standard intensity as defined by ASTM E119. Some refer to this as the assembly's fire endurance. These ratings are typically 20 minutes, 45 minutes, 1 hour, and 1½ hours, 2 hours, and 3 hours. The required rating period is typically tied to the risk associated with the space or usage intended for the space. For example, a stairwell in a hospital used as an emergency exit that is critical to allowing occupants to safely evacuate the building typically requires a 2-hour rating. A smoke resistant corridor wall in a fully fire-sprinkled hospital may only require a 20-minute rating. A fire wall around a storage room usually requires a one-hour rated wall. See Figs. 2.11 and 2.12 for one-hour and two-hour rated wall assemblies.

The manufacturers of these materials, such as USG for drywall, Ruskin for dampers or Weyerhaeuser for doors, pay the independent testing labs to do the actual fire tests on products so they can be certified. Based on these certifications, the building code officials will accept these assemblies for use in the appropriate applications in the building. The designers, architects, and engineers then use the "rated" assemblies to develop the plans and specifications for particular buildings given the fire protection required by the applicable codes. Often they just refer to the testing lab's assembly number. With the critical nature of many patients and the "defend in place" concept discussed earlier, hospitals typically have many fire-rated assemblies. These are used in the "compartmentalization strategy" that is needed in hospitals to provide safe areas to move patients in the event of a fire emergency.

The industry has several independent testing labs that are private companies and universities. These include the National Certified Testing Laboratories (NCTL), Warnock Hersey (WH), Ohio State University (OSU), and the Gypsum Association (GA) fire

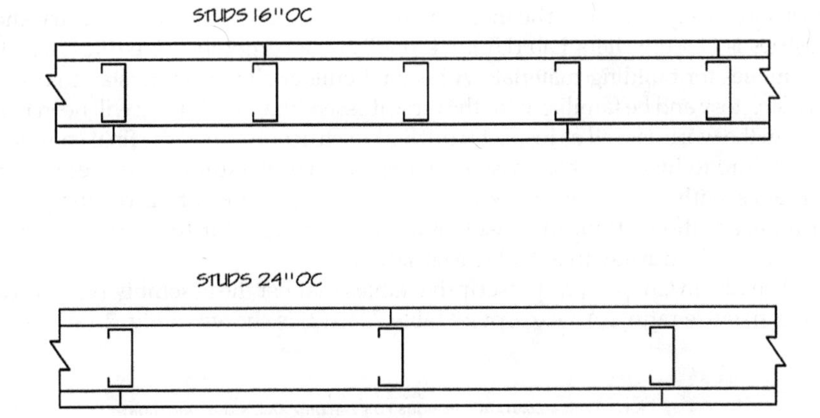

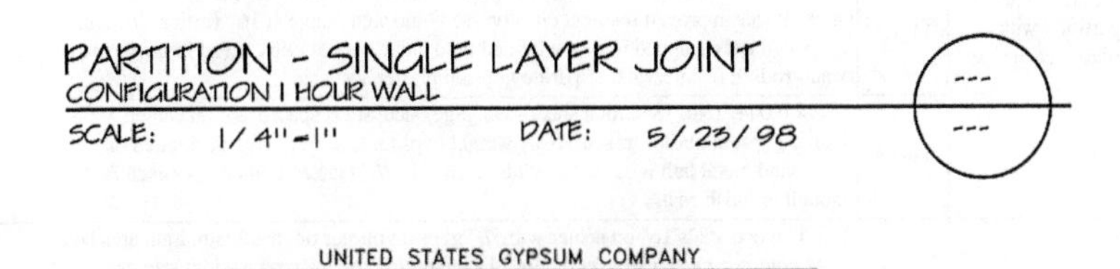

FIGURE 2.11 One-hour rated fire wall. (*Courtesy of Brian Lee, USG Nashville,TN.*)

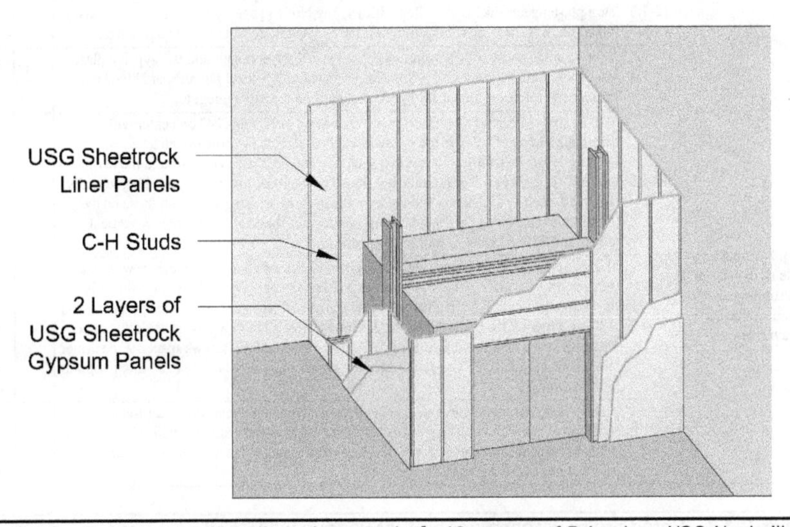

FIGURE 2.12 Two-hour rated wall at elevator shaft. (*Courtesy of Brian Lee, USG Nashville, TN.*)

resistance design manual. The most commonly seen in the hospital industry and other industries is Underwriters Lab (UL). UL publishes a comprehensive directory of fire-rated assemblies for building materials. A hospital builder should understand how to read the UL Directory and be familiar with the typical assemblies used. They will need to know how to plan the work as well as inspect it during construction and at the final completion phase.

Failure to install a rated assembly per the manufacturer's requirements and in accordance with the testing lab's directory voids the fire rating certification. This can result in additional time and cost to redo the work, and further liability in the event the assembly fails during an actual fire situation.

The IBC in Chap. 7 has prescriptive tables that list the assembly type and associated fire-resistance rating. An excerpt of Table 721.1(2) is shown in Fig. 2.13.

TABLE 721.1(2)
RATED FIRE-RESISTANCE PERIODS FOR VARIOUS WALLS AND PARTITIONS [a, o, p]

MATERIAL	ITEM NUMBER	CONSTRUCTION	MINIMUM FINISHED THICKNESS FACE-TO-FACE[b] (inches)			
			4 hours	3 hours	2 hours	1 hour
11. Noncombustible studs-interior partition with plaster each side	11-1.1	$3^1/_4$" × 0.044" (No. 18 carbon sheet steel gage) steel studs spaced 24" on center. $^5/_8$" gypsum plaster on metal lath each side mixed 1:2 by weight, gypsum to sand aggregate.	—	—	—	$4^3/_4$[d]
	11-1.2	$3^3/_8$" × 0.055" (No. 16 carbon sheet steel gage) approved nailable[k] studs spaced 24" on center. $^5/_8$" neat gypsum wood-fibered plaster each side over $^3/_8$" rib metal lath nailed to studs with 6d common nails, 8" on center. Nails driven $1^1/_4$" and bent over.	—	—	$5^5/_8$	—
	11-1.3	4" × 0.044" (No. 18 carbon sheet steel gage) channel-shaped steel studs at 16" on center. On each side approved resilient clips pressed onto stud flange at 16" vertical spacing, $^1/_4$" pencil rods snapped into or wire tied onto outer loop of clips, metal lath wire-tied to pencil rods at 6" intervals, 1" perlite gypsum plaster, each side.	—	$7^5/_8$[d]	—	—
	11-1.4	$2^1/_2$" × 0.044" (No. 18 carbon sheet steel gage) steel studs spaced 16" on center. Wood fibered gypsum plaster mixed 1:1 by weight gypsum to sand aggregate applied on $^3/_4$- pound metal lath wire tied to studs, each side. $^3/_4$" plaster applied over each face, including finish coat.	—	—	$4^1/_4$[d]	—
12. Wood studs interior partition with plaster each side	12-1.1[l, m]	2" × 4" wood studs 16" on center with $^5/_8$" gypsum plaster on metal lath. Lath attached by 4d common nails bent over or No. 14 gage by $1^1/_4$" by $^3/_4$" crown width staples spaced 6" on center. Plaster mixed $1:1^1/_2$ for scratch coat and 1:3 for brown coat, by weight, gypsum to sand aggregate.	—	—	—	$5^1/_8$
	12-1.2[l]	2" × 4" wood studs 16" on center with metal lath and $^7/_8$" neat wood-fibered gypsum plaster each side. Lath attached by 6d common nails, 7" on center. Nails driven $1^1/_4$" and bent over.	—	—	$5^1/_2$[d]	—
	12-1.3[l]	2" × 4" wood studs 16" on center with $^3/_8$" perforated or plain gypsum lath and $^1/_2$" gypsum plaster each side. Lath nailed with $1^1/_8$" by No. 13 gage by $^{19}/_{64}$" head plasterboard blued nails, 4" on center. Plaster mixed 1:2 by weight, gypsum to sand aggregate.	—	—	—	$5^1/_4$
	12-1.4[l]	2" × 4" wood studs 16" on center with $^3/_8$" Type X gypsum lath and $^1/_2$" gypsum plaster each side. Lath nailed with $1^1/_8$" by No. 13 gage by $^{19}/_{64}$" head plasterboard blued nails, 5" on center. Plaster mixed 1:2 by weight, gypsum to sand aggregate.	—	—	—	$5^1/_4$
13. Noncombustible studs-interior partition with gypsum wallboard each side	13-1.1	0.018" (No. 25 carbon sheet steel gage) channel-shaped studs 24" on center with one full-length layer of $^5/_8$" Type X gypsum wallboard[e] applied vertically attached with 1" long No. 6 drywall screws to each stud. Screws are 8" on center around the perimeter and 12" on center on the intermediate stud. The wallboard may be applied horizontally when attached to $3^5/_8$" studs and the horizontal joints are staggered with those on the opposite side. Screws for the horizontal application shall be 8" on center at vertical edges and 12" on center at intermediate studs.	—	—	—	$2^7/_8$[d]
	13-1.2	0.018" (No. 25 carbon sheet steel gage) channel-shaped studs 25" on center with two full-length layers of $^1/_2$" Type X gypsum wallboard[e] applied vertically each side. First layer attached with 1"-long, No. 6 drywall screws, 8" on center around the perimeter and 12" on center on the intermediate stud. Second layer applied with vertical joints offset one stud space from first layer using $1^5/_8$" long, No. 6 drywall screws spaced 9" on center along vertical joints, 12" on center at intermediate studs and 24" on center along top and bottom runners.	—	—	$3^5/_8$[d]	—
	13-1.3	0.055" (No. 16 carbon sheet steel gage) approved nailable metal studs[e] 24" on center with full-length $^5/_8$" Type X gypsum wallboard[e] applied vertically and nailed 7" on center with 6d cement-coated common nails. Approved metal fastener grips used with nails at vertical butt joints along studs.	—	—	—	$4^7/_8$

FIGURE 2.13 Rated fire-resistance periods for various walls and partitions.

Conclusion

With the numerous codes, laws, and regulations that govern the design and construction of hospitals and the risks associated with failure to comply, it is important for the construction manager or builder to be very familiar with the requirements. Copies of the applicable editions of the regulations should be available to the construction staff either in hard copy or with electronic versions. Compliance with the regulations should be incorporated into the quality assurance and quality control process just as with the proper installation of the building materials. Careful attention should be paid to understanding the correct edition or year that is applicable to the specific location of the project as the states and localities are typically not enforcing the latest publication. Training for the management staff and for the workers on the requirements is also a good step in producing a quality project.

Conclusion

CHAPTER 3
Planning and Predesign

All projects are driven by a need that the hospital or system identifies. The reasons can vary widely. Some common ones include

- Recruiting a new physician or group
- Offering a new service line
- Expanding a currently successful service line
- Upgrading medical equipment/technology or an older physical plant
- Addressing the lack of adequate parking
- Building a new facility to replace an antiquated one, or to enter a new market

Projects may be in the planning phase for months or for years. It is very common for capital development efforts to start in a certain direction, or with a defined scope, and to change over time as more information is gathered, the needs change, or funding resources are obtained or reduced. There are often many iterations and the construction manager may be involved in pricing different scenarios, developing different phasing options, and helping find ways to reduce the cost. At some point when the need for the project is confirmed, the project is defined, the funding is available, the regulatory approvals are obtained, and it is approved by the appropriate board or senior management team, the project will proceed. An approval document or funding memo with clearly defined scope, cost, and time should be reviewed with the team, so the goals of the project are shared.

There are two broad classifications of projects

1. Revenue-producing projects
2. Non-revenue-producing projects

Revenue-producing projects often include the following: new services, expansion due to capacity constraints, and upgrading due to new technology. Non-revenue-producing projects include those driven by regulatory and code compliance issues, as was discussed in Chap. 2, the need for central energy and cooling plant capacity, major maintenance repairs and equipment replacements, the need for additional parking, structured or surface, and other facility or infrastructure improvements. Sometimes the need is immediately recognized, such as a failed chiller or corroded waterlines, regulatory changes, or poor inspection results. At times it may take years for a project to develop.

This chapter was written by Tom Koulouris.

One common revenue-producing project is driven by the need to alleviate capacity constraints, which serves the purpose of creating space to treat more patients and increase revenues. An example of this is an emergency department (ED) that is chronically full. Generally speaking, this "full" state is manifested in several different ways: (1) the ED is on diversion or bypass from emergency medical services (EMSs) or the local ambulance system, which means the services are taking patients to other hospitals; (2) there are excessively long door-to-door, door-to-floor, or door-to-disposition experiences (i.e., 3 hours or greater); and (3) complaints from patients or patients leaving before being seen. Sometimes, the problem is not the ED at all but downstream services, such as a shortage of ICU or medical-surgical beds, which causes a logjam of emergency patients. However this is expressed, there is often a real need for improvement, either process driven or physical plant driven. Often in growing areas, expanding the ED is necessary, but not always. Given this, how or who determines what "full" really is? How is this question defined and quantified in clear documented terms? This chapter illustrates how a process-driven approach answers these questions.

Capacity

All service lines have a certain volume of patients based on market share, population in the area served, physician loyalty, insurance contracts, and other factors. The volumes may vary during seasons and can change due to growth or decline in populations, competitive activity, quality issues, and other factors. The facility must match the capacity with the volumes. That is often a difficult balance, as the variations can be significant during different seasons or even during different times of the day. Hospitals in Florida provide a good example, where during the winter flu and "snow birds" (northerners who come to Florida for the winter) season, the volume can be extremely high, while in the summer entire wings of patient rooms may be closed due to lack of patients. Additional capacity usually means more money so hospitals do not want to overbuild, but they must be able to handle the "peak" loads. It is not unusual in some parts of Florida to see patients in the corridors in the ED in the winter.

Capacity is operationally defined as a mathematical representation of how many patients can be treated in a particular department or unit, such as the ED, in a given period of time. Generally speaking, capacity for an ED is expressed in terms of total number of patients seen per year, but more specifically it must be analyzed on a day-to-day or hour-to-hour basis. Continuing with the ED example, the capacity can be determined by utilization rates for the different treatment bays or examination spaces, such as orthopedic, obstetrics and gynecology (ob-gyn), general, trauma, etc. More commonly, it is determined by analyzing peak hour times over the preceding 3 years, regardless of disposition. One important factor is classification of acuity on the standard level 1 to 3 scale, with 1 being the most acute. (Some areas of the United States express this classification as level 1 to 5, with 5 being the least acute.) The most acute patients generally utilize more minutes or hours in the ED. Existing volume analysis can reveal the true numbers of patients currently being seen in the ED, based on the operational procedures in use. If in fact, the existing ED is full, then the volume analysis will reveal that more patients are being seen in the existing ED than what the facility was designed for, either architecturally or operationally. Most hospitals are very resourceful at treating patients, and many routinely operate their ED at a higher capacity than provided for in their original ED design. Continuous process improvement will yield significant results, but at some point a larger facility may become necessary.

The next consideration should be examining why the department or unit is experiencing greater patient volumes than the facility design anticipated. For analysis purposes, assume the downstream services or capacity are not the cause. Where are the patients coming from? Where do they live? What is the predominant demographic? All these questions and more can be answered through demographic studies and volume analysis of exiting patients.

Demographic and Market Study

A demographic study will determine the growth rate in the region, state, county, city, and also in the primary and secondary markets of the hospital in question. The data is often gathered and analyzed based on the ZIP codes of the patients. Market studies can also determine the market share as a percentage of the existing facility, and show reasonable projected increases or decreases due to growth, population shifts, and market share changes. The goal of the demographic and market study is to make realistic projections of the patient volumes that will inform decisions on the need and required size of the capital project. Increases in growth, together with a realistic determination of market share gain, can determine overall projected patient volume. Physician alignment strategies, centers of excellence, marketing, and patient satisfaction efforts are some methods for increasing patient volume.

It is not the intent of this book to explain the marketing analysis of growth and market share, but a quick review will be helpful.

Growth

Growth can be defined as naturally occurring increases in volumes due to a growing population in the area and other factors. The market study will evaluate current and anticipated conditions. This will help the team make realistic projections of the future patient volumes to inform the decisions on the need and required size of the capital project. The same holds true for areas with declining populations, such as parts of West Virginia. While declining patient volumes typically do not lead to capital projects, there may be cases where consolidation of facilities or services may cause the need for capital expenditures. An example would be the closing of an older facility in a city while adding some capacity to another building in the area to handle the volume that will be moved. In this case the net capacity would probably go down.

Market Share

Market share can be defined as the percentage of patients being seen in a facility compared to the total number of the patients in the market. The most common approach is to look at what are called the primary and secondary service areas. These are typically defined by the ZIP codes or counties for the areas that are served by the particular hospital. Primary areas are close to the facility while secondary areas would be those farther away that may also have access to other hospitals. For example in Nashville, Tennessee, the primary service area for Vanderbilt University Medical Center comprises the eight counties closer to the hospital, while the secondary service area is the 36 surrounding counties. The 2012 public disclosure documents indicate that 53 percent of inpatient discharges are from the primary service area, while 23 percent are from the secondary, with the balance from other states, such as northern Alabama and western Kentucky.

Better facilities, specialty services, quicker throughput, better physicians, physician alignment strategies, and better care are all examples of market share determinants that can bring more or fewer patients to a particular facility. Gaining market share in a competitive environment typically requires a concerted effort and sometimes a capital project to improve the facility is part of the strategy.

Volume Analysis

The demographic study (growth and market share) will show anticipated future patient volume expectations. The volume analysis involves evaluating all of the factors that may impact the number of patients that will be serviced by proposed capital project. Once realistic projections are available, the *needs analysis* can be done and planners can develop space programs, one of the first steps toward developing floor plans and preliminary budgets. Care should be taken to not overestimate the percentage increase in market share or growth as it can lead to over-building, while underbuilding can cause capacity constraints too soon after opening a new project. The goal is to *right size* the capital development to meet the projected needs over a practical timeline. A practical approach is to build a reasonably projected growth and include an easy plan to expand if the volumes significantly exceed expectations. This growth and market share should be estimated 5 to 10 years into the future from the occupancy of the new building or expansion. Of course, the farther into the future calculations and predictions are, the more risk there is to their accuracy.

The volume analysis should be conducted by the appropriate staff within the hospital, and in conjunction with the consultants who have prepared the demographic study and market study. This volume analysis will help determine the number of patients, the types of patients, and their profiles.

An important factor is the type of patient, such as pediatric, adult trauma, orthopedics, ob-gyn, general exam, etc., and level of acuity (1 to 5 for the ED example). Patient profiles will provide this type of information. In addition, they will provide an accurate picture of the patient population in terms of economic status, such as Medicare, Medicaid, commercial payer, and other payer types. This is known as the *payer mix*.

All of this data, including projected growth volume, market share volume increases, type of patients, and patient profiles, is then processed to support the next phase which is the needs analysis.

Needs Analysis

The needs analysis is the process of reviewing the data and converting it into quantities of specialized spaces. These spaces in an ED, for example, can be the numbers of general positions, orthopedic rooms, ob-gyn, trauma, pediatrics, behavioral health, etc. Also, the need for services, such as imaging, can be identified depending on the size of the ED and proximity to the radiology department. The spaces identified from the needs analysis are revenue-producing spaces. These are sometimes called key planning units. They may list all of the basic elements, such as number of operating rooms, number of patient rooms, number/type of imaging modalities, etc.

A chart showing key planning units for a new hospital is shown in Fig. 3.1:.

Another approach is called *SMART goals*. These are discussed in the next section.

	COMBINED OPERATIONAL CAPACITY	EXISTING UTILIZATION (2003)	2011 RECOMMENDED CAPACITY	TARGET UTILIZATION	YEAR 3 (2009) UTILIZATION	YEAR 5 (2011) UTILIZATION	Increase (Decrease) in Capacity	NOTES
ACUTE/SUBACUTE BEDS	336	62%	221	*80%*	*80%*	*85%*	(115)	
MED/SURG & TELE	163	72%	156	*85%*	*90%*	*95%*	(7)	
ICU/CCU	25	63%	28	*75%*	*68%*	*72%*	3	Not eqipping 6 rooms initially; Utilization based on 28 total beds
POSTPARTUM	0		16	*70%*	*57%*	*61%*	16	On med/surg floor - space can be cross-utilized
ANTEPARTUM/GYN	26	13%	9	*75%*	*12%*	*12%*	(17)	5 beds for GYN and 4 beds for Antepartum
ANTE/POSTPARTUM	10		0	*70%*	-	-	(10)	
LDRP	18	44%	0	*60%*	-	-	(18)	Moving to LDR Model in new hospital
NICU (LEVEL II & III)	0	na	12	*70%*	*60%*	*64%*	12	
SNF	40	54%	0	*85%*	-	-		
REHAB	30	37%	0	*85%*	-	-	(30)	
PSYCH/GEROPSYCH	24	88%	0	*85%*	-	-	(24)	
OBSTETRICAL SERVICES							0	
LDRs	0	-	7	*40%*	*34%*	*36%*	7	
ANTEPARTUM TESTING/OBSERVATION	1		4				3	Stretcher bays/cubicles within LDR or Postpartum Unit
NORMAL NEWBORN NURSERY	25		16	*70%*			(9)	Size equivalent to Postpartum - max 16 bassinets per Nursery by code
SPECIAL CARE NURSERY (Level II)	4	91%	6	*80%*	*69%*	*72%*	2	Level II NICU within Level III NICU Dep't.
C-SECTION ROOMS	2		2	*50%*			0	
C-Section Pre/Recovery	0		2				2	
SURGICAL SERVICES	15	57%	8	*75%*	*88%*	*91%*	(7)	
GENERAL OR	4		2	*75%*			(2)	2 Additional O.R.s to be shelled
MAJOR OR	8		4	*70%*			(4)	
OPEN HEART OR	1		2	*70%*			1	
CYSTO OR (Dedicated)	2		0	*75%*			(2)	Accommodate w/in General O.R.
Pre-OP/INPATIENT HOLDING	8		4	-			(4)	
OUTPATIENT PREP/RECOVERY	36		30	-			(6)	
PACU	15		15	-			0	1.5 per O.R.
ENDOSCOPY							0	
ENDOSCOPY ROOMS	4	240%	3	*80%*	*70%*	*74%*	(1)	Procedure Rooms; Could use for pain procedure or Bronch
BRONCHOSCOPY (DEDICATED)	1	-		*80%*			(1)	
ENDO STAGING/RECOVERY POSITIONS	10	-		-			(10)	Included in Day Surgery/OP Prep-Recovery
EMERGENCY SERVICES	30	74%	32	*75%*	*73%*	*76%*	2	Includes Observation/Holding Unit
TRAUMA/EMERGENT POSITIONS	9	31%	5	*75%*	*67%*	*70%*	(4)	2 'Trauma' Room with remainder as 'Emergent'
URGENT CARE POSITIONS	13	79%	17	*75%*	*73%*	*77%*	4	
NON-URGENT CARE POSITIONS	8	76%	10	*75%*	*73%*	*77%*	2	
OBSERVATION/HOLDING	0		8	*75%*	*69%*	*73%*	8	Separate Unit from Emergency Department
DIAGNOSTIC IMAGING							0	
GENERAL RAD/TOMO	3			*85%*			(3)	
R/F	5			*75%*			(5)	
TOTAL RAD/FLOURO	8	51%	5	*80%*	*99%*	*103%*	(3)	
ANGIO/SPECIALS	1	24%	1	*75%*	*28%*	*29%*	0	
CT	2	81%	2	*80%*	*94%*	*101%*	0	
MRI	0	47%	1	*75%*	*53%*	*58%*	1	
MAMMOGRAPHY	2	71%	2	*75%*	*83%*	*86%*	0	
MAMMO - STEREOTACTIC	0		1	*75%*	*0%*	*0%*	1	
ULTRASOUND	4	44%	3	*75%*	*68%*	*71%*	(1)	
BONE DENSITY	0		1	*70%*				
NUCLEAR MEDICINE	3	63%	2	*75%*	*87%*	*92%*	(1)	
HEART CENTER							0	
CATH LAB	2	41%	2	*75%*	*48%*	*51%*	0	
CATH PREP/RECOVERY		-	8					Cath/Specials Prep/Recovery w/in Radiology_HeartCenter
EKG/STRESS TEST			3	*75%*			3	
ECHO			3	*75%*			3	

FIGURE 3.1 Example of key planning units for replacement hospital.

SMART Goals

SMART is an acronym for specific, measureable, attainable, relevant, and timely. The term SMART goals describes the process for defining the collective spaces determined through the needs analysis. It is the SMART goal spaces that ultimately will be converted into the functional space program discussed in later chapters.

Specific

This SMART goal identifies the spaces determined from the needs analysis. They are specific to the extent they identify the needs based upon the demographic and market studies. They are revenue producing, and will identify the core purpose of the project (e.g. OR, ICU, imaging modality).

They are documentable in terms of the specific spaces that have been identified, and were developed through analysis. These are the spaces that are truly "what's right" and not a contest of "who's right," as can often happen without analytical measures. It is this process of determining the specifics that hospital operators must utilize to avoid endless committees, meetings, and protracted discussions about what should be built. These specifics are studied, documented, and analyzed so the project is "right sized." All too often when hospitals get caught up in "who's right", the scope of the project becomes too large, resulting in overbuilding. If the scope is not comprehensive enough or the projections are not practical, then the project can be "underbuilt." Both conditions have serious negative implications.

In an "overbuilt," or overscoped, condition, the size of the project based on the numbers of spaces ultimately costs too much to continue the project. Often in an overbuilt situation, new central energy plants, garages, major site work, major renovations, and other costly components become part of the project in order to support the scope determined for revenue-producing spaces. Many times these are necessary add-ons but, nonetheless, they add costs.

In an underbuilt condition, the project is not properly sized and the volumes soon exceed the capacity of the project. This has negative implications simply because the goal of the project was not met and further expansion, or operational "work around," is required. Adding more space shortly after building a project costs far more than what a "right-sized facility" would have cost in the first place.

The SMART goals of the project, properly sized, will also provide a clear basis for the *pro forma*. In an overbuilt situation, a positive pro forma most likely will never be attained, resulting in accountability issues. A clear understanding by the team of how this process works to correctly size health care facilities can help save time and provide better outcomes.

Measureable

This SMART goal has helped properly determine the number of spaces through the *needs analysis*. It is measureable in terms of the method in which the spaces were determined. Demographics, market studies, competitor analysis, physician recruitment strategies, and existing volumes are included in the data for which the numbers of spaces, over a specific period of time, were measured. The financial operating results of the project, after occupancy, can be measured against the original pro forma, which is how the project is measured financially. There can be minimal debate on the numbers of SMART goals since they were developed using measurable data.

Attainable

This SMART goal describes the project in terms of the hospital's ability to reach or attain the specified project goals given the volumes determined through the demographic and market study. The hospital must have sufficient resources, drive, determination, and skill sets within its leadership team to attain this goal. Often, there are challenges on a project related to the cost additional requirements of regulatory agencies, staffing requirements, disruption of services, and political issues. The hospital leadership must be prepared to overcome these issues to attain the originally defined project goals.

Relevant

This SMART goal refers to the overall needs of the hospital. In terms of priorities, the question to be answered is: Are these goals worthy of expending hospital resources of employee energy, political capital, and financial resources? Are the goals relevant in terms of what is happening in the market place? Are these goals at the top of the hospital's priority list, and are they consistent with their long-term strategy? The project must be relevant to the organizational mission statement.

Timely

This SMART goal refers to the overall duration to achieve completion of the project. Will the SMART goals still be "SMART" at the time of project occupancy?

The length of time to deliver the project will affect the pro forma in terms of when revenue will start accruing. Will changes in the health care reimbursement and the managed care landscape affect the pro forma during the first 5 years after occupancy?

All of the SMART goals must be viewed favorably for a hospital operator to recommend moving forward with the project and deploying sometimes scarce capital. If any of the SMART goals are not met, then serious consideration or further analysis of the project may be required, as well as reevaluating the timing.

Provided all the SMART goals are achievable, then the next phase of the project begins with the Project Definition Study (PDS), which includes the functional and space program along with the facility-related due diligence efforts. It is this phase of the project that will begin to transform the planning work into actual measureable scope. The scope can then be accurately budgeted and scheduled to provide further clarity on the potential costs (capital investment), benefits, and returns for the project.

Project Definition Study

As a building is based upon its foundation, so, too, the project plan is based upon the PDS. The PDS becomes the foundation for the entire project from approval to completion. It is ground zero for questions, and is a source for clarification about what was and was not approved, and what was and was not budgeted. Many projects have been delayed, underfunded, and over-designed due to the lack of comprehensive documentation that clearly detailed the goals, scope, and budget. The importance of the PDS cannot be overstressed. The PDS should be used as a comprehensive tool for every single project regardless of the size and/or scope of the project. Once the PDS is understood by all the stakeholders, then "abbreviated" studies may be performed for smaller projects. This will be discussed further later in this chapter.

The PDS process is as follows

1. Iterative
2. Dynamic
3. Predictable
4. Unchanging
5. Documentable
6. Budgeted
7. Approved

Iterative

The PDS is a do–check–act relationship between the owner and design team. After the SMART goals have been delivered to the design team, it is up to them to begin the transformation of the SMART goals into human spaces. These spaces cannot be developed without input from the owner. The architect's response, through drawings and models, is the "do," which is then reviewed with the owner for the "check" phase, and the design team then completes the schematic documents as the "act" phase. This iterative process may occur multiple times. The iterative phase of the PDS occurs somewhat in Stage 1, but mostly in Stage 2.

The PDS typically comprises three separate phases with distinct deliverables. These are Stage 1, Stage 2, and Stage 3.

Stage 1

The Stage 1 deliverable sets forth the general scope of the project in broad terms. It includes a space program, overall schedule, evaluation of unique issues, and key planning units. Its primary objective is to provide enough information for an owner's construction manager to prepare a "desktop" project cost estimate. This estimate is then used to decide quickly if it makes sense to continue pursuing the project.

Stage 2

The Stage 2 deliverables define the project with substantially more specifics in all disciplines, including architecture, mechanical, electrical, plumbing, fire protection, structural, civil, medical equipment, and information systems. In this stage, the single-line drawings prove the space program. Adjustments in design, space, general configuration, and scope are made to provide a functional and efficient project. This is the iterative portion of the PDS discussed earlier.

Stage 3

This last stage of the PDS is confirmation of all the studies, analysis, agreements, budgets, schedules, and scope that have been developed. It is this stage 3 document that is budgeted and becomes the official project document to define scope, schedule, and budget, and ties to the final business case and pro forma.

The formats should be alike for every project, every time, so that anyone familiar with the PDS process can review the package and know immediately where to look for certain pieces of information. Each section should be tabbed separately for easy reference and review.

Dynamic

The PDS is dynamic, in that as new and better information is delivered to the design by other design team members and stakeholders together with the owner, the PDS becomes a living document of the development of the project. It is dynamic in Stages 1 and 2. It is not dynamic in Stage 3. Once the final stage has been completed, the scope and cost should not be changed, as it will require another evaluation of the return on the investment, the strategic goals, and the "right sizing."

Predictable

The machinations of the PDS usually follow a very similar pattern from project to project. This is true whether the design team has followed this process many times or is just following it for the first time. Once the project managers and other stakeholders for the owner learn the predictable nature of the PDS, a comfort level PDS begins to occur. This is the key in developing speed to market for the project, which is discussed later in the text.

Unchanging

The PDS is unchanging as far as the format of the process. It is the same format regardless of project size, scope, or complexity. The key reason is that, within a department, every project manager or senior administrator can review a PDS and know exactly where to find certain key pieces of information. The PDS is a standardized process, fixed in concept and format.

Documentable

The three unique stages of the PDS provide a living document that is historical in its value to document from SMART goals to final scope, schedule, and budget. All of the stakeholders are provided copies of each stage. It is this documentable element that is the key to "buy in" and accountability. Most people that have been involved with hospital development have heard "I didn't know that was part of (or not part of) the project" coming from one of the stakeholders. While the design team is skilled in the design and construction process and vernacular, the hospital administrators and staff are skilled in health care. As a result, there can be an inherent communication gap that exists between the different professions. When the design team is using their unique industry language, many hospital administrators may not follow to the same degree. Therefore, the documentable element of the PDS helps to bridge the communication gap. Other communication tools used are PDS review meetings at each of the three stages, building information modeling, and full-scale mock-ups of key spaces, such as operating rooms and patient rooms.

Budgeted

The PDS must be budgeted throughout the different stages. The Stage 1 budget provides a simple "go/no go" decision on the project cost. The Stage 2 budget is a check to ascertain if the project is still in a "go–no-go" mode and often times picks up additional scope deviations from the Stage 1 pricing. The Stage 3 budget is prepared with input from experienced construction managers and other team members to provide a high degree of accuracy.

Approved

The Stage 3 deliverables are an accurate project budget, a design and construction schedule, the pro forma developed with the project budget, and the business plan. Together, these comprise a sound, accountable approval package. All of the critical elements are clear and the roles and responsibilities of the owner, designer, and builder are well defined.

The PDS components are specific and given here

1. Project information sheet
2. Executive summary
3. SMART goals
4. Space program
5. Single-line drawings
6. Photographs
7. Architectural narrative
8. Interior design narrative
9. Engineering narratives
10. Civil narrative
11. Structural narrative
12. Mechanical narrative
13. Plumbing narrative
14. Fire protection narrative
15. Electrical narrative
16. Information and telecommunication systems narrative
17. Medical equipment and furnishing list

Project Information Sheet

The project information sheet contains a contact list of every stakeholder for the project, such as

Hospital CEO—Chief Executive Officer, COO—Chief Operating Officer, CFO—Chief Financial Officer, CNO—Chief Nursing Officer, and CIO—Chief Information Officer

Vice president of design and construction or capital projects

Facility construction or project manager

Clinical directors involved with the project

Project planner

Project architect and principal

Interior designer

Civil engineer

Structural engineer

Mechanical engineer

Plumbing and fire protection engineer

Electrical engineer

Information systems engineer

Construction manager or general contractor

Medical equipment planner

Geotechnical consultant

Parking consultant

Surveyor

Testing lab

Others

Each contact should contain name, company, title, phone, fax, cell, e-mail information, and/or website. This contact sheet should also identify each individual with their firm or role, such as owner, architect, contractor, and consultant. Further, the contractual relationships between firms should be identified to provide clear direction for communications, payments, and legal responsibilities.

Executive Summary

The executive summary is typically written by the project principal with the architectural firm. It is the architect's understanding of the scope of the project, and the SMART goals. It is written after each of the PDS Stages 1, 2, and 3.

The executive summary should discuss the SMART goals and their implications for the program and design, as well as addressing certain any unique construction issues. This summary is not an exhaustive listing of all aspects of the project but more an overview within a page or two describing the project.

The executive summary should discuss design alternatives or options and methods of obtaining contractor pricing to support the owner in making financial decisions.

SMART Goals

The SMART goals should be reproduced and placed in the PDS. This serves as a single source of the information for which the project is founded.

The SMART goals section should list in detail the supporting documents and data used in producing the program and concept. The SMART goals should be listed in tabular form.

Space Program

The space program is a detailed room-by-room study of all the spaces necessary to achieve the SMART goals. The space program should be departmentalized and include grossing factors for circulation, such as corridors/stairs, mechanical/electrical spaces, canopies/walkways, and other common elements. The space program must consider all spaces and must include SMART goals, spaces required by the Guidelines for the Design and Construction of Health Care Facilities, or other applicable local and state codes. All too often, some of these requirements are overlooked. That is one of the big discrepancies between what the architects might show as the building gross square footage (BGSF) and what the contractors develop during cost-estimating phases. Architects should understand that the contractor's square footage is inclusive of all building components and should not make the error of believing the contractor square footage (SF) pricing includes those items without them being shown on the space program. A good example of this is the use of overhangs and cantilevered canopies, which may not be included in space programs, but nonetheless require money to be built and are part of the constructible square footage.

The space program should be summarized on a single page ultimately showing the departmental and building gross square footages. This is of the utmost importance, so the square footage is stated in clear terms. There should be no room for interpretation by owners, architects, contractors, or other stakeholders as to the exact square footage of the project. The BGSF should include all spaces needed in the building, regardless of interior or exterior usage, whether occupied or not occupied, grade level or above-grade level. The space program should consider all affected areas needing renovation and their downstream effects. For example, if a certain department must relocate to make space available for the project, then the relocated space must be shown in the space program as "renovated" SF. The space program should consider areas needing refurbished finishes. An example would be a new addition being added to the facility. The construction drawings show the interior finishes for the new addition, but may fail to extend the new finishes into the existing facility. This can lead to a glaring problem when you walk into the new project area and, upon entering the existing building, find there is a stark change in finishes without a natural breakpoint. Finishes should be extended into the existing facility at all entry points to natural breakpoints, such as cross-corridor doors or corridor intersections.

The space program is inclusive of all space. For example, shell space must be identified in the space program. Another example is that there should not be spaces without consideration. There can be no spaces left as "unidentified." Unused spaces at a minimum must be labeled "storage" with the appropriate code required for fire-rated construction. The space program should identify any design options that should be priced or considered. These options must show on the single-spaced program summary (Fig. 3.2).

Single-Line Drawings

Single-line drawings typically include the following types of illustrations:

- Existing site and floor plans of the facility for each floor
- Floor plans of the new facility or addition and renovation
- Color-coded plans identifying the SMART goals
- Diagrams illustrating areas of new construction, major and minor renovation, and those spaces that are "existing to remain"
- Diagrams illustrating project phasing
- Plans per discipline (civil, architectural, structural, mechanical, electrical, plumbing, information technology) depending on the scope of work

Department Name	Hospital A					Hospital B (With interventional lab, stereotactic biopsy, and outpatient rehab)a					Hospital B (Without interventional lab, stereotactic biopsy, and outpatient rehab)				
	Key Planning Unit	DGSF/ KPU	Current			Key Planning Unit	DGSF/ KPU	Current			Key Planning Unit	DGSF/ KPU	Current		
			DNSF	Gross Factor	DGSF			DNSF	Gross Factor	DGSF			DNSF	Gross Factor	DGSF
Administrative Services	Beds 42	178	6,560	1.14	7,480	Beds 35	284	8,139	1.22	9,930	Beds 35	284	8,139	1.22	9,930
Public Areas	Beds 42	65	2,345	1.17	2,745	Beds 35	73	2,035	1.25	2,540	Beds 35	73	2,035	1.25	2,540
All Other Administrative Services															
Administration	Staff 4	299	1,015	1.18	1,195	Staff 4	528	1,690	1.25	2,110	Staff 4	528	1,690	1.25	2,110
Business Office	Staff 17	116	1,790	1.10	1,970	Staff 11	114	1,000	1.25	1,250	Staff 11	114	1,000	1.25	1,250
Health Information Management	Staff 9	131	935	1.26	1,175	Staff 7	129	749	1.20	900	Staff 7	129	749	1.20	900
Human Resources/PR/Decision Support	Staff 3	155	375	1.24	485	Staff 2	200	320	1.25	400	Staff 2	200	320	1.25	400
Information Technology/ Communication	Staff 42	20	790	1.04	825	Staff 35	37	1,195	1.10	1,310	Staff 35	37	1,195	1.10	1,310
Medical Affairs	Staff 8	42	310	1.10	340	Staff 35	13	365	1.20	440	Staff 35	13	365	1.20	440
Nurse Administration/Quality	Staff 2	150	275	1.09	300	Staff 9	109	785	1.25	980	Staff 9	109	785	1.25	980
Diagnostic & Treatment Services			25,432	1.49	38,005			25,785	1.47	37,810			22,599	1.49	33,730
Emergency Department	Treatment Rooms 13	703	6,315	1.45	9,135	Treatment Rooms 12	660	5,107	1.55	7,920	Treatment Rooms 12	660	5,107	1.55	7,920
Diagnostic Imaging/Women's	Imaging Rooms 10	872	6,037	1.44	8,715	Imaging Rooms 10	973	6,713	1.45	9,730	Imaging Rooms 10	820	5,653	1.45	8,200
Interventional Services	OR + Endo + C-Sect + Cath 4	3,040	7,680	1.58	12,160	OR + Endo 6	2,935	11,839	1.49	17,610	OR + Endo 6	2,935	11,839	1.49	17,610
Surgical Services	OR & Procedure Rooms 4	2,504	5,645	1.77	10,015	OR & Procedure Rooms 6	2,560	9,884	1.55	15,360	OR & Procedure Rooms 6	2,560	9,884	1.55	15,360
Surgery	Operating Rooms 2	2,668	2,900	1.84	5,335	Operating Rooms 4	2,088	5,566	1.50	8,350	Operating Rooms 4	2,088	5,566	1.50	8,350
Endoscopy	Endo Rooms 2	590	740	1.59	1,180	Procedure Rooms 2	580	775	1.50	1,160	Procedure Rooms 2	580	775	1.50	1,160
Pre-Op/Recovery	Recovery Rooms 12	292	2,005	1.75	3,500	Recovery Rooms 16	366	3,543	1.65	5,850	Recovery Rooms 16	366	3,543	1.65	5,850
Central Sterile Processing	CSP: ORs + C-section 3	468	1,330	1.06	1,405	ORs 4	563	1,955	1.15	2,250	ORs 4	563	1,955	1.15	2,250
Rehabilitation Therapy						Treatment Stations 14	182	2,126	1.20	2,550	Treatment Stations 0	#DIV/0!	0	1.20	0

Figure 3.2 Example of space program for new hospitals.

Department Name	Hospital A						Hospital B (With interventional lab, stereotactic biopsy, and outpatient rehab)							Hospital B (Without interventional lab, stereotactic biopsy, and outpatient rehab)						
	Key Planning Unit	DGSF/ KPU	Current				Key Planning Unit		DGSF/ KPU	Current				Key Planning Unit		DGSF/ KPU	Current			
			DNSF	Gross Factor	DGSF					DNSF	Gross Factor	DGSF					DNSF	Gross Factor	DGSF	
Inpatient Services	Beds + NICU + LDR/P 42	610	17,230	1.49	25,620		Beds+NICU+LDR/P	35	571	13,328	1.50	19,990		Beds+NICU+LDR/P	35	571	13,328	1.50	19,990	
Intensive Care Unit/Step-down	Patient Beds 6	698	2,885	1.45	4,185		Patient Beds	4	1,288	3,433	1.50	5,150		Patient Beds	4	1,288	3,433	1.50	5,150	
Medical Surgical Unit and Observation Unit	Patient Beds 28	501	9,210	1.52	14,025		Patient Beds	31	479	9,895	1.50	14,840		Patient Beds	31	479	9,895	1.50	14,840	
Support Services Group	Beds 42	248	9,630	1.08	10,395		Beds	35	320	9,604	1.17	11,210		Beds	35	320	9,604	1.17	11,210	
Clinical Lab	Beds 42	58	2,260	1.08	2,445		Beds	35	80	2,446	1.15	2,810		Beds	35	80	2,446	1.15	2,810	
Pharmacy	Beds 42	37	1,395	1.12	1,560		Beds	35	42	1,283	1.15	1,480		Beds	35	42	1,283	1.15	1,480	
Dietary Services	Beds 42	74	2,980	1.04	3,110		Beds	35	91	3,030	1.05	3,180		Beds	35	91	3,030	1.05	3,180	
Environment Service & Linen	Beds 42	20	820	1.05	860		Beds	35	24	775	1.10	850		Beds	35	24	775	1.10	850	
Clinical Engineer/Facility Engineer	Beds 42	21	765	1.16	890		Beds	35	29	910	1.10	1,000		Beds	35	29	910	1.10	1,000	
Materials Management	Beds 42	38	1,165	1.08	1,260		Beds	35	37	1,160	1.10	1,280		Beds	35	37	1,160	1.10	1,280	
Central Lockers and Lounges	Beds 42	13	245	1	270		Beds	35	17	554	1.10	610		Beds	35	17	554	1.10	610	
Total Department Square Footage			**58,852**		**81,500**					**56,856**		**78,940**					**53,670**		**74,860**	
Int Mech/Elect/Circ/Elev/Stairs/ Ext Walls	Internal Mech/Elect/Circ/Elev/ Stairs/Ext Walls		% of DGSF	11.2%	9,108		Internal Mech/Elect/Circ/Elev/Stairs/ Ext Walls			% of DGSF	15.0%	11,850		Internal Mech/Elect/Circ/Elev/Stairs/ Ext Walls			% of DGSF	15.0%	11,230	
Penthouse	Penthouse			0.0%	0		Penthouse				0.0%	0		Penthouse				0.0%	0	
Canopies and Soffits@50%	Canopies and Soffits@50%			2.8%	2,268		Canopies and Soffits@50%				2.0%	1,580		Canopies and Soffits@50%				2.0%	1,500	
Central Energy Plant	Central Energy Plant			1.9%	1,580		Central Energy Plant				2.0%	1,580		Central Energy Plant				2.0%	1,500	
Total Building Gross Square Feet (BGSF)	**Total Building Gross Square Feet (BGSF)**			**15.9%**	**94,456**		**Total Building Gross Square Feet (BGSF)**				**19.0%**	**93,950**		**Total Building Gross Square Feet (BGSF)**				**19.0%**	**89,090**	
Area per bed	**Area per bed**				**2,249**		**Area per bed**					**2,684**		**Area per bed**					**2,545**	

FIGURE 3.2 *(Continued)*

Photographs

Photographs are very helpful in explaining existing conditions. Keep in mind that the senior administration or board of directors often may not be familiar with the facility under consideration for capital expansion. This is more likely the case with systems owning many facilities.

A good group of photographs should include

- Aerials.
- Elevations from a distance.
- Potential site photos taken from rooftops looking below.
- Potential problem areas.
- Central energy plants and the equipment, if part of the scope.
- Departments at full capacity, such as patients in hallways, corridors, full waiting areas, etc. Be careful to make sure the photographs taken meet HIPPA privacy requirements.
- Existing site with computer-generated new facility imported into the frame.

Architectural Narrative

The architectural narrative is a detailed discussion on the architecture of the building or renovation. It is in this narrative that the architect should put forth its design considerations, design intent, patient care approach, and overall goal of the design apart from SMART goals. The architectural narrative serves several purposes

- Serves as information for conceptual estimating
- Provides a framework for scope of the project and baseline
- Communicates architectural design intent to other stakeholders
- Provides a detailed overview of the project
- If a stakeholder wanted to read only a portion of the PDS, then the architectural narrative and executive summary would serve that purpose

The document should discuss building types, code criteria for all agencies having jurisdiction (AHJ), site constraints, and complexities due to renovations, logistics, constructability issues, and owner concerns. The architectural narrative also discusses size, height, width, and length in concept. The numbers of floors and other out buildings, such as central energy plants (CEP), parking structures, canopies, gardens, and certain site amenities, also need to be included.

The finish of the building must also be discussed, including exterior openings, windows, doors, storefronts, and facades of all types, roofing, and fenestrations.

Interior Design Narrative

The interior design narrative should include an overview of the concept and an approach to the finishes. Typical wall and floor finishes should be identified, as well as any unique features, such as wood paneling in lobbies or board rooms. The general location of the different types of finishes should be included, so the construction manager can provide conceptual pricing. The finishes should be functional, long lasting, and easy to clean. The overarching goal of the interior design narrative is to not specify colors and finishes, but more to describe the level and coverage of finishes, so that accurate budgets may be prepared.

Engineering Narratives

The following discussions of various engineering disciplines is not intended to be an exhaustive discussion of those disciplines, but more of an introduction into the reasons and importance of including these narratives in the due diligence process. This section is used to provide general information for the owner to understand the scope and for the contractor to provide preliminary pricing, not specific engineering technicalities. A suggested scope sheet is included in each of the following sections. An example of an engineering narrative is included in the Appendix.

Civil Narrative

The civil engineer will be required to produce a narrative for all projects that add square footage to the ground level of a facility, or for projects where systems, such as waste or storm water, water supply, utilities, parking, landscaping, transportation, etc., are affected. The engineer must evaluate the impact of the project on all of these and explain any required supporting systems, both on and off the campus. In addition, it should identify all required agency review requirements, including time impacts of these processes. Many times, the most costly impact on the project might be an item, such as replacing a forced sewer main in the street, that is not owned by the facility. Many municipalities will require the owner to invest in the upgrades required to their systems in order to support the project. These kinds of "unknowns," if not identified in the due diligence, can significantly impact a project budget.

The civil engineer must also look at required water retention on the campus and how the project may be impacted by flooding and/or snow accumulation. In areas where water conservation and control are critical, these issues must be addressed and budgeted. Other issues that must also be addressed include seismic considerations, coastal issues, wetlands, endangered wildlife, etc.

The civil engineer must request all required testing of existing systems in order to confirm capacities. The engineer must also identify all existing systems that will require relocation in order to build the proposed building. Typically, on large projects where building mass is added, a topographical survey must be completed to identify utilities. Sometimes, below-grade investigation such as infrared or magnetic testing must be conducted to verify exact locations of utilities, when critical.

The engineer must also look at parking as it relates to the project. On large campus projects, parking can be one of the most overlooked requirements. Not only must parking requirements be addressed after the project is complete, but interim parking needs during construction must be evaluated and included in the budget. Items, such as off-site parking, can carry heavy price tags that must be factored into the project. Larger projects may also require a traffic study that includes the impact of added traffic caused by the project on the surrounding streets and intersections.

Finally, the civil engineer should coordinate with the other design trades and identify locations of utility services for supporting the project. These include sewer and storm water, domestic water, fire protection water, gas services, electrical services, telecommunications services, etc. All of these must be identified and located, so that they may be properly budgeted. On large projects with substantial site work required, the engineer should provide drawings of the existing conditions and of the proposed locations of new services.

All of these issues must be addressed as required for all projects. The civil narrative should include a summary check list of the required items discussed in the due diligence report.

Structural Narrative

On projects where structural elements are required, the structural engineer should identify all requirements of the systems needed. This can range from confirming that existing floors can support the weight of proposed equipment to specifying the structural system(s) that are to be budgeted for a project. Due to the regional volatility in the steel and concrete markets the engineer may need to provide requirements for different types of systems for cost and time budgeting comparisons. The engineer must identify all required testing in order to determine the proper structural systems to be used. For instance, soil borings and testing must be completed in order to determine the appropriate foundation and construction methodology. The cost of a slab on-grade project compared to one that requires piles is significantly lower. For example, finding out after the budgeting process that piles are required when spread footings were budgeted can impact a project's viability.

The structural engineer must also work with the architect to develop the required spacing of columns, roof construction type, sheer wall locations, etc. In areas where seismic design is required by local codes, the engineer should provide guidelines to meet the seismic needs. Also, in hurricane and tornado prone areas, wind loading must be reviewed and systems designed to meet those requirements. The engineer must also review existing structures and how they relate to current codes related to seismic or wind load needs. The engineer may also be needed to evaluate existing conditions, such as: Can the floor support the weight of a new MRI or chiller? Can an existing 30-year-old building that is stressed to go vertical for additional floors meet the present codes for wind loading and seismic occurrence? These types of issues can all carry significant added cost if not identified during the due diligence phase.

The narrative should include a check list summary of the structural requirements of the project.

Mechanical Narrative

Hospitals are very technologically advanced in heating, ventilation, and air-conditioning (HVAC) systems. Many state codes, the Guidelines, the ASHRAE (American Society of Heating, Refrigeration and Air-Conditioning Engineers) standards, and high-quality patient-focused initiatives will dictate the level and types of HVAC systems required.

Renovation projects of health care facilities are particularly difficult, especially for HVAC. The reasons are many and include such issues as

- Existing systems and proposed tie-ins
- Reduced capacity of older equipment
- Poorly maintained cooling coils
- Fibrous lined ductwork
- Existing ductwork that is broken and open to the ceiling cavity above
- Ductwork that has be "tapped" to provide additional cooling not originally designed or specified

When mechanical engineers are preparing their narrative based on the PDS, extensive research must be undertaken. Testing of existing mechanical equipment is of paramount importance to verify current capacity, performance, and potential problems. Nothing is more frustrating to an owner than occupying a new project and later finding out that new air handling units or chillers are required because the existing equipment was not capable of meeting the new loads. The narrative should provide an overview of the proposed systems, the design approach for energy conservation and infection control, types of chillers and boilers, and any unique issues on the project.

Plumbing Narrative

The plumbing narrative discusses type and quantity of plumbing fixtures, piping types, and systems. Important to the budgeting process is information on how and where to connect sanitary service, size of sanitary, and where increased service size in either water service or sanitary service will be required. Many other components of the plumbing narrative can have significant impacts on projects including forced mains, lift stations, and grease traps. All too often though, especially in code or infrastructure upgrades, "wet" lines above electrical panels or sensitive diagnostic equipment will be found to need relocation. This can be a substantial budget item, especially when attempting to gain access to plumbing lines that are at the uppermost reaches of above ceiling spaces.

Fire Protection Narrative

The fire protection system in a hospital is a critical life safety component. According to the NFPA (National Fire Protection Agency), the fire protection system, above all else, saves lives in the event of fire. Moreover, the fire protection system must synchronize with the fire alarm system (discussed later) including notification of activation, zone identification, and notice to local fire departments.

Electrical Narrative

The electrical, along with the mechanical, are probably the most important narratives. Together these two systems often make up 30 to 40 percent of the construction budget. Extensive research of the existing systems is required on additions and renovations. The constructability of many health care renovation projects depends on when, and how, electrical service can be interrupted for tie-ins to complete the new work.

Within the electrical narrative, the size of service, gear upgrades, code upgrades, grounding, lightning protection, and fire alarm systems all have major budget implications. Special consideration should be given for the power requirements and consumption for medical diagnostic equipment. These pieces of equipment are sophisticated and often require specialized electrical gear and other devices. Large expansion projects, such as new wings, either horizontal or vertical, will generally require new electrical rooms, or central energy plant upgrades, relocations, or other major work. Work, such as this, is often termed an *enabling project* since the work is required before the major portion of a new expansion can be built. Many times these enabling projects, while embedded into the larger expansion project, can be a separate project unto themselves.

In older facilities undergoing renovation, the work often requires the removal of existing ceiling components, such as dropped or lay-in ceilings and drywall or plaster ceilings. In many cases, these above ceiling spaces, over time, have been filled with an enormous amount of cables and wires from previous systems, past renovation projects, and other types of upgrades and additions of new systems. Once the ceilings are removed, much of this "loose" cabling may simply fall out of the corridor ceilings. Obviously, this cabling must be supported, "wrung out"—that is, researching each cable to identify its source, then properly dealt with either by removal or supported by appropriate methods.

Information and Telecommunication Systems Narrative

It is being projected that the cost of information systems (IS) and related health information technologies (HITs) will likely surpass the actual cost for the bricks and mortar of a new health care facility. With that said, and with the current efforts to lower the delivery cost of care, IS and HIT have never been more important to a health care facility than now. Government regulations now are requiring providers to submit reimbursement claims through electronic means, which soon will be the only accepted method for receiving government reimbursement.

Information systems, HITs, telecommunications, handheld devices, bedside charting and reporting, diagnostics, and even apps on an IPhone are now integral components of new hospital designs. Design and operational considerations are integral to the delivery of care, the built environment, and to how care givers will deliver care and report the history, and how providers will be reimbursed for the care of the patient. The IS systems and their interrelation with the operations of the hospital are strategic in nature and have widespread repercussions. Electronic medical records (EMRs) impact many facets of the business and facilities, so these systems must be planned and budgeted in all projects. Generally, EMR includes patient admissions, patient health records, patient discharge and collection, materiel purchasing, supplies inventories, labor costs, and many metrics that are being measured in efforts to improve the patient care quality experience.

Medical Equipment and Furnishing List

All too often, the equipment is considered the "owner's" responsibility. Nothing could be further from the truth. Often termed owner-furnished equipment (OFE), there is a great deal of coordination involved with this portion of the project. OFE may include a wide range of items, ranging from medical equipment to patient beds, and from televisions to office furniture and audio/visual systems. Many factors must be considered, such as who is responsible for the receipt and delivery of the equipment, when it gets delivered, where it gets delivered, who un-crates, installs, and tests it, and who is responsible for security. Every one of these items can have constructor cost impacts.

The equipment and furnishings narrative should include a responsibility matrix that clearly identifies the responsible entity for handling the receipt and delivery of the equipment, when it gets delivered, where it gets delivered, uncrated, installed, or set in place, and when it gets connected, calibrated, tested, and commissioned. Generally, there are three parties to the equipment matrix: the owner, the vendor, and the contractor. Each of the three has various responsibilities with regards to answering those tasks listed above (Fig. 3.3).

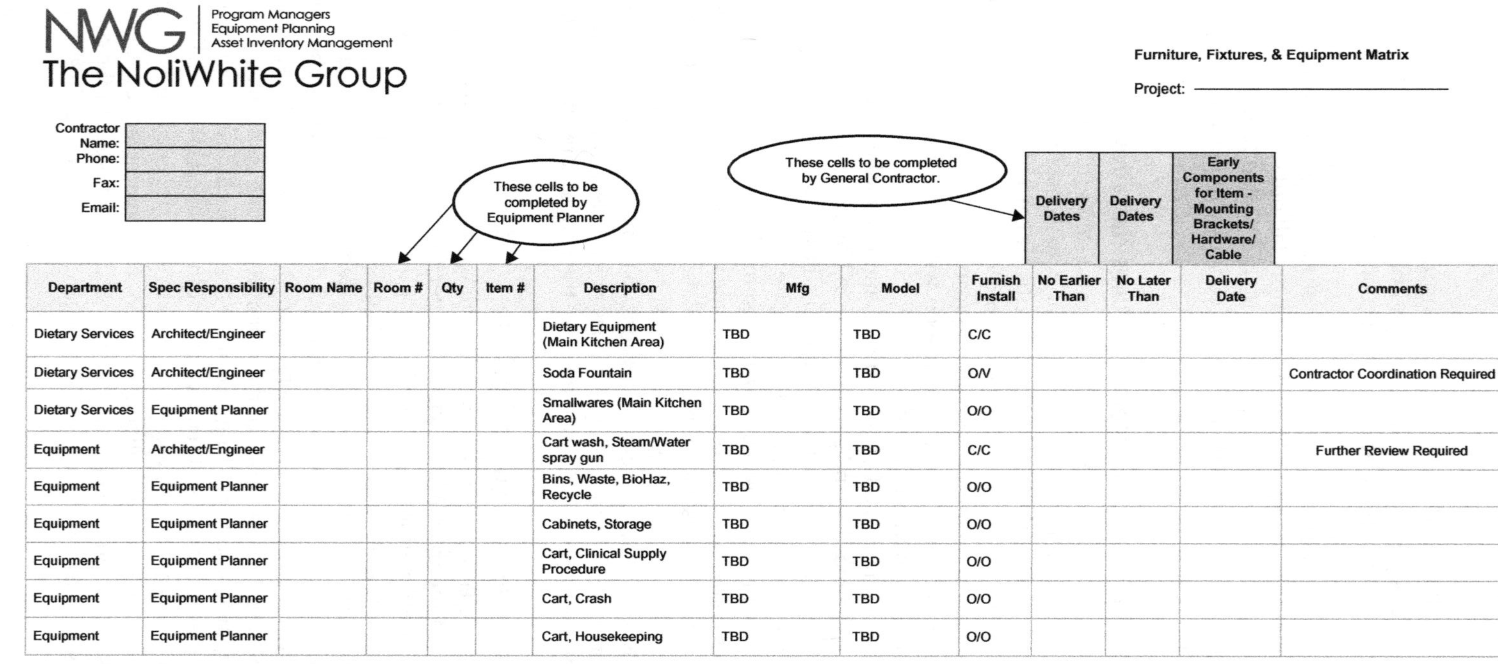

NWG | Program Managers
Equipment Planning
Asset Inventory Management
The NoliWhite Group

Furniture, Fixtures, & Equipment Matrix

Project: ____________________

Contractor Name:
Phone:
Fax:
Email:

										Delivery Dates	Delivery Dates	Early Components for Item - Mounting Brackets/ Hardware/ Cable	
Department	**Spec Responsibility**	**Room Name**	**Room #**	**Qty**	**Item #**	**Description**	**Mfg**	**Model**	**Furnish Install**	**No Earlier Than**	**No Later Than**	**Delivery Date**	**Comments**
Dietary Services	Architect/Engineer					Dietary Equipment (Main Kitchen Area)	TBD	TBD	C/C				
Dietary Services	Architect/Engineer					Soda Fountain	TBD	TBD	O/V				Contractor Coordination Required
Dietary Services	Equipment Planner					Smallwares (Main Kitchen Area)	TBD	TBD	O/O				
Equipment	Architect/Engineer					Cart wash, Steam/Water spray gun	TBD	TBD	C/C				Further Review Required
Equipment	Equipment Planner					Bins, Waste, BioHaz, Recycle	TBD	TBD	O/O				
Equipment	Equipment Planner					Cabinets, Storage	TBD	TBD	O/O				
Equipment	Equipment Planner					Cart, Clinical Supply Procedure	TBD	TBD	O/O				
Equipment	Equipment Planner					Cart, Crash	TBD	TBD	O/O				
Equipment	Equipment Planner					Cart, Housekeeping	TBD	TBD	O/O				

1. All Owner Purchased/Owner Installed units to be assembled and installed by Owner's equipment installation contractor/vendor.
2. At all "Contractor Installed" wall mounted equipment, provide backing/blocking.
3. At all "Contractor Installed" equipment, mtd, free standing, etc., refer to Architect drawings for layout.
4. At all units requiring power and/or plumbing, General Contractor to provide manufacturer required electrical and/or plumbing rough-ins & final connections regardless of installation assignment.
5. At Pharmacy Hoods: Contractor supplies and installs the final connection of all required plumbing, conduit, troughs, special back boxes, AC Wiring/power, shielding as defined by Arch. Owner provides equip. and final certificate.
6. Some items on the above list may not be included in the Project. If there is a discrepancy between the Drawings, Spec and this Matrix, please submit an RFI.

FIGURE 3.3 Example of equipment responsibility matrix.

Conclusion

The completed Project Definition Study serves as the basis and foundation for the project budgeting process and the project approval process, and can be used as an attachment to design and construction management contracts. Using a thorough and well-documented process, the construction manager, in conjunction with the other team members, can help ensure that the scope of the project meets the owner's goals and that the budget is adequate to successfully complete a quality project.

Guidelines for Design and Construction of Hospitals and Outpatient Facilities, FGI, 2010, Dallas, TX.

CHAPTER 4
Project Budgeting

Project Budgeting Process

A budget for a healthcare capital project should be a realistic estimate of the total capital needed that includes all of the costs involved with developing a project, not just construction. Typically, construction is only 60 to 65 percent and medical equipment and information technology are characteristically 25 to 30 percent of the overall expenditures. Additional important cost items include design fees, permits, testing—both geotechnical and materials, land, environmental consulting and/or remediation, program management, and others. Cost data from past projects contains a wealth of information that can be used to more accurately predict budgets for future development. Figure 4.1 shows one approach for a budget summary and key elements for a new hospital and medical office building. Figure 4.2 shows another approach and some key elements for an addition and renovation project for an Emergency Department. The project budget shown in Fig. 4.2 contains 19 items that are commonly seen on healthcare and other projects. In addition to using data from past projects, current pricing in the local market is essential to having an accurate budget. This is one of the benefits of early contractor and subcontractor involvement using a construction management delivery method. If that is not possible, there are estimating services that can be used and local builders will often provide budgeting as part of their marketing services or for compensation. This approach will yield better outcomes at the close of the project, assuming that the scope of the work and the change orders during construction are controlled. A contingency amount is typically included for unknowns that arise on a capital project. It is common to include a design contingency and a construction contingency. There are many approaches to estimating the correct amount of contingency. Some factors to be considered are stage of design (concept, schematic, design development, contract documents), age of the facility, delivery method (design-bid-build, design-build, cost plus, or etc.), project risks, and others. A formal risk analysis process is a good idea for large projects, to define contingencies needed and to develop mitigation strategies. This requires experience and careful consideration for the items that may not be apparent at first.

Realistic numbers should be included that reflect geographical differences, inflation, and other factors, as well as addressing all of the various items in a particular line. As an example, professional services is listed on line item five and might include costs for the following: architectural design, engineering design, program management, interior design, landscape architecture, traffic studies, and services of other professional consultants that may be utilized, along with their reimbursable costs. These might include printing, travel, supplies, models, and other miscellaneous costs. A construction

This chapter was written by Tom Koulouris.

BUDGET

Construction	$ 131,347,844
Medical Equipment	$ 31,106,705
Subtotal	$ 162,454,549
Land	$ 14,657,991
I/S Phones	$ 5,442,951
Interest/Overhead	$ 15,287,375
Subtotal	$ 197,842,866
Total	$ 197,842,866

Design Duration Fast track with construction **Construction Duration** 28 months

Building Area New 434,898 SF MOB 17,308 SF (TI Buildout) Total 452,206 SF

Services

- New acute care (221 bed) Hospital to consolidate and replace Existing Hospital

General Notes

- Site costs for attached third party financed MOB are included in this budget.
- Connectors from MOB to hospital are included in this budget.
- Shell 2 OR's for future.
- Shell 6 ICU rooms for future.

Figure 4.1 Example of capital budget for new hospital and MOB.

manager can develop a database from past projects and use readily available industry sources to yield excellent results by examining similar projects for square footage cost averages ($/SF). The pricing can also be calculated into percentages of construction cost, which is a very valuable and accurate method. Another approach is to look at cost per square foot of different types of spaces, such as patient rooms, operating rooms, etc. By reviewing the percentages of cost items across a large number of projects, it is apparent most projects' percentages will be relatively close. Geographic and market-driven factors must also be evaluated to determine average costs. For example, in 2008–2009, worldwide steel and copper prices spiked due to the high demand from China and India. Armed with this data, a project manager can apply it to the project scope and very quickly have an accurate preliminary or conceptual budget without taking proposals from multiple contractors. This allows users to have an idea of the cost to compare with returns, to make the business decision regarding capital investment with minimal cost invested. On unique projects or those in new geographic areas, additional steps will be required to accurately project the capital costs. One proven way is to obtain estimates from local designers, contractors, and subcontractors. Once again, these should be estimates not "low bids." With these early estimates, there is often not enough

TYPICAL HOSPITAL ADDITION/RENOVATION PROJECT BUDGET						
Date:				**Owner Rep:**		
Facility Name:	Example			**Pgm. Manager:**		
Project Name:	Emergency Dept Addition/renovation			**Architect:**		
Project Duration:	36 Months			**Contractor:**		
Budget Name:	Stage 3			**Equipment:**		
Phase 1: Pre Design	**% of Construction subtotal**					
Planning	0.91%					$ 103,400
Development	0.09%					10,000
Environmental	0.44%					50,087
Testing and Misc.	2.00%					227,106
Total Pre Design	**3.44%**					**$ 390,593**
Phase 2: Design						
Professional Services	14.64%					1,662,581
Building Fees	3.53%					400,692
Total Design	**18.17%**					**$ 2,063,273**
Phase 3: Construction						
New	60.04%	19478	Square feet @	$350 /Sq. Ft.	6,817,300	
Renovation	28.18%	10000	Square feet @	$320 /Sq. Ft.	3,200,000	
Construction Cost		29478			10,017,300	
Site Work	3.52%				400,000	
Tenant Build-out					-	
Other					938,000	
Subtotal Construction					**11,355,300**	11,355,300
Contingency	8.00%					908,424
Total Construction						**$ 12,263,724**
Owner Furnished Equipment						
Equipment	23.85%					2,708,121
Information Systems	3.98%					451,435
Telecommunications	Inc in IS.					See I.S.
Total OFE	**27.82%**					**$ 3,159,556**
Land/Financing						
Land						-
Bond and Financing						-
Overhead						-
Total Land/Financing						-
Total Project Cost						**$ 17,877,146**
Construction cost/Total project cost						64%
Project cost/square foot						$ 606
NOTES:						
1. OTHER; 650 Ton chiller and cooling tower; 650 KW generator; hurricane hardening of CEP; Vertical expansion capability						

FIGURE 4.2 Capital budget.

detail on the documents for firms to effectively "bid for" the work. Bids seem to draw low numbers that do not accurately reflect the scope of a project as they will develop.

Construction Cost versus Project Cost

There is, on occasion, a gap in communication on healthcare projects between the cost of the construction—"bricks and mortar"—and the total capital investment for a project. Contractors often do not get involved with the components of the project budget that are not their responsibility, so they may not consider total capital. Some architects, depending on their experience, will have a better understanding of the total project cost components. Past experience will inform the facility administrators' understanding of all the components.

The costs for a project almost always involve a capital investment unless it is defined as maintenance, such as replacing finishes. Projects can also have significant impacts on operating costs and revenue. In some cases, the movement of existing medical equipment cannot be part of the capital cost, so this will involve operating expenses. Construction activities could impact access to patient rooms, operating rooms, or other services, thus impacting revenue. All of these issues should be discussed and budgeted, so they can be tracked along with all of the other costs as well as the contingency. It is important that all of the project team members have a clear understanding of the entire project budget, how it is prepared, and the meaning of the key financial indicators.

One of the important decisions on a project is which project management and cost control software system to use. There are many web-based computer project management systems on the commercial market for contractors and owners. Large healthcare systems often have their own systems; however, several of the larger systems have selected to outsource and are using a software vendor as a service provider. "Dashboard" capabilities for key financial indicators are valuable, as are summary reports that track change orders, commitments and expenditures, requests for information, and submittals. Detailed reports should be available for a deeper review if required for audits, account payable issues, or other reasons. These systems also can be used for document control and permanent storage of project files.

The Project Budget

Planning

Planning should be one of the first steps on a project so a budget item should be included for this. The costs can vary widely on projects depending on the scope and regulatory agencies involved, especially in Certificate of Need (CON) states. Sometimes the "planning" is included in the architectural fees and is considered the conceptual or schematic phase of an architect's basic services. In those scenarios, the designer may start drawing before the planning work is complete. This can lead to multiple iterations reviewed by many stakeholders in futile efforts to please everyone. This clearly is not possible. Many projects that were well intentioned and seemingly well planned have never been built. The reason for this is often the increased scope that arises with so many drawing iterations. Having a separate planning firm that specializes in this work seems to be the best approach. These professionals can take an analytical view of the true investment needed, given the market and facility conditions. With this clear vision of the project scope, the designers can work to meet the needs while maintaining the desired budget. At some point on many projects, stakeholders perceive that the cost is too much and want to know why the "budget" is so high. Ultimately, many projects do not proceed for financial reasons, so effective planning and managing of expectations is essential. Given this scenario, it makes sense to spend the least amount of money practical, to reach a "go/nogo" decision point. This avoids significant architectural and planning fees that may escalate to hundreds of thousands of dollars and sometimes wind up as "sunk" costs if the project does not proceed. Typical planning costs may include

- Demographic studies
- Market studies
- Financial studies
- Competitor analysis
- Research on physicians in the market
- Certain architectural or engineering efforts, such as the due diligence process and conceptual plans for pricing for financial analysis or regulatory agencies
- Studies such as traffic, codes, parking, utilities, etc.
- Construction estimating services

The sum of these costs is generally in the range of ½ to 1 percent of the project cost. This can vary widely depending on such factors as CON requirements, legal challenges from competitors, complex additions and renovations, and others.

Development

In the planning phase within the due diligence process, scope is often identified that must be completed as part of the project. Some of the items that should be included in the development line of the project budget are

- Rezoning
- Impact fees and other charges from local regulators
- Change of Use and other types of land-related issues or deed restrictions
- Demolition of existing improvements
- Noise and dust abatement issues
- Major utility relocations
- Certain "enabling" projects, such as relocations of people, equipment, and/or utilities/infrastructure

Sometimes, the planning and development fees are distributed among the architectural and construction cost categories. If development fees are included in the design or construction cost per square foot ($/SF), obviously the $/SF line item increases and masks the actual construction costs. Generally, most senior managers, board members, and other reviewing stakeholders will focus on the cost per square foot, as it is easy to compare with other projects. Almost all stakeholders can identify with what this means although it can cause confusion when individuals define the $/SF differently. The best approach is to define the $/SF upfront to include only the construction cost. This means the $/SF is purely the construction cost of the building, operationally defined as the work directly under the footprint of the building and extending, typically, to 5 feet outside the building edge. All other work is carried in other line items. It is best to account for $/SF in this manner to keep an accurate database of costs, as discussed earlier. Often the other costs, such as development, environmental, or site work can be large numbers, skewing the overall cost of the project and yielding confusing numbers, in a database analysis.

Environmental

"Environmental" costs are part of the project directly attributable to several potential issues.

- Hazardous waste products such as asbestos, mold, lead, and other heavy metals, sewage, petroleum products, acids (i.e., battery and lab), and any other substance listed in the federal list of contaminants
- Mitigation of certain issues as required by regulatory agencies such as wetlands, preservation or protected species of animals, birds, fish, and reptiles
- Indoor air quality and monitoring (not the cost associated with keeping negative air in projects)

Many environmental issues require the services of specialized consultants that deal directly with these matters. Certain due diligence procedures are needed to ascertain the magnitude, scope, and cost. Unless a project has a specific environmental focus, such as removing asbestos fireproofing, this line item will likely be a small percentage of overall capital costs.

Some examples of potential environmental costs on a project include

- A renovation project in an older facility, which likely will have some asbestos containing material (ACM) that will have to be mitigated. Most general contractors will not perform this work. An environmental contractor can generally provide an accurate estimate based on SF for items, such as removing vinyl asbestos tile (VAT), sprayed-on fireproofing, pipe insulation, or duct mastic (adhesive).
- A federally protected species is found on land proposed for new construction. For example, one of the authors was involved on a hospital project where eagles built a nest on the site, requiring extensive protection measures.
- The site contains documented wetlands that must be protected or replaced.
- A new addition to an older facility uncovers the remains from previous systems, such as waste systems, septic tanks, fuel oil tanks, and acid pits. Old dry cleaning sites often pose significant problems due to chemical contamination.
- Environmental surveys, remediation, and testing.

Testing and Miscellaneous

Testing is required in many phases of a project. The testing may start early with geotechnical testing of the site conditions for water mitigation and foundation design. It will also be required during construction for a variety of materials including concrete, fireproofing, steel, HVAC–test and balance, and other items, as required by the local building codes. In some states, such as Florida, a special inspector called a "threshold inspector" is required to verify the structural integrity of the facility. Chapter 17 of the International Building Code (IBC) addresses special inspections, structural observations, and tests. The owner is typically responsible for paying these testing costs. The owner often pays the designers to take proposals and administer the testing using third party testing agencies. The owner sometimes hires the testing agency directly. The testing agent should not work directly for the contractor or subcontractor to avoid any conflicts of interest. The IBC requires that "an approved (inspection) agency shall be objective, competent, and independent from the contractor responsible for the work being inspected."

Miscellaneous costs are those that do not fall neatly into the other categories and would skew square footage costs that may affect the database. This budget format is more than just a project-specific budget; it is a tool to manage the database of costs. This allows a desktop budget to be prepared in a matter of hours with minimal information. Some examples of miscellaneous costs include

- Legal fees necessary for land issues, such as rezoning and change of use, contract review fees, filing costs, and notices
- Certificate of Need (CON) costs that may include lawyers and consultants for a new project or service
- Permit expediter fees that can aid the approval process

Professional Services

Professional services typically cover the cost of design services and other consultants that work directly or indirectly for the owner. It is common practice for the owner to contract with the architect, who then contracts with the engineers or consultants. In some cases, the consultant may work directly for the owner, such as a kitchen consultant. These may be contracted as a lump sum, a percentage of construction costs, or on a cost-plus-a-fee basis with hourly rates.

These services would typically include the following:

- Architectural fees for basic design services, which usually consist of schematics, design development, construction documents, and contract administration
- Interior design, which is often done in-house by the architect or can be outsourced
- Landscaping design, which is typically a separate firm
- Specialty consultants, such as acoustics, life safety, thermal, and moisture
- Engineers including mechanical, electrical, fire protection, structural, civil, and others as needed
- Presentation materials of a substantial nature, such as models both actual and electronic
- Reimbursable costs for the designers, such as printing, online plan services, travel, models, special studies, and other miscellaneous costs
- Program management fees for outsourcing the contract administration
- Special consultants for building art, such as a statue or other elements

Building Fees

- Local building permit fees including all miscellaneous permits, such as tap fees, concurrency fees, impact fees, environmental fees, etc.
- State regulatory agency fees
- County environmental fees
- Inspection fees by regulatory agencies

Construction

Construction cost is typically defined as those costs that are the responsibility of the construction manager, subcontractors, and suppliers. These can be summarized as all the costs associated with the labor, materials, equipment, taxes, and fees necessary to construct and make ready for occupancy the work shown and described within the architect's plans and specifications. These are sometimes referred to as the "brick and mortar" costs. Construction costs are usually about 60 to 65 percent of the project cost for hospitals. Variances from this data point include projects that are heavy in diagnostics, those with very long durations (such as multi-phased renovations where the general conditions or project overhead is high), and investments with significant infrastructure (such as power plants or major utilities).

The costs are typically broken down using the Construction Specification Institute's (CSI's) Master Format which contains 49 divisions (www.csinet.org). This is an industry standard and should be used for comparison purposes and analysis. Breakdown of a hospital project in this format is shown in Fig. 4.3.

Site Work

Site work can be and is often a budget "red flag." This is a high-risk area, due to the unknowns underground and the variability in sites. Site work from project to project can range widely and include such risks as rock, water, unstable soils, and other factors. One of the authors saw significant cost issues as a result of prairie dog holes in Oklahoma and a unique rock formation in Las Vegas called caliche. Thorough due diligence and testing as well as talking to local site work subcontractors are the best approaches to help avoid surprises on the site costs.

While the site work is often in the construction contract and part of the contractor's budget, it is often necessary to separate this for a couple of reasons. First, the site work and the land will be depreciated on a different basis than the building. Land typically appreciates, although portions of the work may be handled differently from an accounting standpoint. Second, since the site costs can vary widely depending on the location, it is more meaningful to compare the cost of construction for just the building. This allows for better benchmarking against cost data from previous projects.

Tenant Build-Out

Tenant build-out is often referred to as tenant improvement or TI work. TI costs are generally associated with interior build-outs of medical office buildings (MOBs) or other projects where an ownership or lease of real estate allows for shell and tenant work separately. This can be the case as mentioned above in MOBs or storefronts in retail spaces. Tenant improvements can also be the case when the hospital is providing space

Hospital Budget Using CSI 16 Division Breakdown
Sample project

Area		
Basement	46,932	SF
First Floor	132,975	SF
Second Floor	104,337	SF
Third Floor	90,722	SF
Gross Building Area: (BGSF)	374,966	GSF

16 Division CSI Cost Summary

	CSI			TOTALS:	
	Division 1: General Requirements			$ 0	$ 0.00
13.72%	Division 2: Sitework & Demolition (incl rock piers)			$ 7,414,419	$ 19.77
5.90%	Division 3: Concrete Foundations & Slabs			$ 3,190,516	$ 8.51
4.25%	Division 4: Masonry/Precast			$ 2,817,169	$ 7.51
11.83%	Division 5: Metals			$ 6,393,163	$ 17.05
0.88%	Division 6: Carpentry			$ 473,070	$ 1.26
5.78%	Division 7: Moisture Control			$ 3,121,571	$ 8.32
5.53%	Division 8: Glass & Egress			$ 6,446,215	$ 17.19
2.30%	Division 9: Drywall/Finishes			$ 7,156,522	$ 19.09
1.44%	Division 10: Specialties			$ 1,452,025	$ 3.87
0.00%	Division 11: Equipment			$ 869,021	$ 2.32
0.00%	Division 12: Furnishings			$ 839,507	$ 2.24
0.00%	Division 13: Special Construction			$ 0	$ 0.00
1.88%	Division 14: Conveying			$ 1,015,000	$ 2.71
2.14%	Division 15a: Fire Protection			$ 1,158,807	$ 3.09
4.76%	Division 15b: Plumbing			$ 4,970,309	$ 13.26
25.75%	Division 15: HVAC			$ 13,915,291	$ 37.11
13.86%	Division 16: Electrical Systems			$ 12,772,981	$ 34.06
100.00%	**Estimated Direct Cost:**			**$ 74,005,585**	**$ 197.37**
	Indirect Costs blew:				
	1 LS	Sub Bonding	*applied to trades*	$ 0	$ 0.00
	1 LS	Other-3rd party review		$ 142,812	$ 0.38
	1 LS	Other - LEED/Commissioning		$ 162,000	$ 0.43
	10.86%	Other-Design Costs & Consultants		$ 5,867,000	$ 15.65
	9.73%	General Conditions		$ 5,258,347	$ 14.02
	1 LS	Insurances		$ 010,098	$ 2.69
	1 LS	GC P&P Bond		$ 663,147	$ 1.77
	1 LS	Bldg. Permits Cost		$ 403,542	$ 1.08
	1 LS	Inspection Fee Allow		$ 353,320	$ 0.94
	1 LS	Builder's Risk Insurance		$ 150,000	$ 0.40
	0.65%	Preconstruction Fees		$ 352,000	$ 0.94
	3.48%	Estimating Contingency		$ 1,880,639	$ 5.02
	0.84%	Escalation		$ 453,939	$ 1.21
	7.79%	Fee		$ 5,766,496	$ 15.38
		Total Indirect Costs:		*$ 22,463,340*	*$ 59.91*
	Estimate- Total:			**$ 96,468,925**	**$ 257.27**

FIGURE 4.3 Hospital budget in CSI format.

to a food or retail vendor inside a common lobby or food court area. The most common tenant build-out in healthcare is the construction of physician's offices and outpatient surgery or diagnostic centers.

Other

This section can be used for unique items in a budget that need to be separated for tracking reasons or scope done as part of the enabling work to support or facilitate the main project. In the example in Fig. 4.2, costs are included for an upgrade of the central energy plant (CEP) to support the Emergency Department Addition and Renovation. The "other" costs include a new chiller, a cooling tower, an emergency generator, and hurricane hardening of the CEP for this hospital in Florida.

Owner Contingency

Owner contingency must be dealt with very carefully and the definition clearly communicated to the entire team. Ideally, contingency is included just to cover "unforeseen conditions." The preference is to not use the contingency for scope increases, errors and omissions, stakeholder requests, and items outside the approved scope of the project, although this does happen. There is no simple answer regarding the amount of contingency to be included in a project budget. There are numerous variables, including the amount of due diligence, stage of the plans, geographic location, experience of the design and construction team, age of the facility on an addition/renovation, and the complexity of the project. The average ranges typically seen are 5 to 15 percent, depending on the issues described above. Many owners set aside separate amounts for design and construction contingency.

The reasons for the use of contingency should be cataloged and trends can be further incorporated into the database for use on future projects. For example, if past renovation projects required some "contingency" to pay for additional interior project signage it should signal that more due diligence in the area of project signage should be performed to avoid these overage issues. One approach is to perform a risk analysis where contingency can be assigned to the issues identified. The contingency can then be tracked based on the actual needs and curves can be plotted to compare usage to projections. This will help graphically display the amounts used and contingency remaining to give early indicators of potential budget issues.

Equipment

The equipment budget must be prepared by experts in this field with input from the owner and the vendors. Again, due diligence is required to understand the existing equipment on addition and renovation projects, the clinical needs, and the patient volumes so the proper equipment can be planned and budgeted. The equipment line item is inclusive of the owner cost and vendor cost, but does not typically include the cost of the contractor. Equipment cost should be defined as *fixed equipment* or *movable equipment*. This is an accounting function required due to different depreciation approaches for capital equipment. Moving existing equipment cannot be capitalized. The equipment budget should include medical equipment, such as imaging modalities, furniture, shelving, televisions, kitchen equipment, and other items needed in a healthcare facility.

Information Systems

The information systems (IS) are also sometimes called ITS (information and technology systems). This budget line item should be inclusive of the new hardware and software, as well as of upgrades to existing systems necessary to make the new building operationally ready. This includes the structured cabling, medical devices, computers, monitors, printers, and telecommunications equipment that are specific to the hospital's requirements. This is addressed completely in Chap. 10 of this book.

Land

Land cost is for the acreage that must be purchased for the project. This is typically not used for site work improvements, development, demolition of structures, or other alterations. Realtor fees and transaction costs are commonly included in this category, so the full purchase price of the land is included in this line item.

Bonds and Financing

Bonds and financing costs are those costs associated with selling bonds, the closing costs of financing, debt service over the period of the construction term, and terms associated with the funding of the project. This is not the total long-term interest charges over the life of the bond or loan.

Department Overhead

In some cases, the owner may have internal overhead charges from their design and construction departments. These are cost centers, not revenue producing departments. This cost is sometimes capitalized as part of the project, which is treated differently from an accounting standpoint and provides some advantages.

Project Total

The project total, simply put, is the amount of money that will be considered as the total capital investment required for achieving the operational goals set forth for the project. While funding sources will vary as discussed in Chap. 6, it is always best to view the project from a total capital perspective. It is also important to note that there can also often be some operating costs that must be expended by the hospital on addition and renovation projects that are not in this budget. These costs may include removing or relocating old equipment, temporary services to minimize disruptions, additional staff for cleaning, and other items.

Notes

The Notes section is used to clarify details or to add a few alternates, such as money for shelling an additional floor, equipment upgrades, or other unique items. Clarification should be given if a specific line item or the whole project is outside the cost norms. It could be the site costs are higher due to soil issues, or the $/SF are higher due to the cost of relocating the central energy and central cooling plant. Figure 4.4 provides an example of a new hospital budget broken down by hospital departments to show the relative cost per square foot for the different types of spaces. This also shows the relative size of the different hospital departments. The cost per square foot can range widely for different parts of the United States and for international projects. The market conditions can also impact the costs or materials and labor on a local or national level. Input from experts in the fields of construction and equipment planning for the specific project are essential to developing an accurate capital budget.

Cost Management

Capital project budgeting and management for the owner involve projecting and accounting for all of the costs in a timely fashion that supports the organization's goals, builds the database of cost history, and facilitates effective financial analysis. The project accounting and control system should provide "real time" cost and schedule information to enable the owner and other team members to effectively manage the budget and be

Typical New Hospital Project
October 4, 2013

		Total Capital Budget		Labor and Material Cost	
Licensed Hospital Beds:		100			
Total duration in months:		24			
Allocated Space		*DGSF*	*Total*	*Cost/SF*	*Total*
Cosmetic Renovation		0	0	$ -	$ -
Major Renovation		0	0	$ -	$ -
New Construction		228,770	228,770		
Public Areas		5,000	5,000	$ 190.00	$ 950,000
Administrative Services		10,000	10,000	$ 170.00	$ 1,700,000
Emergency Department		18,150	18,150	$ 250.00	$ 4,537,500
Diagnostic Imaging		20,992	20,992	$ 400.00	$ 8,396,800
Interventional Services		49,308	49,308	$ 375.00	$ 18,490,500
Critical Care Unit		15,450	15,450	$ 350.00	$ 5,407,500
Medical Surgical Unit		42,126	42,126	$ 280.00	$ 11,795,406
Clinical Lab		8,029	8,029	$ 300.00	$ 2,408,571
Pharmacy		4,229	4,229	$ 235.00	$ 993,714
Dietary Services		9,086	9,086	$ 275.00	$ 2,498,571
Environmental Services		2,429	2,429	$ 230.00	$ 558,571
Clinical/Facility Engineering		2,857	2,857	$ 230.00	$ 657,143
Materials Management		3,657	3,657	$ 200.00	$ 731,429
Central Lockers/Lounges		1,743	1,743	$ 210.00	$ 366,000
Mech/Elec/Circulation		29,924	29,924	$ 190.00	$ 5,685,470
Canopies/Soffits @ 50%		1,931	1,931	$ 125.00	$ 241,319
Central Energy Plant		3,861	3,861	$ 250.00	$ 965,275
Shelled Construction		0	0	$ -	$ -
Shell Buildout		0	0	$ -	$ -
Structured Parking		0	0	$ -	$ -
Total Allocated Space		228,770	228,770	$ 290.18	$ 66,383,771
Budget Categories					*Total*
Planning	0.65%	$ 471,882	471,882		
Sitework	17.00%	$ 11,285,241	11,285,241		
Environmental		$ -	0		
Testing	1.50%	$ 1,165,035	1,165,035		
Construction		$ 66,383,771	66,383,771		
Contingency	8.00%	$ 6,213,521	6,213,521		
Arch & Eng Fees	8.00%	$ 6,213,521	6,213,521		
Building Fees	1.00%	$ 776,690	776,690		
Tenant Buildout		$ -	0		
Other (F.A.& Hospital)		$ -	0		
Equipment		$ 29,000,000	29,000,000		
Commissioning	$0.50	$ 114,385	114,385		
I/S & Telecom	7.00%	$ 6,475,676	6,475,676		
PM Fee	2.25%	$ 2,882,244	2,882,244		
					$ -
Total Capital Budget		$ 130,981,967	130,981,967		

Total Capital Cost per Bed	$ 1,309,820
SF per Bed	2,288
Total Capital Cost per SF	$ 573

Figure 4.4 Typical new hospital project budget.

proactive to avoid surprises and budget variations. An owner is always unhappy if the project goes over budget. An owner may be just as unhappy if the contractor completes the project significantly under budget without giving them advance notice so they could add scope or move the budget to other projects. It is best to provide accurate projections of the costs that are not too high or low.

The cost management system must also facilitate accurate entry and tracking of actual commitments and expenditures. At any point in the project, the construction manager should be able to provide the cost to date and a projection of the cost to complete. The system must also allow for auditing of the actual costs, so the owner can verify that the costs billed and paid are in accordance with the contractual requirements. The ability to provide reports in a variety of formats, from executive level reports that may show many projects to a single project report showing much project detail, is essential to effectively manage a hospital project.

Conclusion

Project budgeting is both an art and a science. Careful blending of these two skill sets will result in better, more accurate project cost projections. To a healthcare institution, accurate capital budget forecasting is paramount. All tools available should be utilized with a focus on reducing costs, speed to market, and quality of final product.

CHAPTER 5

Healthcare Facility Justification and Capacity

Normally, a project manager's, contractor's, or architect's involvement in a healthcare construction project begins when he or she is given a description of the project. Usually, the description will include the required number of key planning units (KPUs) and a program. KPUs are primary spaces within a healthcare facility where patient care, treatment, or testing occurs. Examples include operating rooms, inpatient beds, emergency department (ED) treatment beds, and diagnostic imaging modalities. A program is a list of all spaces to be included in the facility or addition/renovation. The program provides the square footage for each space based on the KPUs required, and generally includes grossing factors to address circulation, mechanical/electrical chases, stairs/elevators, and other general areas required.

Unfortunately, what are often not included are the reason or justification for the project and the analysis and assumptions used to derive the number of KPUs. This information can be very valuable to all parties working on the project to ensure alignment of the owner's intent for the project, and the solution developed by the design and construction team.

Healthcare Facility Project Justification

There are many justifications and reasons healthcare entities pursue construction projects. One of the obvious reasons is that a healthcare facility may be no longer appropriate due to age or condition. Another common reason is the need to respond to a change in function or service provided by the healthcare entity. For example, in the 1990s and early 2000s, many small rural healthcare facilities decided (for various reasons) to close their obstetrical units. Upon the closure of those units, a significant amount of space was available for reuse by those facilities. At the same time, utilization in other service lines, such as surgery and cardiology, was on the upswing and many vacant obstetrical units were renovated for use by other service lines.

Often, the exact justification for a healthcare design and construction project will be unclear. Common "mystery" justifications include physician influence, philanthropic influence, and past experience of a hospital administrator. Physicians can exert enormous influence on hospitals through their ability to influence where their patients receive treatment. A hospital's financial success largely depends on attracting as many

This chapter was written by Andrew Collignon.

patients as possible to its facility. Many hospitals attract physicians (and their patients) by providing modern and aesthetically pleasing facilities. For example, many hospitals have built new cardiac catheterization suites for the purpose of attracting a specific cardiologist or cardiology group to come and perform procedures at their hospitals, as opposed to a competitor's hospital that might have older, less attractive facilities.

Philanthropic influence manifests itself when a donor wishes to contribute money for use on a specific project or in a specific area of a healthcare facility. Cancer centers, children's hospitals, and rehabilitation facilities are common beneficiaries of philanthropy.

The movement of healthcare administrators from one facility to another is quite common, especially among those at the beginning of their careers. Often, at some point, an administrator will be involved with or be responsible for a successful service line. When they pursue a new opportunity at another facility and are charged with improving the bottom line at their new facility, they will naturally look back to the successful service line they were involved with at their previous employer.

Patient Volume Change

All of the aforementioned reasons are common and can serve as justification initially, but alone they are not necessarily legitimate reasons for a healthcare design and construction project. If they are the sole reasons for a project, it is more likely that the project will be inappropriate in purpose and scope. To ensure success, a project should promote or facilitate continued or additional patient volume and meet a need in the community.

A healthcare facility can realize increased patient volume from two primary sources. The first is from additional patients in the facility's service area; the second is by taking patients from a competitor. A service area defines where a majority of a healthcare facility's patients originate. The service area is usually defined using ZIP codes and can include a primary service area (PSA), a secondary service area (SSA), and occasionally, a tertiary service area (TSA).

Population Change

The number of patients in a service area can change for two reasons: first, due to a change in population, and second, due to a change in utilization of healthcare services. Table 5.1 shows the 13 fastest growing cities during the period from July 2011 to June 2012.[1] Note that cities in boldface are located in Texas.

For an example of the impact population growth can have on a healthcare market, consider Cedar Park, Texas. If the population of Cedar Park continues to grow as it has for 5 years starting in June 2012, its population will increase by 36,462. According to data from the Henry J. Kaiser Family Foundation for 2011,[2] Texas residents require hospitalization at a rate of 99 per 1000 people. Applying that utilization rate, approximately 3610 more people would require hospitalization in 2017 than did in 2012. To address the healthcare facility impact of those additional inpatients, let's assume that on average, patients in Texas stay in the hospital for approximately 5.2 days. Assuming that average length of stay (ALOS), the patients would spend a total of 18,770 days in a

---

[1]Source: United States Census Bureau. http://www.census.gov.

[2]Henry J. Kaiser Family Foundation. Inpatient admissions per 1000 in 2011 in the State of Texas. http://kff.org/other/state-indicator/admissions.

Rank	Location	Annual Growth (%)
1	**San Marcos**	4.91
2	South Jordan, Utah.	4.87
3	**Midland**	4.87
4	**Cedar Park**	4.67
5	Clarksville, Tenn.	4.43
6	Alpharetta, Ga.	4.37
7	**Georgetown**	4.21
8	Irvine, Calif.	4.21
9	Buckeye, Ariz.	4.14
10	**Conroe**	4.01
11	**McKinney**	3.95
12	**Frisco**	3.92
13	**Odessa**	3.83

Source: http://www.census.gov/newsroom/releases/archives/population/cb13-94.html.

Table 5.1 Thirteen Fastest Growing Cities during the Period from July 2011 to June 2012

hospital. Divide that number by 365 days each year and the average number of those patients in a hospital on an average day, or the average daily census (ADC), is 51.4.

You could stop there and assume that you would need 51.4 beds for those patients, but that would not account for the fact that the actual patient census can fluctuate significantly. To account for that fluctuation, the ADC is usually divided by a target occupancy percentage in order to derive a number of additional beds above those that are required to accommodate the average number of patients expected. In this example, dividing the ADC (51.4) by a target occupancy factor of 75 percent would produce a target number of 69 beds. Described another way, if you had 69 beds and your ADC was 51.4, you would have a 25 percent cushion, or 17 additional beds, to be able to accommodate those days when you have more patients than average. The choice of 75 percent for the target occupancy percentage is customary for general medical/surgical beds, but can vary depending on many factors, including the acuity of the patients, staffing patterns, all private or semi-private beds, and the overall number of beds in the healthcare facility.

Healthcare Utilization Change

Utilization was mentioned and used in the example above. It is basically the rate of use of healthcare services expressed in per capita terms. Changing utilization of healthcare services is inevitable and ongoing due to advances in technology, changes in the national healthcare laws, and the way patient care is delivered. A University of Michigan Health System study from 2011 found that 3.2 percent of emergency patients received CT scans in 1996, while 13.9 percent of emergency patients seen in 2007 received them.[3]

[3]Kocher, KE, Meurer, WJ, Fazel, R, et al. "National Trends in Use of Computed Tomography in the Emergency Department," *Annals of Emergency Medicine*. doi:10.1016/j.annemergmed.2011.05.020.

Description	2010	2012	2017	Volume Change	Percent Change (%)
Total population	218,900	228,334	264,796	36,462	15.97
Median age	34.81	34.99	35.24	0	0.71
Total population by age					
00–14	56,161	58,556	68,156	9,600	16.39
15–24	24,404	24,638	26,879	2,241	9.10
25–34	30,138	31,845	37,248	5,403	16.97
35–44	39,748	40,538	46,095	5,557	13.71
45–54	32,805	33,174	35,606	2,432	7.33
55–64	20,510	22,589	27,590	5,001	22.14
65+	15,138	16,994	23,222	6,228	36.65

TABLE 5.2 Population of Cedar Park, Texas, Broken Down by Age Cohort

This is an example of a change in utilization. The facility impact is a proportional increase in the need for CT scanners for ED patients. In some facilities, this has resulted in the need for a CT dedicated for use in the ED.

Changing population demographics can also have a significant impact on utilization. Anecdotally, older populations have higher utilization of healthcare services. Refer to Table 5.2 for the population of Cedar Park, Texas, broken down by age cohort. Not only is the overall population increasing but some of the most significant growth is also occurring in the 55 to 64 and 65+ age cohorts.

A recent study of how the aging of the U.S. population would affect demand for ED services and hospitalizations found the following:

> With US emergency care characterized as "at the breaking point," we studied how the aging of the US population would affect demand for emergency department (ED) services and hospitalizations in the coming decades. We applied current age-specific ED visit rates to the population structure anticipated by the Census Bureau to exist through 2050. Our results indicate that the aging of the population will not cause the number of ED visits to increase any more than would be expected from population growth. However, the data does predict increases in visit lengths and the likelihood of hospitalization. As a result, the aggregate amount of time patients spend in EDs nationwide will increase 10 percent faster than population growth. This means that ED capacity will have to increase by 10 percent, even without an increase in the number of visits. Hospital admissions from the ED will increase 23 percent faster than population growth, which will require hospitals to expand capacity faster than required by raw population growth alone.[4]

[4]"The Care Span: Population Aging and Emergency Departments: Visits Will Not Increase, Lengths-of-Stay and Hospitalizations Will," *Health Affairs*, July 2013, 32(7): 1306–1312.

Competition for Healthcare Services

Aside from the growth of patient volumes occurring in the market due to population changes and utilization changes, competitive healthcare organizations generally won't hesitate to take patients from another local healthcare provider. This practice also isn't limited to for-profit healthcare providers; not-for-profit entities also compete vigorously for patients. A common saying in the sometimes mission- or charity-oriented not-for-profit world is "No margin, no mission."

Here is an example of a competitive maneuver and the impact it had on facility requirements. A large multispecialty physician practice was recently purchased by a healthcare provider (hospital A). There is one other acute care provider in the market (hospital B). The physician practice was composed of approximately 90 physicians and several mid-level providers. Before the acquisition, the majority of the physician practice's patients went to hospital B. After the acquisition, the physicians were free to start sending their patients to hospital A. Before that could happen, the volume of patients that would be sent to hospital A instead of hospital B needed to be identified, to determine the facilities required to accommodate those patients.

It was estimated that 85 percent of the physician group's patients would go to hospital A instead of hospital B. It was assumed that 15 percent would not or could not go to hospital A for various reasons, including personal preference, loyalties, travel distances, and, most importantly, restrictions contained in some private pay insurance programs. Using the estimated capture rate of the physician group's patients, the anticipated swing in patients from hospital B to hospital A was 32,000 patient days, 7500 surgical cases, 4000 ED visits, 1230 cardiac catheterizations, and 850 baby deliveries. The corresponding impact on hospital A was the net need for 113 additional patient beds, 9 additional operating rooms, no change in the number of ED beds, 1 additional cardiac catheterization lab, 4 additional labor-delivery rooms (LDRs) and 12 additional postpartum rooms. No additional rooms were required in the ED because it already had available capacity to absorb the additional patients. Existing capacity for all existing departments was determined and then compared with the demand anticipated from the physician practice's patients to arrive a net number of additional diagnostic and treatment spaces.

Another factor to consider in the example above is the timing. As soon as the agreement between hospital A and the physician group was reached, the physicians could literally begin sending their patients to hospital A instead of hospital B. This necessitated a design and construction or capital project where the sooner the project was finished, the sooner hospital A could start seeing and treating those patients.

Healthcare Fraud and Abuse Laws

It is important to note that there are three federal laws that should be considered when interactions between physicians and healthcare entities occur. They are the Anti-Kickback Statute, the Ethics in Patient Referrals Act (Stark Law), and the False Claims Act. The Federal Anti-Kickback Statute, 42 U.S.C. § 1320a-7(b), prohibits providers of services or goods covered by a federal healthcare program (Medicare, Medicaid, etc.) from knowingly and willingly soliciting or receiving or providing any remuneration, directly or indirectly, in cash or in kind, to induce either the referral of an individual, or furnishing or arranging for a good or service for which payment may be made under a federal healthcare program. A common example of this is a hospital offering a physician below-market rent rates in medical office space in order to persuade them to send their patients to the hospital for surgery, diagnostic imaging, or other services. The Stark Law, 42 U.S.C. § 1395nn, generally prohibits physician self-referrals for certain health services that may be paid for by a federal healthcare program. An example would be a physician who has an ownership interest in a surgery center and refers all of his patients to the center. There are exceptions called "safe harbors" to this law that would allow certain instances of self-referral. Finally, the False Claims Act, 31 U.S.C. § 3729 (FCA), imposes civil liability on persons who knowingly submit a false or fraudulent claim to a federal healthcare program. The most common example of this would be fraudulent billing to a federal healthcare program for services not rendered or supplies not used.

"Build It and They Will Come"

There is an assumption that if you build a new healthcare facility or an addition to an existing healthcare facility, that alone will be enough to generate additional patient volume. As the analysis in Table 5.3 shows, that isn't always the case. Only two of three recent replacement hospitals had positive outcomes when their volumes from 1 year before opening and 1 year after opening were compared. There are always unforeseen circumstances and occurrences that can prove assumptions wrong.

Past Experiences

Often, when a hospital administrator relocates to a new healthcare provider, they will look back at previous experiences when evaluating the prospects for growth in certain service lines at their new facility. Often, that evaluation includes past experience with return on investment (ROI) among different service line construction projects. Table 5.4 shows relative ROIs for several service line construction projects. If you are a hospital administrator trying to make the difficult decision of allocating capital between two service lines, this comparative analysis may help with your decision. However, with two of the service lines, other factors should be considered along with the ROI.

	2011–2013	
Replacement Hospital A Opened 8/2012	Admissions	10.7%
	Surgeries	2.1%
	ER Visits	0.7%
	Deliveries	3.6%
Replacement Hospital B Opened 10/2012	Admissions	3.2%
	Surgeries	11.9%
	ER Visits	8.2%
	Deliveries	35.5%
Replacement Hospital C Opened 4/2012	Admissions	−6.4%
	Surgeries	−8.4%
	ER Visits	20.7%
	Deliveries	−8.8%

Table 5.3 Replacement Hospital Volume Analysis

Project Type	n-value	Relative ROI
Cardiovascular	7	37.3
Radiology/Diagnostic	7	28.5
Beds Expansion	12	23.5
Surgery	13	21.6
Emergency Department	15	20.1
Women's Services/NICU	6	18.0
Facility Renovations	10	16.4
Total/Average	**70**	**23.6**

Table 5.4 ROI for Service Line Construction Projects

The ROI considered in this analysis is a direct ROI, meaning that the "return" portion of the ROI equation only includes revenue directly from the service line itself. So, in looking at the emergency department, revenue from other downstream ancillary services including diagnostic imaging, surgery, lab, and inpatient departments would not be considered. This obviously works in favor of the other areas, as ED patients usually make up a significant portion of patient volume for those areas. For example, at some facilities, up to 60 percent of inpatient admissions come from the emergency department.

Another department that has a low relative ROI in Table 5.4 is Women's Services and NICU. According to the Bureau of Labor Statistics, women make approximately 80 percent of healthcare decisions for their families and are more likely to be the caregivers when a family member falls ill.[5] So, anecdotally, if women have a good impression of a healthcare facility when they are there giving birth, it is assumed that, when the need arises for other healthcare services in their family, they would choose and direct those other family members to the healthcare facility where they had that positive experience during birth. Therefore, while the direct ROI for Women's Services and NICU is low, the potential downstream impact for other healthcare service lines at the facility could be significant.

Appropriate Facility Capacity

After an appropriate justification for a project has been established and the expected volume of patients has been determined, it is important to test for appropriate capacity. A capacity analysis can be used to determine the number of KPUs required to accommodate anticipated service line patient volumes (patients, cases, procedures, patient days, etc.) given certain operational parameters (average length of stay, average case time, operational hours, etc.). For example, the "capacity" of an inpatient OR may be 700 to 1000 inpatient cases per year, depending on the types of procedures being performed.

[5]United States Department of Labor website, http://www.dol.gov/ebsa/newsroom/fshlth5.html.

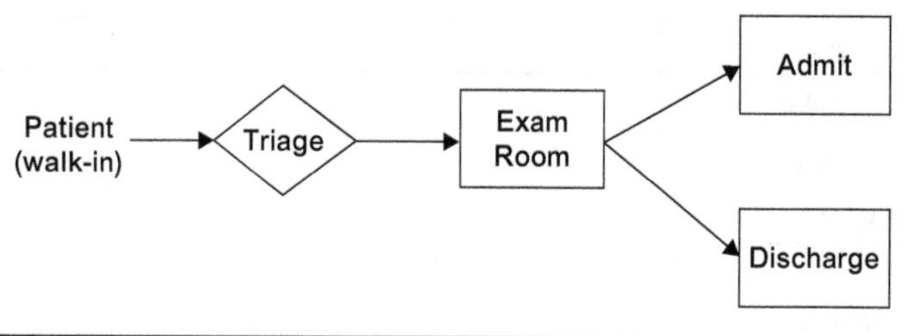

FIGURE 5.1 Patient flow for a traditional emergency department.

If you anticipate 1400 to 2000 inpatient surgery cases per year, you would need two ORs to accommodate that volume.

Service Line Operational Models

A service line operational model describes how patients flow through a healthcare diagnostic or treatment area. Understanding the operational model being used to care for patients in those areas is one of several factors that should be considered when determining appropriate service line capacity. The flow chart in Fig. 5.1 shows the basic patient flow for a traditional ED. First, an ambulatory or "walk-in" patient arrives at an ED and is seen in a triage area where their acuity or the seriousness of their injuries or illness is determined. Unless they have an emergent condition or need immediate intervention, they are then taken to an exam/treatment room. After any necessary tests are made and they are treated for their illness or injury, they are either discharged or admitted to the hospital as an inpatient.

Recently, many hospital emergency departments have been investigating ways to improve their efficiency and patient satisfaction. One operational model that has been employed to that end includes the use of a *results waiting area* (RWA) or internal disposition area (IDA). As you can see in Fig. 5.2, an RWA or IDA produces an alternative path for ED walk-in patients. Usually, a patient will follow the same path as a patient in a traditional ED until they are in the exam/treatment room. While in the exam room, the patient will have blood drawn or other samples taken for lab tests and have diagnostic-imaging (general radiology, computed tomography [CT], EKG, ultrasound, etc.) tests made. In the traditional model, the patient would wait in the exam room while the results from the tests are determined. Alternatively, a low-acuity or relatively healthy patient that will likely be discharged could be moved to an RWA/IDA while waiting on the test results. The RWA/IDA is a sub-waiting area for patients and can be nicely appointed with comfortable seating, TVs, and other amenities that would make it preferable to an exam/treatment room.

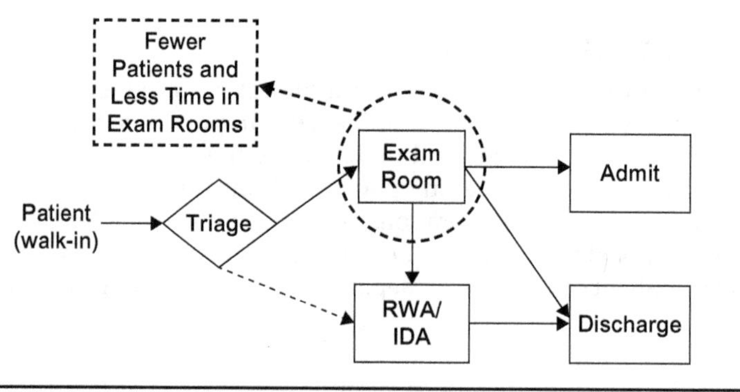

FIGURE 5.2 An alternative path for emergency department walk-in patients.

Making this transfer can accomplish two important things. First, it decreases the time a patient spends in an exam/treatment room, making that room available for the next patient sooner or effectively increasing the capacity of that room. Second, it can also increase patient satisfaction by allowing the patient to wait in a more comfortable environment. Studies have also shown that this type of movement from one location to another generally gives a patient a sense that he or she is making progress through the healthcare system compared to spending the same amount of time in a single location.

Service Line Operational Parameters

Service line operational parameters greatly affect capacity requirements. Examples of operational parameters and variables that are considered in a capacity analysis include target occupancy rates for inpatient beds, average case times and operational hours for different surgical service lines, room turn times for diagnostic imaging modalities, and C-section rates for OB services. Three important operational parameters for an ED include peak patient volume times, the patient acuity mix, and ALOS.

Peak Patient Volume Times

Continuing with the emergency department example, peak patient volume times should be considered as part of a capacity analysis for an ED. This involves analyzing the volume of patients that arrive or present at an ED throughout a typical day. The chart in Fig. 5.3[6] illustrates a pattern of arrival times that can generally be applied to most emergency departments across the United States.

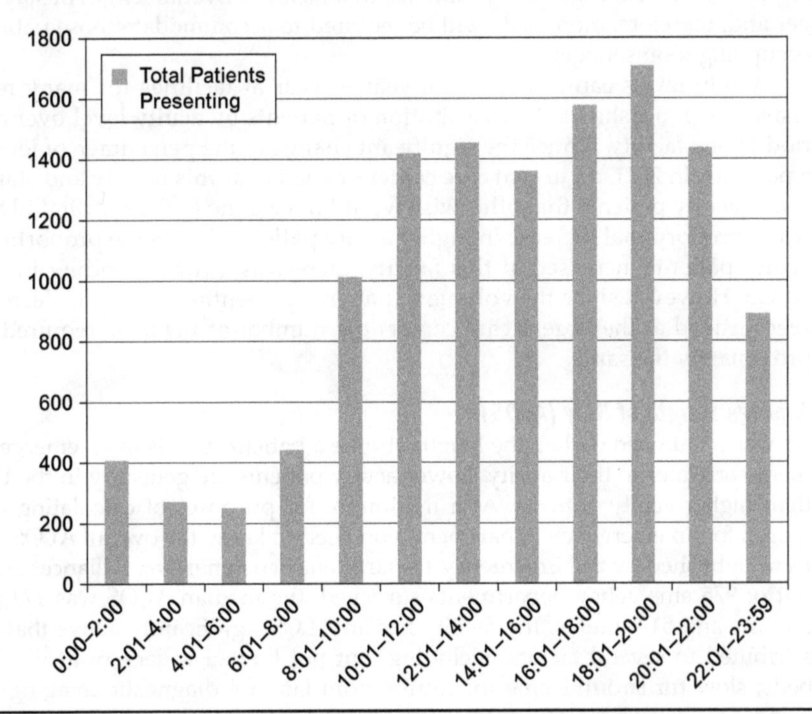

FIGURE 5.3 Pattern of arrival time applied in most emergency departments.

[6]Source: Dr. James Augustine, Emergency Department Benchmarking Alliance (EDBA), 2012.

The volume of patients presenting at a typical ED will be lowest in the early morning hours then increase significantly around mid-morning. Volumes of patients will peak sometime during midday and continue at high levels before dropping down around the end of the day. The dilemma for facility planners is trying to determine an appropriate number of exam/treatment beds to accommodate a fluctuating patient volume. If you plan for enough beds to accommodate the peak volume of patients, you will waste money by building spaces that are significantly underutilized during the low-volume periods of the day. Conversely, if you build only enough beds to accommodate the patients presenting during the low-volume periods of the day, you will have patients waiting for beds during the high-volume periods of the day. The ideal number of patient rooms will *always* be a compromise between the two extremes.

Patient Acuity Mix

When a patient arrives at an ED, most facilities will triage or classify them into different categories based on their acuity (how soon they need medical attention). One such triage system in common use is the Emergency Severity Index (ESI) created by the Agency for Healthcare Research and Quality.[7] In this system, patients are grouped into five categories: immediate (level 1), emergent (level 2), urgent (level 3), semi-urgent (level 4), and non-urgent (level 5). Generally, lower acuity patients will have shorter ALOS than higher acuity patients. Knowing the allocation of ED patients by their acuity levels is useful in the overall calculation of bed need. For example, if you are at a facility that has high proportion of high acuity patients, that facility's overall length of stay will be longer and, therefore, more beds will be required to accommodate those patients that are occupying rooms longer.

Acuity levels can change from year to year at facilities for many reasons. The chart in Fig. 5.4 shows the distribution of patients by acuity level over a 3-year period at one facility. Notice the significant change in the percentage of levels 3, 4, and 5 patients. In 2011, an urgent care center opened near this facility and started to treat lower acuity patients that otherwise would have gone to the hospital ED. The result was a proportional increase in higher acuity patients. When the proportion of higher acuity patients increased at this facility, there was a corresponding increase in the ALOS. However, since the volume of patients presenting at the ED decreased (some seen instead at the urgent care center) the number of ED beds required stayed approximately the same.

Average Length of Stay (ALOS)

As was mentioned earlier, the length of time a patient spends in an emergency department correlates to their acuity. Lower acuity patients are generally in the ED less time than higher acuity patients. At a minimum, for purposes of calculating the beds required for an emergency department, you need to know the overall ALOS for patients. Data published by the Emergency Department Benchmarking Alliance[8] suggests that, out of 975 emergency departments surveyed, the median ALOS was 171 minutes, or 2 hours and 51 minutes. If a facility has an ALOS significantly above that, it could be attributed to several factors, including, but not limited to lack of available inpatient beds; slow turnaround time for results from lab and diagnostic imaging; the facility being a teaching hospital/children's hospital; inefficient design or department layout;

---

[7]http://www.ahrq.gov/professionals/systems/hospital/esi/esi1.html.
[8]http://www.edbenchmarking.org.

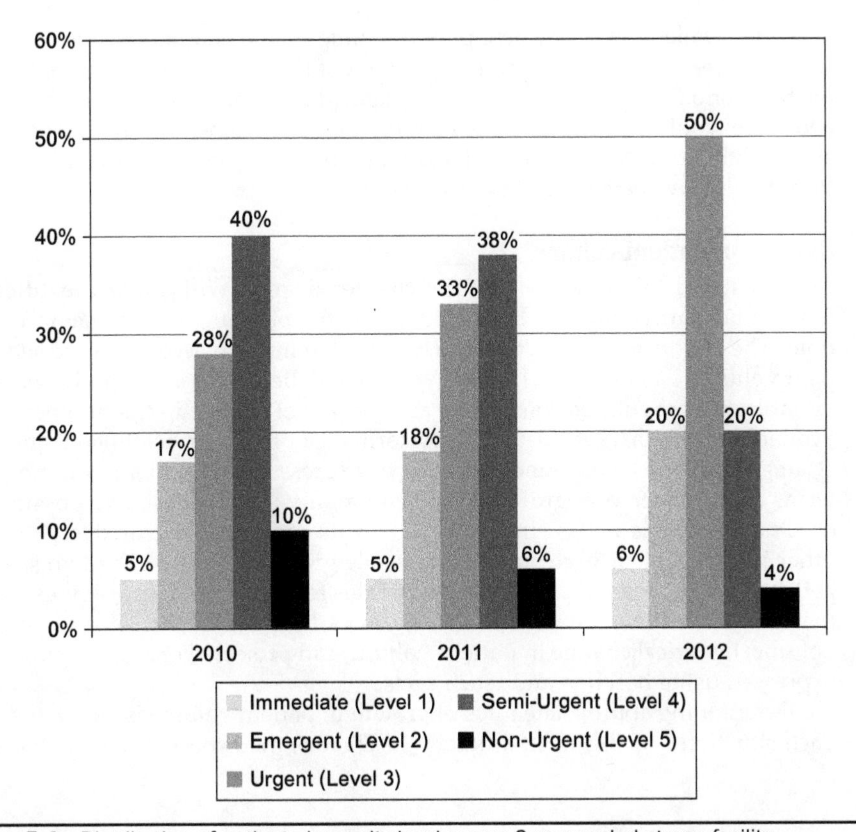

FIGURE 5.4 Distribution of patients by acuity level over a 3-year period at one facility.

higher than average patient acuity; or, possibly, inefficient staff operations. Before calculating bed need for an ED, it is critical to identify the source of an above average ALOS. In some cases, facility changes can alleviate the problem. In those cases, an improved ALOS can be used for planning purposes. But, in other cases, you will need to assume that the above-average ALOS is unavoidable and plan for it to continue.

Statistical Probability

One method of quantifying bed utilization in this case is through the use of statistical probability modeling. One such probability model is call the Poisson distribution, and is named after French mathematician Siméon Denis Poisson.[9] This probability model has been shown to be particularly effective at expressing the probability of a given number of events occurring during a fixed interval of time when the average rate of occurrence is known. In this case, the events are the arrivals of patients to an emergency department. The results can also be expressed in terms of the likelihood of events in the form of a sensitivity analysis. For example, the analysis can be set up to calculate the number of beds required so that 99 percent of the time there will be a bed available for arriving patients. By modifying a variable in the analysis, the number of beds required

[9]http://en.wikipedia.org/wiki/Poisson_distribution.

to ensure availability at any give period of time can be determined. Expressed another way, with the time frame in years, 99 percent of the time would mean that 1 percent of the time, or 3.65 days out of a year, the planned number of beds would be inadequate, and there would be other patients waiting for a bed. Some facilities might be comfortable with building fewer beds and accepting that 5 percent, or 18.25 days out of a year, they would have patients waiting for a bed.

Change in Patient Volume

As was stated in the section of this chapter dealing with project justification, the most important variable to be considered is the potential for change in patient volume. The majority of projects are driven by the need to accommodate *additional* patient volume. Calculating what that volume will be is both an art and a science. Some variables that should be considered are the rate of change in the number of patients available in the market, historical performance of the service line in question, and strategic plans. I have found it useful to express future patient volume growth in terms of year-over-year growth, and I commonly will propose two possible scenarios. One scenario is a "baseline" or the low side of a possible growth scenario and the other is a "reach" or high side of a possible growth scenario. The two scenarios together create a range of possible growth rates for future patient volumes. See Fig. 5.5 for part of an ED analysis that incorporates the market rate of change in patient volume; historical change in patient volume and projected change in patient volume expressed using baseline and reach rates.

Determining appropriate rates of growth in patient volume is an art and a science. Each situation is going to be different. In Fig. 5.5, the expected rate of change for ED

ED Facility Growth Rates

	ED	
Market	0.94%	
2010	0.47%	
2011	–2.46%	
2012	1.17%	
	Baseline	Reach
2013	0.94%	1.88%
2014	0.94%	1.88%
2015	0.94%	1.88%
2016	0.94%	1.88%
2017	0.94%	0.94%
2018	0.94%	0.94%
2019	0.94%	0.94%
2020	0.94%	0.94%

Figure 5.5 Emergency department analysis that incorporates the market rate of change in patient volume.

patients over the 8 years starting in 2013 is 0.94 percent annually. From 2010 through 2012, this facility's patient volume fluctuated up and down. The baseline rate of future change was set to match the forecast market rate of change, 0.94 percent. The reach rate of change was set to equal two times the market rate of change from 2013 through 2016 and then revert to the market rate of change thereafter. In this situation, the rate of growth seemed reasonable when all available information about the market and strategic plans was taken into consideration.

Capacity Analysis Model

Knowledge about an emergency department's operational model, peak patient times, patient acuity mix, average length of stay, and projected patient volumes allows for the creation of a mathematical model that will predict the number of beds that will be required at some point in the future. Table 5.5 shows a summary of all parameters except the future patient volumes.

Note that, in this analysis, the facility has decided to include a results waiting area in their emergency department. For that reason, two separate lengths of stay are included. One for the time a patient would spend in an exam room and the other for the time a patient would spend in a results waiting area.

Peak Volume Hours	14.00				
Percent during Peak	75.0%				
PATIENT TYPE	**Percent by Type (%)**	**Percent to Results Waiting (%)**	**Exam Room ALOS(HH:MM)**	**Results ALOS(HH:MM)**	**Total ALOS (HH:MM)**
Non-urgent (Level 5)	**2.0**	0.0	0:30	0:30	**1:00**
Semi-urgent (Level 4)	**20.0**	0.0	0:45	0:45	**1:30**
Urgent (Level 3)	**40.0**	0.0	1:15	1:00	**2:15**
Emergent (Level 2)	**25.0**	0.0	3:45	0:00	**3:45**
Immediate (Level 1)	**7.0**	0.0	4:00	0:00	**4:00**
Behavioral health	**6.0**	0.0	6:00	0:00	**6:00**
Total/Average	100.0	0.0	2h 14m	0h 34m	2h 48m
				* ESI Average =	2h 40m

Table 5.5 Summary of All Parameters Except Future Patient Volumes

Room Need

Pulling all of the variables and parameters together allows for the calculation of key planning units (KPUs) needed, in this case for the number of ED exam/treatment beds required. The final calculation is actually quite simple.

1. Begin with the time a single KPU is available for use to get average KPU availability.
2. Determine the average time a patient would be in the KPU for examination or treatment.
3. Determine the total number of expected patients in a certain time frame, such as a year.
4. Multiply the average time per patient by the total number of expected patients to get total patient time.
5. Divide the total patient time by the KPU availability to get raw number of KPUs required.

However, in most cases, this basic calculation will underestimate the actual number of KPUs required. For example, let's assume that an ED exam/treatment room is available 365 days per year, 24 hours per day. That equals 8760 hours and equates to KPU availability. Next, calculate total patient time. Let's assume 50,000 ED patient visits are expected and that the ALOS for those patients is 2 hours. That equals 100,000 hours. So, 100,000 divided by 8760 equals 11.42 KPUs. For reference, divide 11.42 KPUs into 50,000 patients and you will get 4378 patients per KPU per year.

Experience has shown that an ED exam/treatment room can accommodate between 1250 and 2000 patients per exam/treatment room per year. So what's missing? The fact is that other variables already discussed have not been taken into account. The basic analysis above assumes an average number of patients per hour. We know that the volume of patients arriving at an ED varies depending on the time of day. That variable and the others must be accounted for in the final analysis. See Fig. 5.6 for a sample chart that shows the final results of a sample capacity analysis to determine the number of

Non-urgent/ Semi-urgent Beds	Urgent/emergent Beds	Trauma/cardiac Beds	Behavioral Health	Patients per Bed
3.1–3.1	17.3–17.3	3.4–3.4	4.0–4.0	1,892–1,892
3.1–3.1	17.4–17.4	3.4–3.4	4.0–4.0	1,893–1,893
3.0–3.0	17.0–17.0	3.3–3.3	3.9–3.9	1,888–1,888
3.1–3.1	17.2–17.2	3.3–3.3	4.0–4.0	1,891–1,891
3.1–3.1	17.3–17.5	3.4–3.4	4.0–4.0	1,893–1,895
3.1–3.2	17.5–17.8	3.4–3.4	4.0–4.1	1,895–1,898
3.1–3.2	17.6–18.1	3.4–3.5	4.1–4.2	1,896–1,902
3.2–3.3	17.8–18.4	3.4–3.5	4.1–4.2	1,898–1,906
3.2–3.3	17.9–18.5	3.5–3.6	4.1–4.3	1,900–1,908
3.2–3.3	18.1–18.7	3.5–3.6	4.2–4.3	1,902–1,910
3.2–3.4	18.2–18.9	3.5–3.6	4.2–4.3	1,904–1,912
3.3–3.4	18.4–19.0	3.5–3.7	4.2–4.4	1,906–1,914
				Total Beds
0	0	0	0	30
4	18	4	4	30
2017 Patients per Bed (Without Results Waiting Chairs)				1,820–1,889

Figure 5.6 Sample chart showing the final results of a capacity analysis to determine the number of exam/treatment rooms.

exam/treatment rooms needed, given all of the variables and parameters discussed in this chapter. Note that this was a very efficient ED with around 1900 patients per exam/treatment room per year.

Conclusion

The design and construction of healthcare facilities has evolved over time to a highly refined process that many project managers, contractors, and architects spend their entire careers perfecting. Because the industry is in such a highly evolved state, the opportunities for value engineering and budget reduction after a project has been designed are minimal, as indicated by Fig. 5.7.

The greatest opportunity to impact the cost of a project is at the earliest point in the life of a project. Given that, accurately determining the appropriate number of KPUs early will have a much greater impact on a project's budget than squeezing out a few square feet during design or using a less expensive finish during construction, for example. Project managers, contractors, and architects that understand how a project was justified and can reinforce that justification through their own analysis can ensure alignment of the owner's intent for the project and the solution developed by the design and construction team.

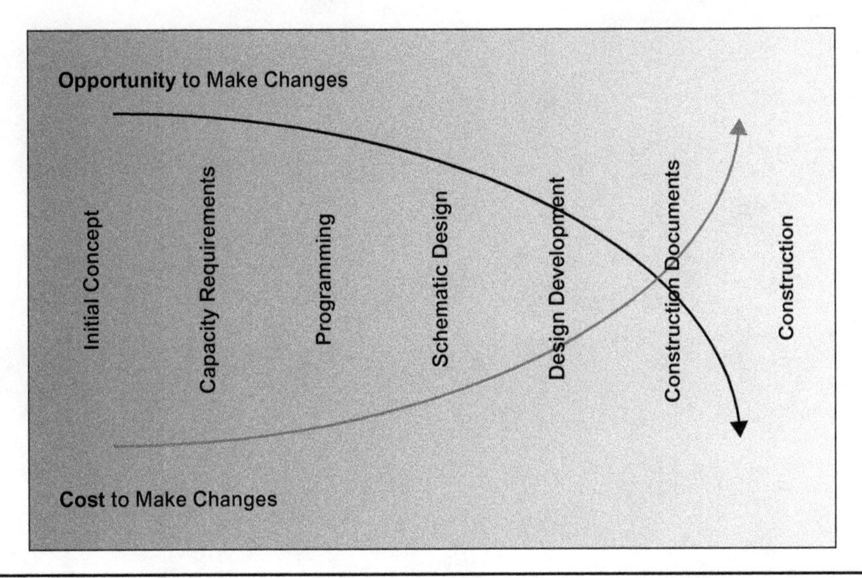

Figure 5.7 Opportunities for value engineering and budget reduction diminish after a project has been designed.

CHAPTER 6

Healthcare Project Financing

Construction projects, like all assets, are funded with capital. The term *capital* is generally considered to refer to the combined total of long-term debt plus fund balance (equity) of an organization. The not-for-profit version of equity is called *fund balance*, but it represents the same thing that equity, or stock, represents to a for-profit company, i.e., investment. Investment, in the not-for-profit healthcare industry, is built up through donations and the excess of revenues over expenses (surplus) over a period of time. For-profit healthcare organizations have the ability to sell shares of stock to build investment (equity), in addition to generating profits and receiving donations. Few hospitals have the ability to generate enough capital through earnings and donations to fund all of the necessary capital projects to remain competitive in the industry. As a result, most hospitals have turned to debt to fund some of their capital needs.

Alternative Sources of Capital and Their Relative Costs

The funding of every construction project should be consistent with the organization's capital plan, and every capital plan should begin with an evaluation of the alternative forms of capital that are available to the organization. The stronger the financial condition of the organization, the more alternative sources of capital will be available, and at a lower cost. A generally accepted premise is that it is best to use the cheapest (least expensive) capital available to fund a project. The cheaper the capital used to fund a project, the lower the break-even point for that project becomes, which leads to greater profitability. There are four categories of capital that should be investigated by most healthcare organizations. The first is an evaluation of the funds that have been generated internally by the organization, namely *cash reserves*. A second source of capital would be *donations* (philanthropy). A third potential source of capital is external *equity*, where investors actually participate with an ownership interest. For-profit healthcare organizations are able to take advantage of this source of capital, while not-for-profit organizations must be very careful on how they make use of equity so they do not jeopardize their tax-exempt status. The fourth source of capital available to healthcare organizations is *debt*. It is important for a healthcare organization to be able to

---

This chapter was written by Chris Payne.

discriminate between these sources of capital, especially in terms of cost, to be able to generate the most cost-effective capital plan.

Debt

Debt issued on behalf of not-for-profit healthcare organizations can be taxable or tax exempt. That is, the interest paid to bondholders (investors) can be taxable income to the investor or tax-exempt income to the investor. The key distinction between taxable and tax-exempt debt is that borrowers pay lower interest rates on tax-exempt debt than on taxable debt with comparable terms. This is because investors demand a higher interest rate on taxable debt to make up for the added tax liability on the interest received. As a result, tax-exempt debt usually should be a cheaper source of capital than comparable taxable debt. In many states, debt that is exempt from federal income taxes is also exempt from state income taxes, making it that much more attractive.

Internally Generated Funds and Philanthropy

The cost of utilizing capital generated internally or through philanthropy is the opportunity cost that is incurred if these funds are spent on projects rather than retained and invested for use on future projects. Not-for-profit healthcare organizations have the ability to borrow capital for specific projects in the tax-exempt markets while retaining and investing their own cash reserves at higher rates in the taxable markets for later use. As a result, the overall cost of a project that is eligible for tax-exempt financing may be higher if internal funds are used and the investment opportunity is lost. Although this opportunity cost never shows up on an income statement, it represents a real economic loss to the healthcare organization.

This is a very important concept. *Properly structured tax-exempt debt can be a cheaper source of capital than a healthcare organization's own cash reserves.* In this case, "properly structured" means tax-exempt debt that is issued efficiently, and with terms that are favorable to the borrowing organization, resulting in the lowest possible cost.

It is important to emphasize that the proceeds of tax-exempt debt must be used to fund the projects for which the bonds are issued, and may not be invested and used to hedge bonds or earn profits. The cash that may be invested is the healthcare organization's own cash reserves, which are not bond proceeds. Proceeds of taxable debt may be used in any way a healthcare organization chooses, but it is unlikely that a healthcare organization could consistently invest its cash at a rate of return higher than the interest rate paid on its taxable debt. Because of this, taxable debt is usually a more expensive form of capital than either a healthcare organization's cash reserves or tax-exempt debt.

Equity

The equity markets generally require a rate of return that is greater than the return on taxable debt. This is because equity carries more risk, in that debt holders (creditors) get paid prior to equity shareholders in the event of a bankruptcy. This means that the use of equity to fund a project typically will cost a healthcare organization more in the long run than using debt or its own cash reserves. The additional cost may be warranted if other benefits are generated through the use of equity, such as firming up relationships with doctors or risk sharing. Of course, the primary advantage of outside equity is the freedom from a contractual obligation to repay the amount invested or to generate any set return for the investor. The use of equity may also allow a healthcare organization to tap a larger pool of capital, resulting in the ability to fund more projects, albeit at a higher cost.

The use of equity by not-for-profit healthcare organizations has become increasingly difficult as more and more scrutiny is being placed on interactions with for-profit entities. As a result, funding a project with outside equity may not only result in utilizing more expensive capital, but it may end up causing complications relative to the tax-exempt status of the organization, if it is not done very carefully. As government regulations become increasingly burdensome to not-for-profit healthcare organizations, there may be increasing incentive for these organizations to give up their not-for-profit status and turn to the less restrictive but more expensive taxable capital markets, especially when interest rates are low and the difference in cost is less significant.

In summary, properly structured tax-exempt debt usually is the least expensive form of capital available to not-for-profit healthcare organizations. Consequently, a not-for-profit healthcare organization should use tax-exempt debt whenever possible and maintain its cash reserves for projects that are ineligible for financing with tax-exempt debt, or for future projects when tax-exempt debt may no longer be available. Taxable debt and equity are relatively more expensive forms of capital and are more appropriately used when tax-exempt debt is not available and cash reserves are down to the minimum required for operations. Because some form of debt is the most likely source of capital to fund construction projects, the remainder of this chapter will focus on debt.

Financing with Debt

Debt is the general term used to describe a variety of obligations to repay borrowed money. Debt is usually evidenced by instruments such as bonds, notes, or mortgages. In each case, a lender or group of lenders allows a borrower the use of a specified amount of money, which is the *principal* amount of the loan, in return for a rental fee for use of that money, known as *interest*. Both principal and interest must be paid to the lender at specified times pursuant to the terms and conditions of the loan. The required payments of principal and interest are referred to as *debt service*. The required payments of principal and interest for any given year are referred to as *annual debt service*. The sum of all debt service payments is referred to as the *total debt service* under the obligation.

Taxable Debt versus Tax-Exempt Debt

Once a healthcare organization has determined that it will use debt to finance all, or a portion of, its capital projects, it must determine whether taxable or tax-exempt debt is more appropriate. For every form of tax-exempt debt, there is a comparable version of taxable debt that could be used if tax-exempt debt is not available or desirable to the borrower. Although tax-exempt debt is usually less expensive for long-term financing, federal law severely restricts its use, making it unavailable for a variety of projects. The Tax Reform Act of 1986 specifies that, with very few exceptions, any project that directly or indirectly benefits any person or entity other than a tax-exempt entity cannot be financed with tax-exempt debt. Examples of projects that normally cannot be financed with tax-exempt debt under current tax law include office buildings for physicians who are not employees of the healthcare organization, parking garages used by people visiting non-employed doctor's offices, athletic facilities used for purposes other than rehabilitation, or any joint venture with individuals or for-profit corporations. As such, for-profit healthcare organizations currently are prevented from using tax-exempt debt, and not-for-profit healthcare organizations are prevented from using tax-exempt debt for projects that involve almost any direct benefit to individuals or for-profit entities.

For interest on debt to be exempt from federal income tax, the debt must be issued by a state, or a political subdivision of a state, such as a county or city. Some states (and some state subdivisions) have legislatively established agencies, known as *authorities*, to issue tax-exempt bonds for specific public purposes such as housing, education, or healthcare. Private not-for-profit healthcare organizations must use a conduit issuer to make their bonds tax-exempt. A *conduit issuer*, such as an authority, a city, or a county, issues tax-exempt bonds on behalf of a borrower, and lends the proceeds of the bond issue to that borrower. The availability of tax-exempt conduit issuers varies state by state. Some states severely limit the number of entities that can issue bonds for private not-for-profit healthcare organizations. Public healthcare institutions, which are owned by a city, county, hospital district, authority, or other political subdivision, may be able to issue their own tax-exempt bonds without the need for a conduit issuer.

The key drawbacks to tax-exempt debt are the limitations for which it can be used and under which it can be issued. A not-for-profit healthcare organization can greatly offset these drawbacks by financing more eligible projects with tax-exempt debt, and building and maintaining its own cash reserves for later use on projects not eligible for tax-exempt financing. Taxable debt carries no such restrictions and can be issued for any purpose, which may change at any time without the borrower having to redeem the debt. As a result, taxable debt and equity offer a degree of flexibility that is not available with the proceeds of tax-exempt bonds. For example, pension shortfalls are not an eligible use of tax-exempt proceeds, but taxable financings can be used to fund the shortfall in a pension plan. Additional advantages of taxable financings would include the lack of a need to use a conduit issuer or to comply with tax laws for the useful life and use of bond proceeds.

There are two primary drawbacks to taxable financings besides higher cost. First, if done on a variable rate basis, the rates may seem attractive while interest rates are low; but when rates rise, taxable financings will become proportionately more expensive than tax-exempt debt. This implies that it is most appropriate to issue fixed rate bonds in a taxable financing unless it is short term. The second drawback to taxable financings is that the most efficient market is for issues in excess of $250 million. This limits the most significant benefit to very large systems. Smaller issues can get done, but at slightly higher cost than the larger issues.

Publicly Issued versus Privately Issued Debt

When representatives of a healthcare organization go to borrow money directly from a bank, they sign a note to evidence the obligation to repay the debt with interest. The note will detail the terms and conditions under which the debt obligation must be repaid. If the terms and conditions tend to be lengthy, they are often contained in a separate loan agreement. When an institution borrows from the *market*, the document evidencing the obligation usually is some sort of a *bond*. A bond form contains some detail on the relevant terms and conditions of the obligation, but the complete detail is contained in supporting documentation that typically is held by a bond trustee. The bond is what is held by the depository on behalf of the bondholder as the primary evidence of the debt.

In some cases, the bonds are *publicly offered*, which means that they are offered to many potential purchasers. In other cases, bonds are sold in *private offerings*, in which case there are a limited number of potential sophisticated investors (usually institutions) that are given the option to purchase the bonds. Bond documents for publicly offered financing are drawn up before the actual lender is known. The bonds are sold after the documentation is either fully or mostly, completed. Purchasers of bonds in a public offering typically are not given the chance to comment on the bond documentation, other than in details that are finalized during pricing such as call provisions, interest rates, or the amount of original issue discount or premium. When bonds are being sold in a private placement, the purchaser(s) of the bonds is(are) frequently given the ability to make final comments on the documentation to customize the bonds.

Credit Enhancement

The term "credit" generally deals with an entity's ability to repay its debt obligations and other liabilities in a timely manner. In some situations it is desirable, or necessary, to obtain *credit enhancement* on a bond issue—which means to increase the credit standing (or credit rating) of the bond issue above what it would otherwise be. When a bond issue is credit enhanced, the credit risk becomes that of the credit facility provider rather than that of the underlying borrower. This means that the bond issue will be given a rating based on the credit strength of the credit facility provider rather than on the credit strength of the healthcare institution incurring the debt.

Credit facilities on healthcare bond issues almost always are provided by one of two sources: letters of credit from banks, or insurance from FHA. Bond insurance from municipal bond insurers was an option until the financial crash in 2008. After that the bond insurers were downgraded to the point where their insurance provided little or no value, so bond insurance is no longer a viable alternative. A guaranty of the debt of an institution by a related entity with a higher credit rating, such as a parent corporation, could also be considered credit enhancement.

Bank Facilities

The most common form of credit facility provided by banks is a letter of credit. A *letter of credit* (LOC) is a document that requires the bank by which it is issued to make payment, according to its terms, no matter what happens to the underlying borrower. In other words, the bank is obligated to pay even if the borrower goes bankrupt. With a *line of credit*, the bank is obligated to pay unless the borrower is bankrupt, or in default of the line of credit agreement in some other way. An LOC provides both credit and liquidity to a bond issue, meaning that the bond trustee can draw funds under the LOC regardless of the financial condition of the borrower. A line of credit provides liquidity but does not change the credit, meaning that a bond trustee can draw funds under the line of credit only if the borrower is not in default of the line of credit agreement. When a line of credit is used, the rating of the bank is not transferred to the bonds.

With an LOC backing a bond issue, the rating on the bonds reflects either the rating of the bank or the rating of the underlying borrower, whichever is higher. There is little point in paying for an LOC if the rating of the borrower is higher than that of the bank, so it would be unusual to structure a bond issue with an LOC unless the intent was to obtain the rating of the bank on the bonds.

Fixed Rate versus Variable Rate Debt

When the interest rate on a debt instrument is set to maturity, the debt is said to have a fixed interest rate and is referred to as *fixed rate debt*. If the interest rate changes prior to maturity, it is *variable rate debt*, which is also referred to as *floating rate debt*. Fixed and variable rate financings have their appropriate uses by healthcare organizations.

The primary advantage of fixed rate debt is that the interest cost is a known factor throughout the life of the debt. The primary disadvantage is that fixed rate debt generally costs more than variable rate debt at the time of issuance, both in terms of issuance costs and interest rates. At any point in time, current long-term fixed interest rates typically will be higher than current variable interest rates. If variable interest rates rise and remain relatively higher over the life of the debt, the fixed rate may prove to be the less expensive option in the long run. Conversely, if interest rates fall or remain relatively constant over the life of the debt, the variable rate option will prove to be relatively cheaper. Fixed rate debt is generally most appropriate for funding projects where the borrowing organization does not have cash reserves to offset the risk of variable rates. It should be used when long-term rates are relatively low and acceptable to the borrower.

Variable rate tax-exempt debt has been a popular choice of financing for healthcare organizations since the 1970s. It became especially popular in 1981, when interest rates were reaching all-time highs and borrowers did not want to lock into very high long-term interest rates. Since that time, many borrowers have chosen to utilize fixed rate debt with the assumption that long-term rates would be low at the time of issuance and with the intent of protecting themselves from higher rates in the future. This is especially true today as long-term rates are nearly the lowest they have ever been. However, variable rates have been so low that any borrower that issued tax-exempt variable rate debt to date has enjoyed a lower overall cost of debt than had they issued comparable fixed rate debt. Of course this will change if rates do go up significantly, but variable rate financing has proven to be a very sound form of financing for the last 30 years.

Direct Bank Loans

While direct loans from banks to any private organization are taxable, not-for-profit private healthcare organizations can borrow on a tax-exempt basis from banks through a conduit issuer, thereby lowering the interest cost. Like other long-term tax-exempt healthcare revenue bonds, tax-exempt bank loans generally amortize over long-term periods (20 to 30 years or possibly longer) tied to the useful life of the assets financed. However, the bank purchaser will require a mandatory tender of the bonds after a shorter period of time—usually in 3, 5, 7, or 10 years. The interest rate on these loans can be fixed for that period or variable. At the end of the agreed-on time period, the bank has the option to extend the financing to a new mandatory tender date, up to and including the maturity date of the debt. In the event the bank is not willing to extend the tender date, the healthcare borrower can remarket the debt to another bank or different type of investor, or refund it with a different bond offering.

Direct bank tax-exempt debt bearing interest at a variable rate typically is pre-payable without penalty at any time. Fixed rate debt is pre-payable either with a make whole provision at any time (similar to the structure present on taxable corporate bonds) or non-callable for a certain period of time with a call premium after the initial non-call period. Under the make-whole prepayment scenario, the borrower has the right to make a lump sum payment to redeem the debt derived from a formula based on the net present value of interest payments that will not be paid because of the call.

One advantage of bank debt over other types of debt is that banks can provide a draw-down feature for the debt proceeds so that a healthcare borrower can obtain the borrowed funds from the bank as needed, rather than all at once. This feature not only is convenient but can save a significant amount of money, as the borrower does not have to reinvest the construction funds in short-term, lower yielding investments during the construction period. It also avoids the need to capitalize a portion of interest payments on the debt during the construction period.

Even though most tax-exempt bank loans will be on *parity* with the borrower's other long-term debt (and thus share in the security interest in hospital revenues and any other pledged property), a bank will require a few of its own financial covenants and different events of default and default remedies than a traditional fixed rate debt issue sold to the investing public. Any additional covenants will be similar in scope and level to the additional covenants found in the same bank's letters of credit and standby bond purchase agreements.

Many banks also expect the healthcare borrower to purchase noncredit banking services now or in the future as part of the overall compensation to the bank for extending credit. This additional business requirement can make the financing more complicated, but often leads to the conclusion that a healthcare organization should be buying all or a portion of its noncredit banking services from a bank which "lends its balance sheet" in support.

Variable Rate Demand Bonds

Variable rate demand bonds (VRDBs) are issued with a long maturity, but the bondholders have the right to *put* (which means the right to sell) these bonds to the trustee, at *par* (face value), on short notice. While the put period typically is 7 days, VRDBs can be set up with the put option daily, weekly, monthly, or any period desired. This put feature is what allows these bonds to sell at short-term interest rates rather than at long-term rates, even though the bonds have a long-term maturity. An underwriting firm serves as a *remarketing agent* to resell the bonds if they are ever put. The remarketing agent also serves as a *pricing agent* for VRDBs, with a rate determined by that agent for each pricing period.

A letter of credit typically secures the VRDBs to provide cash to buy them back from a bondholder if they were ever put and the remarketing agent is unable to resell them. The letter of credit provides both credit strength and liquidity. In other words, the letter of credit could be drawn upon to make regularly scheduled interest payments (thereby providing credit) in the event that the underlying borrower is not able to do so, in addition to providing the liquidity needed in the event that the bonds are put and the remarketing agent is unable to resell them. Purchasers of these bonds generally require a rating of A1/A+ or better. The use of a letter of credit allows the VRDBs to obtain the ratings of the bank rather than the ratings of the borrower. This limits the number of banks that can provide these letters of credit.

Most versions of VRDBs provide additional flexibility by including a *multimode* format where a 7-day repricing feature is only one alternative. Some VRDBs include a *commercial paper mode,* which allows different bonds within one series to be priced at fixed rates for different periods. This mimics the performance of true commercial paper, where bonds are sold at fixed rates for differing short-term maturities ranging from one day to 270 days, and then *rolled over*, or reissued for additional short-term periods. The net effect of a commercial paper program is similar to a VRDB. The difference is that, with commercial paper, the bond actually matures on each roll-over date and is reissued for the next period. With a VRDB, the bond does not mature until the final maturity, but is repriced periodically.

Floating Rate Notes

Floating rate notes (FRNs) are long-term bonds sold to institutional buyers for periods of 3 to 15 years. Unlike VRDBs, the bondholder has no right to put the bonds until the mandatory tender. This means that no separate liquidity facility is needed. FRN's price at a spread off of an index such as SIFMA or a percentage of LIBOR, rather than having rates set by a remarketing agent. The rates on FRNs tend to be higher than the rates on bank loans or VRDBs, but offer another alternative to healthcare borrowers. Unlike most other forms of variable rate debt, FRNs are typically not callable until 6 months prior to the tender date.

Total Return Swaps

Another option to achieve variable rate exposure is to issue long-term, fixed rate bonds that are sold to a financial institution and later swap the effective interest rate back to a variable rate for 3 to 5 years, at rates comparable to VRDBs, with a vehicle called a total return swap (TRS). Different bankers offer slight variations on this theme. It is a structure that can result in fixed rate debt on the balance sheet, but with a net variable rate exposure. There are market risks associated with most versions of these structures when the TRS matures but, in some cases, they are not any greater than the risks posed with other variable rate structures. Currently, the IRS is investigating a few TRS deals, so bond counsel and/or your financial advisor should be consulted before too much time and effort is expended considering one of these transactions to make sure the proposed structure does not run afoul of tax law.

Developer Financing

In some cases, the healthcare organization may find it desirable to let the developer provide the financing for a construction project. These financing vehicles can take a number of forms, such as construction loans, permanent financing, sale-leaseback, or leases. In almost all cases, the sources of capital used by the developer will be taxable, so it would be unusual for the all-in cost of developer financing to be lower than what a creditworthy not-for-profit healthcare organization should be able to get on its own tax-exempt financing. The developer can take advantage of depreciation expense offsetting some of the cost differential. Developer financing is probably most appropriate for smaller transactions that do not have the economies of scale to warrant a bond issue. This approach had been used often for development of medical office buildings in the past, but the current accounting requirements to show master leases as capital commitments has greatly reduced the use of this funding vehicle.

Appropriate Uses of Variable Rate Debt

Variable rate debt is most appropriate in two situations: first, when long-term interest rates are at unacceptable levels and it is desirable to avoid being locked into a long-term fixed rate until rates drop; and, second, when a healthcare organization has sufficient cash reserves that will, in effect, offset the risk of the variable rates.

If a healthcare organization issues fixed rate debt while maintaining significant cash reserves, and rates decline in the future, it may find itself in a situation where its cash reserves are earning a lower rate than what is being paid on the debt. The healthcare organization typically will invest its cash reserves at short- to intermediate-term maturities to maintain liquidity and avoid market risk. The rate on its fixed rate debt would be set for as long as 30 years or more. Consequently, when interest rates drop, the healthcare organization may not be able to invest at the desired shorter maturities and still earn enough return (or yield) to equal or exceed the rate it is paying on its tax-exempt bonds. In this case, the cost of capital for the fixed rate debt would be higher than the cost of using the cash reserves.

This risk can be avoided by issuing variable rate debt instead of long-term, fixed rate bonds. When variable rate debt is issued, the cost of the debt moves in tandem with changes in the return on any cash reserves invested in the short-term markets. In most markets, the healthcare organization should be able to invest its cash reserves at rates higher than the rates on the variable rate debt, thereby generating a net gain in income. This is especially true when rates are higher in general, because the net gain comes from the spread between income earned on the taxable investments and the interest cost of the tax-exempt bonds, rather than from the absolute level of the interest rates.

It is important to point out that the cash reserves must not be tied to a tax-exempt issue in any way lest they be considered to be an invested sinking fund. An *invested sinking fund* is a cash reserve that can be reasonably expected to be used to pay debt service on the tax-exempt bonds. The determination that the cash reserves are an invested sinking fund would result in a *yield restriction,* which is a limitation placed on the yield that can be earned on the cash reserves, eliminating part of the advantage of such a program.

Benefits of Variable Rate Debt and Cash Reserves

Using variable rate debt to finance projects while a healthcare organization maintains cash reserves has several significant benefits. The first is the considerable flexibility that the cash reserves provide to the healthcare organization. There are no restrictions on these reserves, other than those self-imposed by the organization itself. These reserves can be used to fund projects that are not eligible for tax-exempt financing. They may also be retained and invested for use on future projects. The healthcare organization retains total control of these reserves, which results in a greater ability to take advantage of future opportunities for expansion or acquisitions when its competition may not be able to do so. Cash reserves provide a not-for-profit healthcare organization with capital that has flexible uses, comparable to the flexibility of taxable debt and equity used by for-profit healthcare organizations, but at a lower cost.

A second benefit is that the market and the rating agencies place a strong emphasis on the liquidity of healthcare organizations when evaluating credit strength. A healthcare organization with significant cash reserves is more likely to be perceived as being successful, well run, and able to withstand swings in the business environment. Therefore, a healthcare organization with $100 million in debt and $125 million in cash reserves is likely to be perceived as being stronger than a healthcare organization with no debt and $25 million in cash reserves.

Third, as discussed above, it should be possible to invest the cash reserves at a return that is greater than the cost of the variable rate debt until the reserves are used. This results in a lower overall cost of capital for the borrower than if its own cash reserves were spent on the project.

A fourth benefit is that most forms of variable rate debt can be converted to a fixed rate when the healthcare organization decides to spend the cash reserves that are serving to offset the risk of the variable rates. This ability to convert to a fixed rate might prove to be valuable if Congress ever eliminates the ability of healthcare organizations to issue new tax-exempt debt, or if the healthcare organization has a significant capital need that is not eligible for tax-exempt financing. In either case, a healthcare organization might choose to use its cash reserves to fund projects, rather than use the more expensive alternative of taxable debt, and find it beneficial to convert its variable rate debt to fixed interest rates.

A fifth benefit is that most forms of variable rate debt can be paid off at par, with no prepayment penalty, on short (30 days or less) notice. This ability to call on short notice removes any risk of not being able to pay off the bonds whenever the healthcare organization decides it wants to.

Typical Characteristics of Healthcare Bonds

Bonds, like other types of debt, may be fixed rate or variable rate, long term or short term, secured or unsecured, taxable or tax exempt, or contain any of the characteristics described previously. Every bond has a set maturity, and interest and principal are payable to the bondholder according to the terms spelled out in the bond documentation.

Bonds are usually *dated* as of a certain date. The *dated date* is most commonly the first day of a month, although it is not unusual to see bonds dated as of the 15th of a month. The dated date is usually a date prior to the closing date of the bond issue, and represents the date from which interest will accrue. On the day the bond issue is *closed*, all of the documents have been (or will be) executed, and the borrower usually receives the bond proceeds and the purchaser of the bonds receives the bonds. The day on which bonds are released to the purchaser is referred to as the *delivery date* of the bonds. While it is most typical for the closing date and the delivery date to coincide, it is possible to close in escrow where the delivery of the bonds and the release of cash to the borrower happen after some period of time or after some event occurs.

On the day of closing, the purchasers of the bonds will pay what is called *accrued interest* in addition to the purchase price (par value less any original issue discount or plus any original issue premium) of the bonds. This is because the borrower will be required to pay interest according to the schedule on the bonds from, and including, the dated date on the bonds. The accrued interest represents the amount of interest that will be paid on the bonds from the dated date until the closing date. The net effect is that the purchaser of the bonds pays the accrued interest at closing and then those funds are returned to the bondholder at the first interest payment date. In some types of bond issues, especially VRDB issues, it is common to have the bonds dated as of the date of closing so that the bondholder will not have to pay any accrued interest when the bonds are purchased.

Long-Term versus Short-Term Maturities

A generally accepted premise in the business world is that the lives of assets should match corresponding liabilities; that is, long-term assets should be funded with long-term debt, and short-term assets should be funded with short-term debt. An exception to this rule may be taken in the case of tax-exempt debt because long-term fixed rate tax-exempt debt generally contains provisions that allow it to be called (paid off) in 10 years or less, while variable rate debt can, in most cases, be called in 30 days or less. As a result, it usually is prudent to stretch the maturity of tax-exempt debt as long as possible. This gives a healthcare organization the ability to make efficient use of the cheaper capital to fund additional projects in the future, with cash generated through the depreciation of the original assets, even if the ability to issue new tax-exempt debt is eliminated. If the healthcare organization chooses, it can always pay off the debt on a call date, if it is no longer beneficial. Current tax law limits the ability to stretch the maturity of tax-exempt debt by requiring that the average maturity of a tax-exempt bond issue not exceed 120 percent of the useful life of the project being financed. The average life of a 30-year bond issue with level debt service is typically in the range of 18 to 20 years. This means the average life of the project financed with such a bond issue must be at least 14 to 16 years. If the average life of the project is less than this, then the maturities of the bond issue will have to be shortened to comply with this rule (Fig. 6.1).

Serial Bonds and Term Bonds

Long-term bond issues usually are structured like a home mortgage, with a growing amount of principal maturing each year and a decreasing amount of interest due each year, so that the net effect to the borrower is level payments. This is referred to as having *level debt service*. A long-term fixed rate bond issue usually consists of two types of bonds, serial bonds and term bonds. Serial bonds typically mature each year, creating a succession of principal payments to bondholders by the borrower. Term bonds mature on a given date several years from the date the bonds were issued, and the borrower typically makes periodic payments into a separate fund known as a *sinking fund* to provide for their repayment. With some types of debt instruments, sinking funds accumulate money over a period of time and then pay off the debt. In a tax-exempt bond issue, the sinking fund will usually accumulate funds for 1 year, and then make a partial call on the next maturing term bond and redeem that amount of bonds. As a result, sinking funds in a tax-exempt bond issue typically are cleaned out once a year and do not accumulate for longer than that. The typical healthcare organization long-term bond issue includes several serial maturities and one or more term maturities. Usually, the shorter the maturity of the bond is, the lower the interest rate will be. By using serial bonds in the early years of a bond issue, the overall interest cost is reduced.

Ideally, a healthcare organization would issue serial bonds for each year of the entire bond issue, thereby keeping its interest rates as low as possible. Unfortunately,

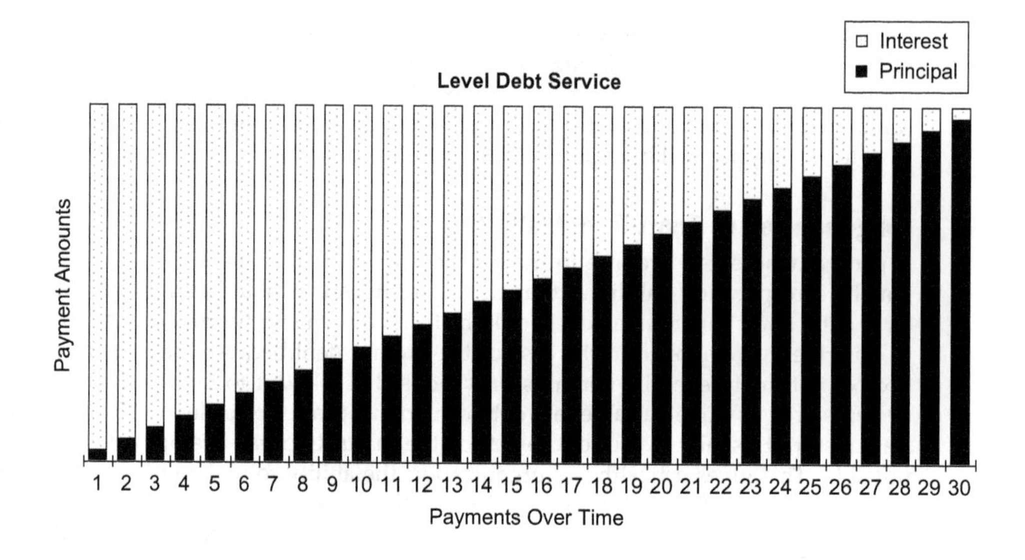

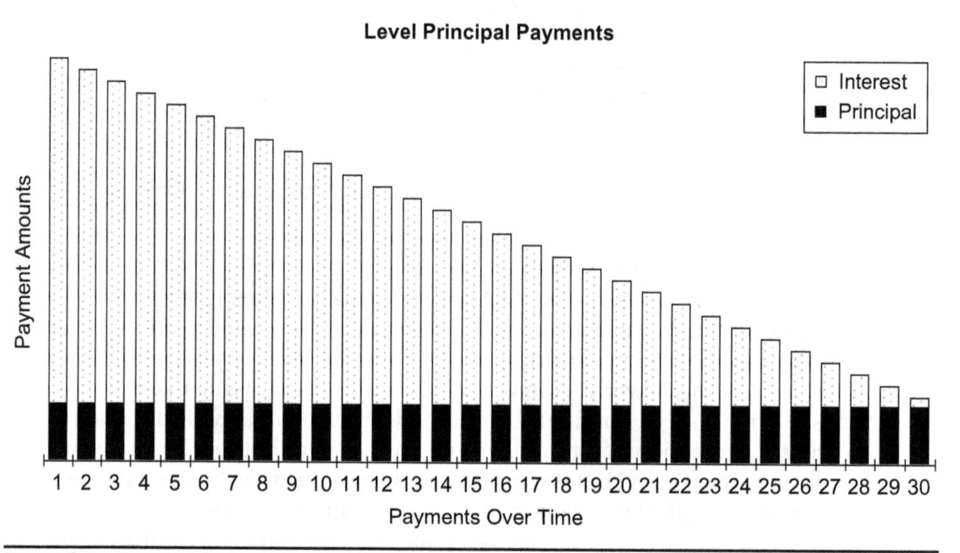

FIGURE 6.1 Level of debt service and level of principal payment for a 30-year bond.

there are usually no buyers for the middle maturities, such as 13 to 20 years. As a result, a typical long-term structure might include serial bonds maturing each year for as many as 10 to 12 years, and then one or more term bonds maturing in 25 or 30 years, or both. In the early years, the principal component of the borrower's loan payment goes to pay off the maturing serial bonds. After the final serial bond has been paid off, the next year's principal payment will go into a sinking fund and will be used to redeem, or call, a portion of the next maturing term bond. More than one term bond might be used when the additional term maturity can be sold at a lower interest rate than the final maturity.

Average Life

The *average life* of a bond issue is a weighted average of the period of time that the debt will be outstanding. This calculation takes into account the principal amount of each maturity as it becomes due. Because there is a large amount of principal outstanding in the early years of a bond issue with level debt service, and a small amount of principal outstanding in the last few years, the average life of a bond issue set up with level debt service is usually a little more than half of the actual life. For example, the average life of a 30-year bond issue with level debt service typically will be in the range of 17 to 19 years. If the principal amortization of a bond issue is weighted more heavily in the later years, the average life lengthens. Similarly, if the amortization is weighted more in the early years, the average life of the issue is reduced.

Call Features

When a borrower retains the ability to pay off a debt prior to maturity, that debt is said to be *callable*. Most bank debt is callable at any time, but most fixed rate bonds have some restrictions on the ability of the borrower to call them. Most variable rate bond issues are callable at any time, upon notice to the trustee of 30 days or less, because the bondholder is receiving a short-term rate and is treating the bond like a short-term security. In a typical long-term, tax-exempt, fixed rate bond issue, the term bonds will not be callable solely at the option of the borrower for roughly 10 years, although bonds sold in the retail market can sometimes have 5-year call provisions. Bonds that mature in less than 10 years are typically non-callable.

The ability to call bonds prior to maturity solely at the option of the borrower is considered to be a benefit to the borrower, and a detriment to the bondholder. Because of this, including optional call provisions in the bond issue generally results in a higher interest rate than if the bonds are not callable at all. There may be certain situations where a healthcare institution chooses to take advantage of a lower interest rate by eliminating optional call provisions in its bond issue. These considerations should be evaluated very carefully by the borrower and its financial advisor to determine the advisability of deviating from the standard call provisions that the market is accustomed to seeing.

Original Issue Discount and Premium

When bonds are offered at a price below face value (par) at the time of issuance, they are said to contain an *original issue discount* (OID). The OID can range from a very small amount, such as 0.25 percent, to a deep discount of 5 to 50 percent, or more. The bonds with the greatest OID are zero coupon bonds, where no interest is paid until the final maturity. Original issue discount bonds are sought by many institutional investors. On a typical long-term fixed rate healthcare financing, institutional investors might be willing to buy the term bonds at a yield that is 10 to 20 basis points (0.10 to 0.20 percent) lower if an OID is offered, compared to term bonds offered at par. An OID usually has little effect on the retail market.

A disadvantage to the borrower of bonds issued with an OID is that the yield paid on the bonds will be greater if the bonds are called prior to maturity, because they will be paid off at par or at a premium. The borrower only gets the benefit of the lower coupon for the number of years until the bonds are called, rather than until the final

Sample Maturity Schedule and Interest Scale for a 30-Year, Fixed Rate, Tax-Exempt Healthcare Bond Issue Issue Size $30,000,000			
Maturity Date (August 15)	Principal Amount ($)	Interest Rate (%)	Price (%)
2014	535,000	2.000	101.257
2015	545,000	3.000	103.664
2016	560,000	4.000	107.807
2017	585,000	4.000	109.301
2018	605,000	4.000	109.838
2019	630,000	2.250	99.322
2020	645,000	2.750	99.934
2021	660,000	5.000	114.194
2022	695,000	3.125	99.332
2023	715,000	4.000	105.612
2024	745,000	5.000	112.748
$ 3,320,000 4.0% Term Bonds Due August 15, 2028, Priced at 100.409% $ 5,060,000 5.0% Term Bonds Due August 15, 2033, Priced at 108.839% $ 14,700,000 5.0% Term Bonds Due August 15, 2043, Priced at 106.014%			

FIGURE 6.2 Sample maturity schedule for 30-year tax-exempt bond.

maturity. Because of this, the likelihood of a call prior to maturity should be evaluated before a healthcare organization decides to use a significant OID on its bonds. Similarly, a healthcare organization should determine whether its bonds are likely to be sold in the retail market or the institutional market, for an OID typically will not help lower interest rates as much in a retail offering.

In periods of very low interest rates, institutional investors often demand an *original issue premium* (OIP) on the bonds, where the coupon will be higher than the yield to the bondholder because the bondholder pays more than the face value for the bonds when they are issued. The problem for the borrower with OIP is that the bondholder will not pay the premium beyond the first call date of the bonds, typically 10 years. If the bonds remain outstanding beyond the first call date, the effective interest rate to the borrower goes up every day the bonds stay outstanding beyond the call date (Fig. 6.2).

The Process of Issuing Tax-Exempt Bonds

Once a healthcare institution has determined the amount of tax-exempt financing they will need, the process of issuing that debt begins. The capital plan should have determined the amount of debt to be incurred, and the nature of the debt to be issued, i.e., long term versus short term, fixed rate versus variable rate, conventional versus derivatives, etc. In addition, the capital plan should recommend whether the bonds will be offered through a competitive sale method or a negotiated sale method and whether the offering should be public or private in the event the negotiated sale method will be used.

Public Offering versus Private Placement

A key decision to be made when putting together a bond issue is whether to sell the bonds through a public offering or a private placement. In a *public offering*, the minimum denomination of a tax-exempt fixed rate healthcare bond is typically $5,000, and the bonds are offered to both the *retail market* (consisting of individuals) and the *institutional market* (consisting of large, "sophisticated" investors such as banks, insurance companies, bond mutual funds, and sometimes very wealthy individuals). A *private placement*, or *institutional offering*, is implemented when bonds are offered to a limited number of sophisticated potential buyers, as defined in the securities laws, and the minimum denomination often is $100,000 or more. In these offerings, bonds are not offered to the retail market, which consists of the general public.

The lowest interest rates usually are obtained in a public offering because there are more buyers competing for the bonds. However, in a public offering, securities law requires more extensive disclosure about the borrower than is required in a private placement. Another potential drawback to a public offering is that there usually will end up being many bondholders, so the likelihood of obtaining bondholder consent for a future amendment to the bond documents is greatly reduced. A private placement often allows a closer working relationship between the borrower and the lenders, and the lenders frequently have some input as to the final structure of the deal. With a limited number of lenders, a healthcare institution may have a greater ability to amend its documents in the future if the need arises.

Selecting the Working Group

The first step in actually implementing the financing is selecting the remainder of the financing team. It is preferable for the financial advisor to be selected prior to the development of the capital plan, and to be instrumental in the development of that plan. If a financial advisor was not used to develop the capital plan, then this should be the first member of the financing team to be selected, as one of the key roles of the financial advisor is the coordination of the financing team.

The financial advisor should review the capital plan, and then assist in the process of selecting the remaining members of the working group who will best be able to execute the capital plan. Some conduit issuers of tax-exempt debt require input as to the selection of certain members of the financing team, so this should be a consideration in the selection of an issuer, if more than one entity is able to serve as issuer. Once the financial advisor and issuer have been retained, the remaining members of the team can be selected. Other members would include an underwriter, bond counsel, borrower counsel, and bond trustee.

Working Group Meetings

Once the financing team has been assembled, an *organizational meeting* is scheduled to bring each member of the team up to speed on the financing plan. During that meeting, the borrower and its financial advisor should explain the nature of the project, the types of bonds they expect to be issued, and the time frame under which they expect all of the necessary requirements to be fulfilled for the issuance of the bonds. By the end of the organizational meeting, each member of the financing team should have a good understanding of their "marching orders."

During this meeting, the borrower and its financial advisor should be prepared to provide the group with information on any relevant topics, including the following:

The borrower

- The identity of the borrower
- The legal organizational structure of the borrower and its related affiliates
- Will a master trust indenture be used?
- Will there be an obligated group or a single borrower?

The issuer

- Who will issue the bonds?
- What is the process required by the issuer (e.g., number of meetings, timing requirements, uncommon policies, etc.)?

The project

- Will the project consist of new money (i.e., financing a project that has not been financed in the past), refunding, or both?
- Description and cost of any new money project, or project for which proceeds of any debt to be refunded were used (e.g., construction, renovation, equipment, etc.)
- Total amount to be financed
- Is any portion of the project not eligible for tax-exempt financing?
- Is a certificate of need or other permit required?
- Description of all approvals required and the process for obtaining them

Structure of the financing

- Fixed rate or variable rate?
- Will the bond offering be competitive or negotiated?
- Anticipated security for the bonds
- Will there be credit enhancement or not?
- Will the issue be rated or not?
- Will a debt service reserve fund be included or not?
- The term (final maturity) of the bonds
- Amortization of the bonds
- Types of bonds to be offered (e.g., conventional, derivatives, etc.)
- Is the financing *interest rate sensitive* (i.e., if interest rates go up by 100 or 200 basis points, is the borrower prepared to proceed with the financing)?

During the organizational meeting, a preliminary distribution list should be passed around the table and additional information and corrections for a master distribution list obtained from all parties. The distribution list will contain the address, telephone, and e-mail information for each person who is to receive any documents, and should be prepared and distributed to the working group by the financial advisor or underwriter shortly after this meeting. A schedule for the financing should be agreed upon at the organizational meeting, providing for two or three drafting sessions, due diligence meetings, meetings with rating agencies and/or bond insurers, specific meeting dates of the conduit issuer, a mailing date for the preliminary official statement, the pricing date, closing dates, and any other items that are necessary for the issuance of the bonds.

Prior to the first drafting session for the documents, bond counsel will prepare and distribute a first draft of the necessary bond documents. This would usually include a bond trust indenture and a loan agreement (or their equivalents) as a minimum. Prior to any drafting session, each member of the team should have reviewed the documents distributed by various counsel so that comments on the documents can be given at the meeting. Prior the second drafting session, underwriter's counsel should have prepared a first draft of the preliminary official statement.

The Official Statement

An *official statement* (generally referred to as an "OS") is the offering document whose purpose is to disclose all relevant information about the bonds and the borrower to prospective purchasers of the bonds in a publicly sold tax-exempt bond issue. A *preliminary official statement* (generally referred to as a "POS") is prepared prior to the sale of the bonds with all of the final information except for those details that will not be known until the bonds have been sold, such as the final interest rates, the final size of the issue, and the amounts of bonds that will mature each year. The POS is sometimes referred to as a "red herring" because it displays a statement printed in red ink along the left margin and top of its cover warning that the document is not a final OS.

Both the OS and the POS contain complete disclosure of relevant information on the bonds, the borrower, any credit enhancement, the issuer, potential risks to bondholders, and any other information that the working group feels would be relevant to a potential purchaser of the bonds. A legal opinion from one or more of the counsel in the working group is required before the bonds are issued, to the effect that all relevant information has been included in the OS, and that none of the information is misleading. The bonds are sold based on the information in the POS, and the final OS must be delivered to each purchaser prior to, or at the time the bonds are delivered.

The Trust Indenture

A *trust indenture, bond resolution*, or similar document is used to instruct the trustee on the terms, conditions, and procedures to be followed on a bond financing. It describes the security for the bonds and most of the covenants that the trustee is expected to monitor. There may be other documents that contain additional legal covenants, or actually grant any security interest to the trustee, but the trust indenture is essentially an instruction manual for the trustee on how to oversee the bond issue. A standard bond trust indenture is designed to govern an issue of bonds, and sometimes additional issues of bonds that share in all of the rights granted to bondholders under the indenture. A *master trust indenture* is a document that contains covenants and security like a bond indenture, but is designed to govern multiple obligations.

Any obligation issued pursuant to a master trust indenture shares a parity (equal) interest in whatever security and covenants are contained in the master trust indenture. This becomes particularly helpful when a healthcare organization wishes to incur debt from several different sources. For example, it may issue bonds through a tax-exempt issuer, and then wish to obtain a bank loan at a later date, as well as wish to participate in some other financing program from another issuer. Each of these lenders can receive a parity security interest each time debt is issued by securing their notes with an obligation issued under the borrower's master trust indenture, without modifying the underlying documents that grant the security to the master trust indenture.

A master trust indenture often includes an *obligated group* structure to combine the credit of several entities. The obligated group structure works like a set of cross-guarantees so that each member of the group is obligated to pay on any debt issued by any member of the group, if that debt is secured by a master note. This has become desirable for many healthcare organizations since the trend began in the early 1980s to reorganize into multi-corporate organizations. Some healthcare organizations found that they had reorganized to the point that the credit quality of the remaining healthcare corporation itself was no longer sufficient to obtain attractive financing. The use of an obligated group allows a multi-corporate organizational structure while preserving the ability to present a unified credit to the financial markets.

Obtaining Necessary Approvals

Unless a healthcare institution is owned by a governmental body, it will require the use of a conduit issuer to make its bonds tax exempt. Each issuing body has its own system and approval process. It is important for the borrower to investigate this process early on, so the schedule for the financing can be worked around the schedule of the issuing body.

In addition to the issuer, approvals may be necessary from local planning and zoning departments. Some states also require a certificate of need for certain projects, which can require a lengthy process to obtain. Bond counsel typically will require all necessary approvals to be in place by the time the bonds are issued, or at least to have evidence that the necessary approvals will be forthcoming.

Obtaining Bond Ratings and/or Credit Enhancement

The first step in obtaining bond ratings or credit enhancement is to prepare a package of information and pertinent materials to send to the rating agencies and/or letter of credit providers. If a rating is being sought, meetings with the various rating agencies should be scheduled, either at their offices in New York or at the facility of the borrower. Presentations to the rating agencies typically involve a more formal presentation than required by letter of credit providers. The materials should be sent to the agencies at least one week prior to any meeting. Final approval is not likely to be received until one or two weeks after the meetings. Meetings with credit enhancement providers may, or may not, be necessary. If they are needed, they will be requested by the providers.

Selling the Bonds in a Negotiated Offering

When the working team has determined that the financing documents are in final form, all necessary approvals have been received, ratings and/or credit enhancement are in place, and the POS is in final form, the borrower is ready to enter the market to sell the bonds. The POS typically will contain financial information on the borrower. This information can become stale if an extended period of time goes by before the bonds are sold. Usually, the market likes to see financial information current to within 135 days from the time bonds are sold and within 150 days from the time the bond issue closes. There may be some flexibility on this timing, but it makes sense to include the most current financial information available in the POS, to build some leeway in the time frame by which the bonds must be sold.

In a negotiated offering, the POS typically will be distributed to prospective bondholders roughly 1 week before the bonds are expected to be offered for sale. There are certain situations where an underwriter will be comfortable offering bonds for sale when the POS has been in the hands of prospective bondholders for as little as a day. Sometimes they want the bondholders to review the information for two or more weeks. The actual timing is determined by several factors, such as expectations for market movements in the near future, the complexity of the issue, and the need to tell a story about either the bonds or the underlying credit. Bonds that have some unusual characteristics, or are issued by a credit that requires a special sales effort, are referred to as *story bonds* by the underwriters. Conventional bonds issued by a strong credit that is well known to the market will require a relatively short period of time for the POS to be in the hands of bondholders before they can be sold effectively. Conversely, a bond issue for a difficult credit that is new to the market (i.e., story bonds) may require the underwriter to provide additional information besides what is contained in the POS to prospective bondholders before they are willing to purchase the bonds, thereby requiring the POS to be distributed earlier.

Once the POS has been distributed and held by prospective bondholders for the desired amount of time, and the market appears to be in a favorable condition for the sale of bonds, the bonds will be priced. *Pricing* bonds is the process of negotiating interest rates at which the borrower is willing to sell, and the underwriter is willing to buy, the bonds. It generally is advisable to schedule a preliminary pricing call on the afternoon prior to the day the bonds actually are priced. This call should include the borrower, the borrower's financial advisor, and the underwriters. The purposes of this call are for the underwriters and the financial advisor to discuss the current market conditions and the conditions expected for the following day, to agree upon a scale of interest rates for the bonds, and to develop a marketing strategy for the bonds.

The *scale* of interest rates is a listing of the interest rates (coupon rates) and prices for each maturity of the bonds. The price and the coupon rate determine the yield at which the bonds in each maturity will be issued. The shorter the maturity, the lower the yield will be in most market situations. The scale typically will contain interest rates for maturities in each year from 1 to 10 (sometimes out to year 15), and then one or two term bonds in the later maturities.

The bonds may be offered with, or without, original issue discount or premium, and any other aspects of marketing strategy should be discussed on this telephone call. In most cases, the parties on that call will agree upon a scale of interest rates so the underwriter can offer the bonds for sale first thing the following morning, without the need to contact the borrower and its financial advisor. In the event there is any change in the market and the underwriter determines there needs to be some revision in the scale of interest rates, another call would be placed the morning of pricing. Most underwriters will not offer to underwrite (purchase) the bonds unless they have obtained verbal orders from their customers for almost all of the bonds at the designated scale of interest rates.

On the day of pricing, the underwriter will broadcast the proposed scale of interest rates to the market on the various wire services. The scale is set at levels that are expected to result in offers to purchase the amount of bonds that are offered. If interest rates are set too high, the bonds are likely to be *oversubscribed*. This means the underwriter receives orders for more bonds than they have to sell in a given maturity. Conversely, if interest rates on the scale are set too low, sufficient orders for the bonds may not be received.

The inverse relationship of price to yield can cause some confusion when the pricing of bonds is discussed with underwriters. The underwriters usually are the initial purchasers of the bonds. Their interests are opposed to the borrower in that they will always prefer a higher yield on the bonds while the borrower will always prefer a lower yield. When underwriters refer to bonds as being *cheap*, they mean that the yield is relatively high for the current market and the price is, therefore, relatively low. Conversely, when they refer to the bonds as *rich*, they mean that the yields are relatively low for the current market, and the price is relatively high. When the bond market is *up*, it means that prices on outstanding bonds have risen and interest rates moved lower. When the market is *down*, prices on outstanding bonds have lowered and interest rates (yields) have risen. Because they are selling the bonds, borrowers prefer the market to be up and their bonds to be sold rich.

When the sale is broadcast over the wire services, an *order period* is established by which time all orders must be received by the lead underwriter. Usually, it will run from the morning of the pricing until early afternoon. If enough orders for the bonds have not been received by the end of the order period, the borrower either must raise the yields on the bonds and again attempt to sell them or pull the issue from the market. When an issue has entered the market and the market has not been receptive due to the interest scale being too aggressive, it usually takes a fairly good bump in yield to get the market to seriously reconsider purchasing the bonds. In other words, it is usually not sufficient to raise the yields on the scale by only five basis points (0.05 percent) or so. In many cases, it will take increasing the rates anywhere from 10 to 25 basis points for the market to regain interest in the issue. Because of this, it generally is advisable to enter the market with an interest rate scale at which the underwriter and financial advisor are fairly comfortable that sufficient orders will be received to sell the bonds. If the bond issue is oversubscribed, it is fairly easy to reduce the interest scale and then verify which orders will hold at the new scale. In order words, it generally is more effective to set the interest scale a little too high and then reduce it than it is to set the interest rate scale too low and then have to increase it.

During the initial pricing call, a follow-up pricing call should be scheduled for shortly after the end of the order period. On this call, the underwriter will explain to the borrower and its financial advisor the results of the sales effort during the order period. The ideal situation is for the underwriter to have orders for roughly 75 to 90 percent of the bonds. This means the interest rate scale was not so high that the bonds were oversubscribed, but high enough so the majority of the bonds were sold. If a sufficient number of the bonds have been sold, the underwriter should agree to underwrite the remaining portion of the bond issue at the interest scale that was proposed. If sufficient orders were not obtained, the underwriter still might offer to underwrite all of the bonds, but at a slightly higher interest scale than was initially proposed. The borrower's financial advisor should have conducted research throughout the day to develop an educated opinion as to whether or not a change in the interest scale is appropriate. The only options open to the borrower at this point are either to agree to sell the bonds at the underwriter's offer or to pull the issue from the market and try to sell the bonds on another day.

Bond issues are usually priced on either Tuesday, Wednesday, or Thursday of any given week. The rationale for this is that Monday is used to ascertain whether or not there has been a shift in the market over the weekend. If there is a need to price the bonds on a Monday, the order period would normally not begin until afternoon for this reason as well. In situations where pricings begin in the afternoon, the order period often will flow over to the next morning. Once the terms of the pricing have been agreed upon, a verbal award is made, and a purchase contract is prepared. The *purchase contract* is the binding written contract that details the terms and conditions under which the borrower agrees to sell, and the underwriter, or other purchaser, agrees to purchase the bonds. The purchase contract normally is executed a day or two after the verbal award is made, due to the number of specific details that have to be finalized. Bond issues rarely are priced on Fridays because underwriters are hesitant to hold a bond issue over a weekend on a verbal award alone.

Closing the Bond Issue

Once the purchase contract has been signed, and all approvals obtained by the borrower and issuer, the lawyers finalize all of the documents for *closing* when, under normal circumstances, the money from the proceeds of the bond issue is received by the borrower's trustee and the bonds are released for distribution. At closing, the legal counsel responsible for the disclosure will be asked to give an opinion that everything in the final OS is true and correct, and that no relevant information which should have been included was omitted. Frequently, different legal counsel representing different parties will opine to the sections of the official statement that are relevant to their respective clients. Before the bond issue closes, all sections of the official statement will be covered by one or more legal opinions confirming the adequacy of disclosure.

The actual closing of a bond issue is usually a two-day process, and sometimes longer than that. The day prior to the actual closing is usually referred to as the *pre-closing*, and most documents are finalized and executed on that day. In a well-executed closing, all of the documents will be signed and executed on the day of the pre-closing, except perhaps for the cross-receipts, which evidence receipt of the actual funds and the bonds. In this situation, the underwriter will initiate wire transfers for the bond proceeds first thing in the morning on the day of the closing. When all of the people assemble for the closing, the only remaining step necessary is for the trustee to verify that the funds have arrived. At that point, the bond issue is closed and the bonds are released to the underwriters. In many situations, however, many of the documents are not finalized until the day of closing. This can cause some excitement, especially in situations where the bonds are going to be issued in book entry fashion. This is due to the fact that the depository companies that hold the bonds have cut-off times by which the issue must close, or else they will not accept bonds until the next day. Currently, the cut-off time used by most of these institutions is 1:00 p.m. Eastern time (although they have been known to accept bonds as late as 1:15 p.m. Eastern time.).

Key Covenants Found in Healthcare Bond Issues

Covenants are legal obligations of the borrower found in bond documents and other debt instruments, designed to protect the lender's ability to receive timely payment of principal and interest. A borrower violating a covenant is in default of the agreement containing the covenant. A default can result in different consequences depending on the terms spelled out in the document, the most severe of which may be an *acceleration* of the debt, requiring immediate repayment of the remaining outstanding principal.

As a rule, the stronger the credit, the fewer covenants the market requires. For example, a healthcare organization rated AA/Aa can issue debt with few covenants, while an organization rated BBB/Baa is likely to be required to comply with significantly more. Bond insurers and banks typically require a more restrictive set of covenants than the public market in general. In addition, tax-exempt bond issues contain certain tax-related covenants that are not necessary for taxable issues.

Security Alternatives for a Bond Issue

Prior to the early 1980s, it was common for healthcare organizations to secure their bond issues with mortgages on their primary healthcare facilities. In addition to a mortgage, they usually were required to pledge their revenues or gross receivables as additional security. A pledge on gross revenues and a pledge on gross receivables are considered equivalent by the market. Bond counsel firms differ as to which form of security is more appropriate or enforceable. Both attempt to achieve the same purpose of attaching any revenues of an organization after a default on the bonds.

Mortgages rarely are required by the market for healthcare financings with debt rated in the A category or better. A typical healthcare bond issue without a mortgage will still be secured by a pledge of revenues where all revenues of the healthcare organization would first be available to pay down bonds in the event of a default. In this case, there would also be a *negative pledge* on the borrower's healthcare facilities to prevent the borrower from giving a lien on the real estate to any other creditor. Weaker credits (BBB/Baa and below) may be required to secure their financings with mortgages if they are to obtain the best possible interest rates.

Debt Service Coverage

Debt service coverage is a concept that is used in many of the covenants and tests found in healthcare bond documents. It measures the number of times that annual debt service on the bonds (principal and interest payments that are due that year) could be covered from the revenues that are available to pay debt service. The formula is typically stated as income available for debt service divided by annual debt service. Income available for debt service includes net income (profit) plus depreciation and interest expense. Income available for debt service is intended to include all of the cash flow that the borrower generates, and is available to pay debt service after the normal operations of the borrower are covered. Debt service coverage can be calculated for any period of time, but most tests evaluate either coverage for the most recent fiscal year for which an audit is available, or test the income available for debt service relative to the maximum annual debt service in any future year, given the current debt outstanding and any proposed debt.

Rate Covenant

A *rate covenant* requires a healthcare organization to charge sufficient rates and fees to produce enough revenue to cover the debt service on its bonds by a certain amount, in addition to the revenue needed to operate the healthcare facilities. This coverage requirement is frequently set at 125 percent of annual debt service, but has been reduced on many financings to 110 percent. In the event that a borrower does not meet the required coverage, it may be required to retain a qualified consultant to review the situation and make recommendations to the healthcare organization that are intended to allow it to reach the required coverage in the following fiscal year. The borrower can avoid a default on its bond issue if it retains the consultant within a specified period of time and follows the consultant's recommendations. In the event that the coverage is not increased to the required level, the borrower usually is required to retain a consultant every 2 years. Typically, there is a provision that allows the required debt service coverage to drop to 100 percent if a consultant's report is delivered to the trustee stating that government regulation has prevented the healthcare organization from obtaining the required coverage, and that the healthcare organization has operated so as to maximize its revenues.

Liquidity

A *liquidity covenant,* which requires that the borrower maintain a certain level of liquid assets such as cash or marketable securities, is intended to make sure that the resources are available to allow the borrower to continue operations for a longer period of time in the event of a problem, so the recommendations of the consultant will have time to take effect. The most common form of liquidity covenant is *days cash,* which is the total of cash and liquid investments divided by one day of expenses for the organization.

Additional Debt Restrictions

Additional debt provisions restrict the ability of the healthcare organization to incur additional indebtedness in the future. There normally are different covenants for long-term and short-term debt.

The ability to incur long-term indebtedness is usually restricted by several alternative sets of tests. There is typically one alternative that looks at historical debt service coverage of the projected maximum annual debt service of all long-term debt, assuming issuance of the proposed debt. If a healthcare organization can meet this test, then it can proceed to issue debt. As an alternative, the healthcare organization is allowed to retain a consultant to project expected coverage of maximum annual debt service for the two years following the completion of the project being financed.

In addition to the tests for additional debt described above, bond covenants should also allow the healthcare organization to issue a restricted amount of additional long-term or short-term indebtedness without meeting any tests. A common limitation is an amount not to exceed 25 percent of the prior year's total operating revenue. This provision is commonly referred to as a *basket* for additional debt.

Transfers of Assets

Covenants restricting transfers of the borrower's assets are intended to protect the bondholders' interest in the healthcare organization remaining a viable, revenue producing healthcare facility. These covenants usually allow transfers of obsolete assets, sale of assets on an arm's-length (fair market value) basis, and transfers of assets that total a maximum of 5 to 10 percent of the total book value of assets in any year. Frequently, an additional provision allows transfers of a greater amount of assets if it can be demonstrated that the borrower could have passed the additional debt coverage tests for the incurrence of $1 of additional debt, assuming the assets had been transferred prior to such calculation. In other words, more assets can be given away if the revenue producing capability of the healthcare organization is not reduced below a specified amount. In addition, there are tax law restrictions on the transfer of property financed with tax-exempt debt.

Consolidation or Merger

These covenants typically allow a consolidation or merger if it can be demonstrated that the consolidated entity could pass the coverage tests for the incurrence of $1 of additional debt after the merger. Again, this test allows a merger if the revenue producing capability of the new entity is not below a certain amount. Evidence is also required that such a merger will not create a default under the terms of the bonds and will not adversely affect the tax exemption on any outstanding tax-exempt bonds.

Corporate Existence and Maintenance of Properties

The borrower is required to maintain its legal existence and to maintain its property in good working order. These covenants generally contain a series of common sense requirements such as complying with laws, paying taxes and assessments, making timely payments on debt, maintaining licenses, and maintaining its tax-exempt status (if applicable).

Insurance

The borrower typically is required to maintain insurance in form and amount customary for similar corporations and activities. There typically is a requirement for a review by an insurance consultant at least once every 2 or 3 years for commercial insurance, and more frequently for self-insurance.

Use of Insurance or Condemnation Proceeds

Use of insurance or condemnation proceeds covenants require that if the proceeds of insurance or condemnation exceed a certain specified amount, then the healthcare organization will either repair or replace the facility that was affected, or use the proceeds to repay the debt that was issued to finance the facility. There may be exceptions to these requirements if certain coverage ratios can be demonstrated.

Limitation on Liens

Limitation on lien covenants limit the ability of any other lender to obtain a security interest that is prior (senior) to the interest of the bondholders in any of the borrower's property. It is customary to have certain exceptions to these restrictions that allow a limited amount of senior indebtedness and for the exclusion of property not necessary for the operation of the healthcare organization.

Joining or Leaving the Obligated Group

In the event the financing contains an obligated group structure, there will be restrictions limiting the ability of entities either to join, or to be removed from, the obligated group. These tests usually conform to the coverage tests for mergers and additional debt.

Conclusion

The financing of any capital expenditures should be consistent with the capital plan of the organization. Most creditworthy healthcare organizations are fortunate to have access to many alternative forms of financing. The overall success of any significant construction project depends largely on the cost and terms of the capital used to finance the project. Proper planning and execution can greatly enhance the success of these endeavors.

CHAPTER 7

Project Delivery Methods for Healthcare Projects

Hospitals, by their very nature, are often large, complex buildings, and hence so is the process of planning, designing, and *managing the construction* of hospitals. There are also some smaller hospitals, such as critical access and primary care facilities with 30 or fewer beds. These too pose challenges during construction and especially renovations.

For a typical project of moderate size and complexity, the owner often employs a designer [an architectural engineering (AE) firm] who prepares detailed drawings (plans) and specifications (specs) for the constructor (a general contractor). On smaller projects, the designer may also act on behalf of the owner to oversee the project implementation during construction. On larger projects the owner may have in-house staff that performs this role, or the owner may hire an independent owner representative or program manager. The general contractor or construction manager is typically responsible for the construction itself, even though the work may actually be undertaken by a number of specialty subcontractors.

The owner or their representative usually negotiates a fee for services of the AE firms. In some cases, the owner may take bids or have some competitive process for selecting the design team. In addition to the responsibilities of designing the facility, the AE firm also exercises supervision, to some extent, of the construction as stipulated by the owner. Field inspectors working for an AE firm (or directly for the owner) usually follow through the implementation of a project after the design is completed. Because of the litigious environment over the last two decades, most AE firms only provide observers, rather than inspectors, in the field. Even the shop drawings of fabrication or construction schemes submitted by the contractors for approval are reviewed with a disclaimer of responsibility by the AE firms.

The owner may select a general contractor or construction manager either through competitive bidding or negotiation. Public agencies often rely on the competitive bidding, while private organizations may choose either mode of operation. There are many variations of competitive selections from "hard bids" with lump sum fixed prices to submission of just fee and general conditions with the team, time, and relevant experience as other criteria. In using competitive bidding, the owner is forced to use the designer-constructor sequence since detailed plans and specifications must be ready before inviting bidders to submit their bids. If the owner chooses to use a negotiated

---

This chapter was written by Sanjiv Gokhale.

contract, the owner is free to use phased construction if the owner so desires. Breaking down a project into small logical construction/permit phases allows construction while finalizing the design requirements. The typical phases for a building project are foundation, structural frame, shell, and core. The other benefit of early selection of the contractor is to have their input during the design process on cost, constructability, and phasing.

The general contractor may choose to perform all or part of the construction work, or to act only as a manager by subcontracting all the construction to subcontractors. The general contractor may also select the subcontractors through competitive bidding or negotiated contracts. According to the 2012 U.S. Construction Industry Productivity report prepared by FMI, only 23 percent of the general contractors self-perform a significant portion of the work (FMI, 2012).

Integrated project delivery (IPD) is a new and growing trend in the healthcare industry. The definition of IPD, as described by the American Institute of Architects (AIA, 2007), "a project delivery approach that integrates people, systems, business structures, and practices into a process that collaboratively harnesses the talents and insights of all participants to optimize project results, increase value to the owner, reduce waste, and maximize efficiency through all phases of design, fabrication, and construction."

IPD principles can be applied to a variety of contractual arrangements and IPD teams can include members well beyond the basic triad of owner, architect, and contractor. In all cases, integrated projects are uniquely distinguished by highly effective collaboration among the owner, the prime designer, and the prime constructor, commencing at early design and continuing through to project handover.

Due to long-term ownership and to aggressive goals of constructing high-performing buildings with lower energy expenditures, healthcare building owners are constantly seeking new avenues to streamline the risk and opportunities associated with standard contractual relationships. Each contractual relationship offers a variety of risks and rewards for the entire project team; however, the advent of LEED (leadership in energy and environmental design) (USGBC, 2009), has created an additional layer of complexity and potential risks for the client and project team.

The goal of the LEED for healthcare rating system is to help design, build, and operate high-performance healing environments. The needs of healthcare facilities are unique. Healthcare buildings often have strict regulatory requirements, 24/7 operations, and specific programmatic demands *not* covered in LEED for new construction. The LEED for healthcare rating system acknowledges these differences both by modifying existing credits and by creating new, healthcare-specific credits. The goal is to help promote healthful, durable, affordable, and environmentally sound practices in these projects. LEED for healthcare is geared toward inpatient and outpatient care facilities and licensed long-term care facilities. It can also be used for medical offices, assisted living facilities, and medical education and research centers.

Projects that meet certain criteria are required to use LEED for healthcare. These include licensed and federal inpatient and outpatient care facilities and licensed long-term care facilities. Another possible alternative for the design and construction of high performance buildings is compliance with the International Green Construction Code (IgCC) as published by the International Code Council. The IgCC is an overlay document to be used in conjunction with other codes and standards adopted by the jurisdiction. The IgCC was developed in collaboration with cooperating sponsors such as the American Institute of Architects (AIA), ASTM International, the American Society of Heating, Refrigeration and Air-Conditioning Engineers (ASHRAE), the Illuminating Engineers Society (IES), and the U.S. Green Building Council (USGBC).

With additional potential risks threatening the project objectives, the IPD method provides a refreshing and radical alternative to standard construction contracts—one that results in mutual risk and reward sharing among all major team members. IPD offers an alternative in design and construction management in which obstacles between project team members are removed, encouraging open dialogue, and reducing interdisciplinary conflicts by using BIM (building information modeling) technology and an open-book approach. This process is still fairly new, so there is little or no case law that defines the potential legal risks with this delivery method.

This chapter is devoted to the discussion of the more traditional methods of project delivery, such as design-bid-build (DBB), that are evidenced in healthcare projects, as well as a discussion of some of the more modern methods that lay claim to improvements in design, construction cost, speed, and quality of healthcare projects. The chapter will include case studies, where necessary, to better illustrate the application of a project delivery method.

Project Delivery Systems

Project delivery systems (PDSs) refer to the overall processes by which a project is designed, procured, constructed, and maintained. In the public sector, this has traditionally entailed the use of the DBB system, involving the separation of design and construction services and sequential performance of design and construction. In recent years, however, the public sector has begun experimenting with alternative methods to improve the speed and efficiency of the project delivery process.

These alternative systems move closer to the integrated services approach to project delivery favored in the private sector. To illustrate this concept, the commonly utilized delivery systems have been arranged below on a continuum (Fig. 7.1), with the traditional DBB approach appearing on the top and the more innovative systems arranged from top-down according to increasing similarity to the private sector model in terms of shifting greater responsibility and risk to the constructor, and less separation between design and construction services.

Design-Bid-Build

The design-bid-build, or design then bid then build, project delivery method is thought of as the traditional method by most people in the construction industry and related professions. Although various alternatives to this traditional method have come into greater use in recent years, the traditional DBB method is still preferred by many owners,

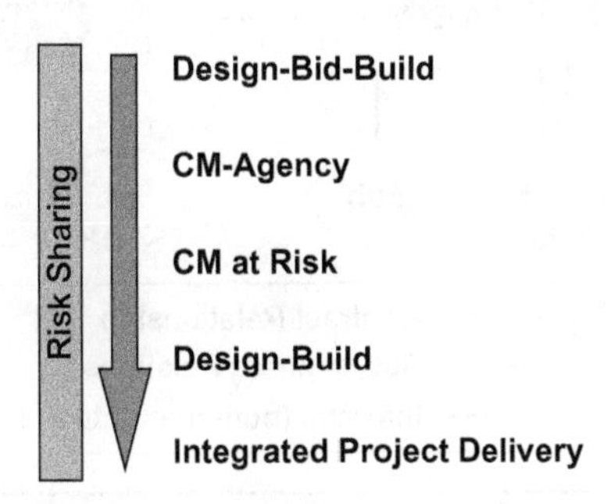

FIGURE 7.1 Risk sharing in project delivery systems.

particularly in the public sector. The DBB method is characterized by an owner (or agency) utilizing in-house staff (or, alternatively consultants) to prepare fully completed plans and specifications that are then incorporated into a bid package. Contractors competitively bid the project based on these completed plans and specifications. The agency evaluates the bids received, awards the contract to the lowest responsible and responsive bidder, uses prescriptive or method specifications for construction, and retains significant responsibility for quality, cost, and time performance.

The traditional method has a number of fundamentally sound aspects. It is a logical and orderly method that is well understood throughout the United States by all parties. There is significant case law that provides insight into how the court may rule in the event of disputes. It easily meets all procurement procedure requirements, being free of conflicts of interest. It provides a clear and transparent method for obtaining direct "apples-for-apples" competition for a fully described and illustrated end product before construction starts. It also provides for the highly desirable direct professional relationship between the owner/user and the architect/engineers for the project. Additionally, it is felt that this method of competitive bidding results in the best (lowest) price for the owner. Figure 7.2 illustrates the basic organization of a typical DBB project.

The architect and the architect's consulting engineers first carry out schematic design (SD). With the owner's review(s) and approval, the AE then more fully develops the drawings and preliminary specifications in the design development (DD) phase of services. With the owner's further approval, the AE prepares the final "working drawings," and detailed specifications and contractual requirements referred to as the construction documents (CDs). Figure 7.3 illustrates the main phases of DBB project. Figure 7.4 shows the main elements of the construction document (CD) prepared by the design team. At that point, the owner requests and receives competitive bids from general contractors, or a price may be negotiated with a general contractor who might be selected at that point, based on past-relationship or pre-qualification. During the construction phase, the AE typically works with the owner's program manager (internal or external) to observe the work in progress, review shop drawings, approve progress,

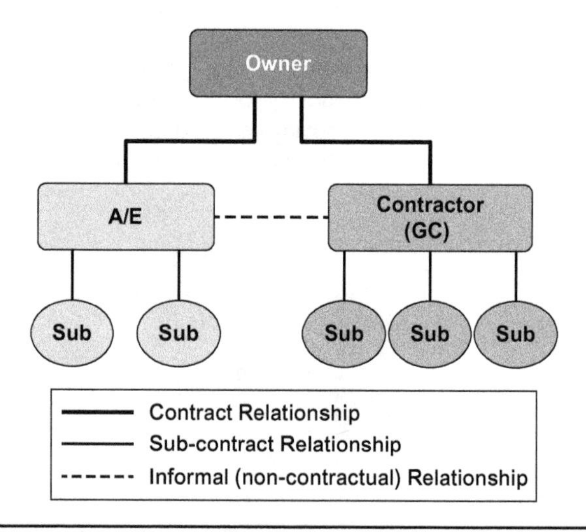

FIGURE 7.2 Organization of design-bid-build project.

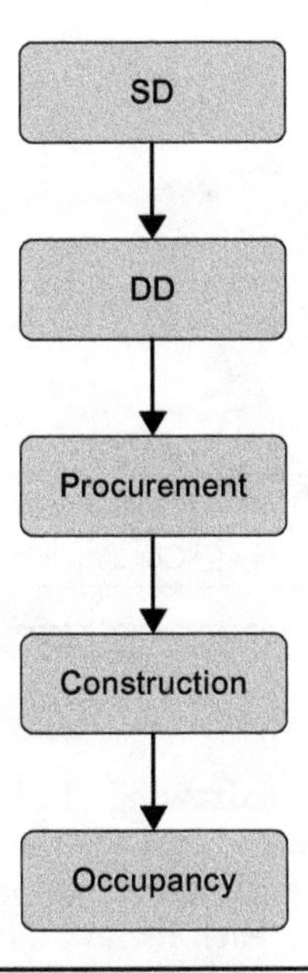

Figure 7.3 Main phases of design-bid-build project.

Bidding Requirements	Contract Forms	Contract Conditions	Specifications	Drawings	Addendum
• Invitation to Bid • Instructions to Bidders • Information Available to Bidders • Bid Forms • Bid Security Forms	• Contract Agreement • Performance Bond • Payment Bond • Certificates	• General Conditions • Supplementary Conditions	• CSI Master Format (2004) • Divisions 01 Through 50	• Architectural • Civil/Site • Structural • Electrical • Mechanical • Plumbing • Interior • Equipment • --------	• Supplementary documentation issued prior to the contract award that changes or clarifies information in the bid documents

Figure 7.4 Elements of construction document (CD) prepared by the design team.

and final payments, process any change orders, and generally act as an advisor to the program manager and the owner. This set of services is in accordance with a standard form of agreement between owner and architect as published by the American Institute of Architects (AIA), as well as similar forms of agreement used by many public and private sector owner organizations.

The distribution of the effort in the various phases for a typical DBB healthcare project by the AE is shown in Fig. 7.5. For a DBB project, the owner's program manager coordinates and oversees reviews at the design criteria, conceptual, preliminary engineering, 30, 60, 90, 100 percent, and bid document stages. The percentage refers to the approximate ratio of design budget spent over total design budget. Reviews at these points in design are key control points in the design management process. A design review is a detailed, analytical, and unbiased approach used to verify that the appropriate

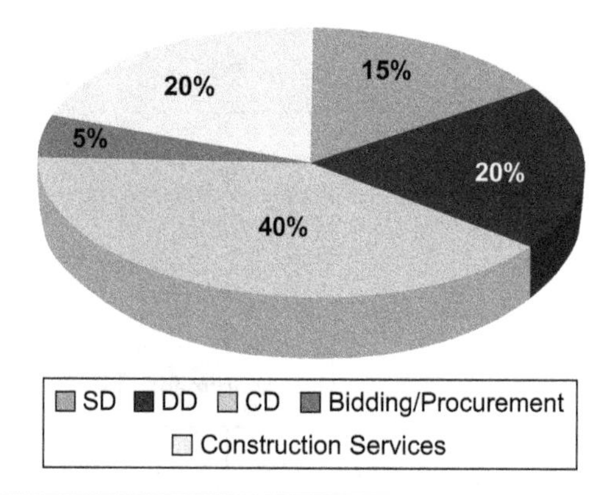

Figure 7.5 AE distribution of effort for a typical healthcare project.

deliverables (e.g., studies, final drawings, technical specifications, and construction bid documents) are being prepared, that the design is meeting the owner's program or scope, and that the design consultant is maintaining pace with the budget and project schedule. In addition, during each review, a current construction cost estimate can be developed and compared to the anticipated cost.

Following are some key considerations in conducting design reviews:

- Adherence to design criteria and environmental documents
- Quality of the design
- Identification of errors and omissions
- Building codes compliance
- Operational and functional objectives are met
- Coordination between engineering disciplines
- Adherence of cost estimates to the budget
- Designers' feedback before progressing further
- Design is biddable, constructible, and cost effective
- Interface compatibility: adjacent project elements and the existing transit system
- Final construction contract documents comply with the design criteria, environmental document, codes, and regulations

Issues with the traditional process that have caused some owners to change to alternative methods include the fact that the pre-construction phase is fairly lengthy with the owner having a good bit of the project funds at risk before obtaining a firm price on construction. The method also leaves the owner somewhat vulnerable to contractor-initiated changes orders and claims, as well as delays and additional costs for the owner in correcting post-construction problems discovered after occupancy. Furthermore, some of the best sources of knowledge on material pricing, labor availability, and constructability (specialty subcontractors and building product manufacturers) are not involved in the design process in this method. Advantages and disadvantages with DBB project delivery system are summarized in Fig. 7.6.

Design-Bid-Build Advantages and Disadvantages	
Advantages • Applicable to a wide range of projects • Well established and easily understood • Clearly defined roles for all parties • Provides the lowest initial price that responsible, competitive bidders can offer • No legal barriers in procurement and licensing • Insurance and bonding are well defined • Discourages favoritism in spending public funds while stimulating competition in the private sector	***Disadvantages*** • Tends to yield lower level quality • Least-cost approach requires higher level of inspection to assure quality • Initial low bid might not result in ultimate lowest cost or final best value • Lack of input from the construction industry during the design stage exposes the agency to claims related to design and constructability issues • Tends to create an adversarial relationship among the contracting parties, rather than foster a cooperative atmosphere in which issues can be resolved efficiently and effectively • Owner bears design adequacy risk • No built-in incentives for contractors to provide enhanced performance (cost, time, quality, or combination thereof) • Greatest potential for cost/time growth (in comparison to other delivery methods) • Prone to adversarial positions that lead to disputes, claims, and litigation

Figure 7.6 Advantages and disadvantages with DBB project delivery system.

Case Study: Design-Bid-Build

(http://www.lasvegas.va.gov/new_hospital/new_las_vegas_hospital_nursing_home.asp)
(http://www.parsons.com/projects/pages/las-vegas-va-medical-center.aspx)

In May 2004, the Department of Veteran Affairs (VA) established a program to improve access to and quality of care for veterans throughout the United States. This program, known as CARES (Capital Asset Realignment for Enhanced Services), called for building new medical centers in Las Vegas, NV, and Orlando, FL, as well as

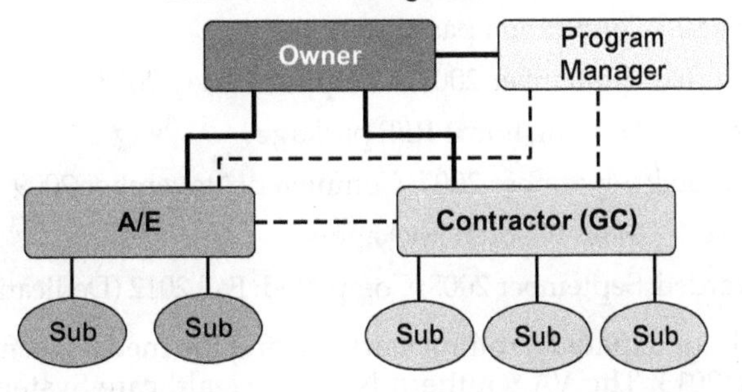

Figure 7.7 Organization of the Veterans Administration (VA) Medical Center project,Las Vegas, NV.

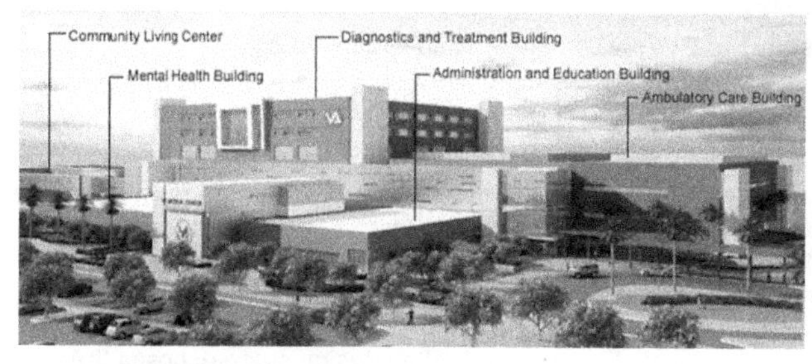

FIGURE 7.8 VA Medical Center, Las Vegas, NV—architectural rendering (*left*) and completed project (*right*).

more than 150 new community-based outpatient clinics. In August 2006, the VA selected Parsons (http://www.parsons.com) to provide program and construction management (PM/CM) services for the new medical center near Nellis Air Force Base in North Las Vegas at Pecos Road and the Woodbury Beltway. The estimated construction cost for this 150-acre greenfield hospital, the *Crown Jewel of the VA*, was $650 million. Major components of the project include a 790,000-square-foot medical center tower, 120-bed nursing home care unit, 20-bed mental health facility, and a 47,000-square-foot energy center that will house all major infrastructure equipment and system utilities. Structures vary in height from one to seven stories. Figure 7.7 depicts the Organization of the Veterans Administration (VA) Medical Center project, while Fig. 7.8 shows the architectural rendering of the project and the completed project. Due to the large size and complexity of the project, the Las Vegas VA Medical Center was undertaken in four major phases:

- Phase I—Design-bid-build: construct energy center and site utilities
- Phase II—Design-bid-build: construct foundations for hospital
- Phase III—Design-build: construct 120-bed nursing home care unit
- Phase IV—Design-bid-build: construct main hospital and build out heavy equipment within energy center

Construction progress—target award dates

- Phase 1—Central plant and off-site package

  Awarded: September 2006; Completed: April 2008
- Phase 2—Foundation package

  Awarded: September 2007; Completed: May 2008
- Phase 3—Nursing home RFP package

  Awarded: September 2007; Completed: November 2009
- Phase 4—Main hospital package

  Awarded: September 2008; Completed: Fall 2012 (Dedication on August 14, 2012)

Officials anticipated a rolling start to bring the medical center into full operation by January 2013. The VA Southern Nevada Healthcare System serves about 46,000 veterans. When the hospital becomes fully functional—with 90 inpatient beds and the 120-bed community living center, or "skilled nursing home care facility"—it will serve about 60,000 veterans out of some 400,000 who live in Nevada. The cost of building and staffing the medical center with 1800 healthcare professionals and support staff escalated over the years as the recession hit and the cost of steel rose. "Construction costs are in the neighborhood of $600 million. By the time you add the equipment and furniture and pay for the staff, we're going to be bumping up close to $1 billion," said John B. Bright, director of the VA Southern Nevada Healthcare System. The figure includes four primary care clinics that have opened around the Las Vegas Valley and two that will be established in Laughlin and Pahrump.

Case Study: Integrated-Design-Bid-Build (IDBB) Project Delivery

(http://enr.construction.com/bonus_regions/midatlantic/2012/1210-fort-belvoir-hospital-uses-new-delivery-method.asp)

Upon completion of the Fort Belvoir Community Hospital in August 2011, the project team not only delivered a significant healthcare facility, it also tested a new procurement method for the U.S. Army Corps of Engineers (USACE).

Developed as part of the Department of Defense's Base Realignment and Closure (BRAC) program, the new $958 million hospital replaced the 50-plus-year-old DeWitt Army Hospital at Fort Belvoir and greatly expanded its facilities. In total, more than 2.3 million square feet of structures were built in less than 5 years. The main facility consists of a 9-story main hospital building, two 3-story clinical buildings, and two 2-story clinical buildings. The complex also includes an ambulance shelter, helipad, two parking structures, a central utility plant and a utility tunnel. Figure 7.9 shows the new hospital facade and a typical patient room inside the new Fort Belvoir Community Hospital.

To meet its fast-track schedule, with a fixed end date established under BRAC, USACE used a new integrated DBB delivery method. As USACE had not previously used IDBB on a hospital project, "there was no Army playbook on how IDBB should work," says Victor Mudryk, project executive, Gilbane Building, Laurel, MD.

The joint venture of Turner Construction, Arlington, and Gilbane worked with the design team of HDR, Alexandria, Virginia, and Dewberry, Fairfax, Virginia, to develop new procedures to expedite paperwork flow, submittal procedures, reporting, and decision making.

Early contractor involvement was a key component of the plan. Mudryk says the team was brought on board when designs were at 12 percent, including a group of key trades such as site work, structural steel, mechanical,

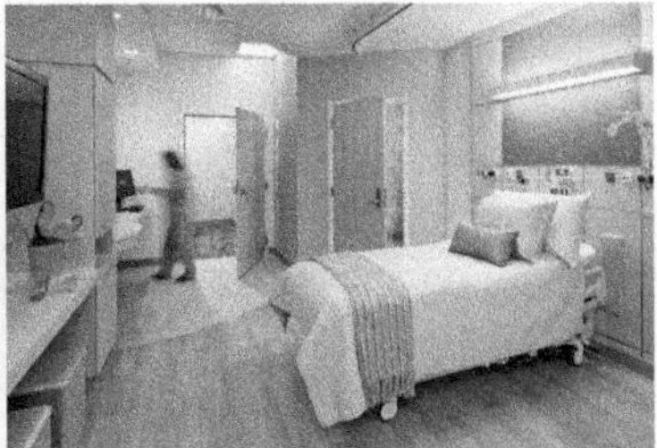

Figure 7.9 Fort Belvoir Community Hospital, VA, operated by Department of Defense—new hospital facade (*left*) and patient room (*right*).

electrical, fire protection, roofing, and waterproofing contractors. By participating in early design collaboration, the team quickly translated design decisions into direct implementation in the field. Up to 1500 trade workers were on site at peak construction. The team coordinated with 90 key subcontractors and vendors.

"[IDBB] proved to be the right process given the schedule," Mudryk says. "It opened up communications, and everyone knew where everyone else was at all times."

Owner: U.S. Army Corps of Engineers, Norfolk District, Norfolk, Virginia

General contractor: Turner Gilbane Joint Venture, Arlington, Virginia

Lead design, structural/civil/MEP engineer: HDR, Alexandria, Virginia/Dewberry, Fairfax, Virginia

Construction Manager–Agent (CM-Agency)

Construction manager (CM)–agent (also known as CM-agency or CM-fee) is a fee-based service in which the CM is exclusively responsible to the agency (owner) and acts as the agency's representative at every stage of the project. The CM is selected based on qualifications and experience, similar to the selection process for design services. Best suited for larger or complex projects, including new or renovation healthcare facilities, the CM is responsible exclusively to the owner and acts in the owner's interests at every stage of the project. Comprehensive management of every stage of the project, beginning with the original concept and project definition, optimizes the level of services and advocacy to the benefit of owners.

Figure 7.10 represents the organization of a CM-agency project. A project using this approach typically has two phases: pre-construction (concurrent with the AE's design) and construction. In the pre-construction phase, the CM provides input to the architect and owner regarding cost estimates, scheduling, constructability, value engineering, and general insight. CM responsibilities may include evaluating bids from prime and trade contractors. During construction, the CM oversees various trade contractors on behalf of the owner—but the various trade contracts are actually held by the owner, or there may be an at risk contractor that holds the subcontracts. The CM oversees the construction and manages project cost, schedule, and quality. The CM does not guarantee a price for the project or take on the contractual responsibility for design and construction.

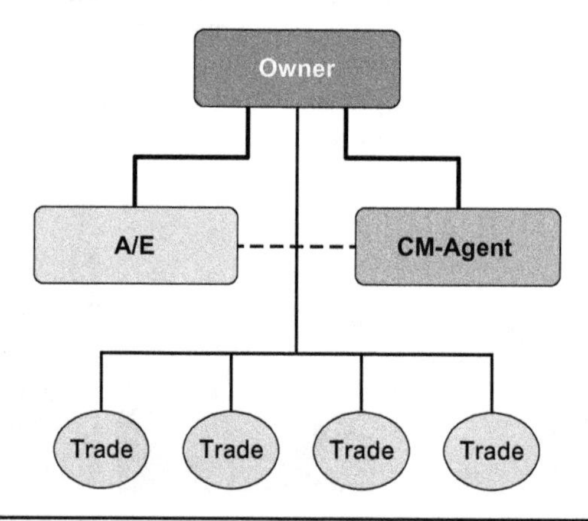

Figure 7.10 Organization of construction manager–agent (CM-agency) project.

Turner Construction Company

JOB: ABC OFFICE BUILDING
TRADE: ROOFING
SPEC SECTION: 07530
BID DUE: 6/1/99 - 12:00 NOON

	SUB / SCOPE	TCCO ESTIMATE				OK ROOFING	GOOD ROOFING	BETTER ROOFING	BEST ROOFING
		QUANTITY	UNIT	UNIT COST	TOTAL	Joe Ballast	Joe Ballast, Jr.	Tom Reroof	Will Leak
						708-555-0001	312-555-0001	312-555-1001	708-555-0101
1	BASE BID:					810,000	675,000	920,000	780,000
2	DRAWINGS PER PLANS AND SPECS					✓	✓	✓	✓
3	FULLY ADHERED ROOFING SYSTEM	220000	SF	3,25	$ 715,000	219,750*	220,000*	221,050*	219,600*
4	FLASHING	600	LF	10,00	$ 6,000	✓	✓	✓	(S) 5,800
5	PRECAST ROOF PAVERS	100	LF	6,00	$ 600,00	✓ (T) 600	(T) 600	✓	✓
6	ROOF HATCHES	2	EA	500,00	$ 1,000	✓	✓	✓	(T) 500
7	10 YEAR WARRANTY					✓	✓	✓	✓
8	60 MIL MEMBRANE					✓	✓	✓	✓
9	MANUFACTURED BY GOODYEAR/FIRESTONE/CARLISLE					✓ GOODYEAR	(X) JOE'S MEMBRANE	✓ FIRESTONE	✓ CARLISLE
10	INCLUDE INSTALLATION					✓	✓	✓	✓
11	INCLUDE DELIVERY AND HOISTING					✓	✓	✓	✓
12	INCLUDE SALES TAX					✓	(S) 600	✓	✓
13									
14									
15									
16									
17									
18									
19									
20									
21									
22									
23									
24									
25	SUBTOTAL WITHOUT BOND:				$ 722,600	810,600	676,200	920,000	786,300
26	BOND:				$ 7,300	*INCL.*	8,100	*INCL.*	7,000
27	TOTAL INCLUDING BOND:				$ 729,900	810,600	684,300	920,000	793,300
28									
29									
30	ADDENDUM NO 1-Add Roof Curbs				$ 200	300	(T) 200	✓	400
31									
32	ADDENDUM NO 2					N/A	N/A	N/A	N/A
33									
34									
35									
36	TOTAL INCLUDING BOND:				$ 730,100	810,900	684,500 (LOW BID)	920,000	743,700

Figure 7.11 Tabulating bids received from prequalified subcontractors in CM-agency. (*Courtesy: John Gromos, Turner Construction Company.*)

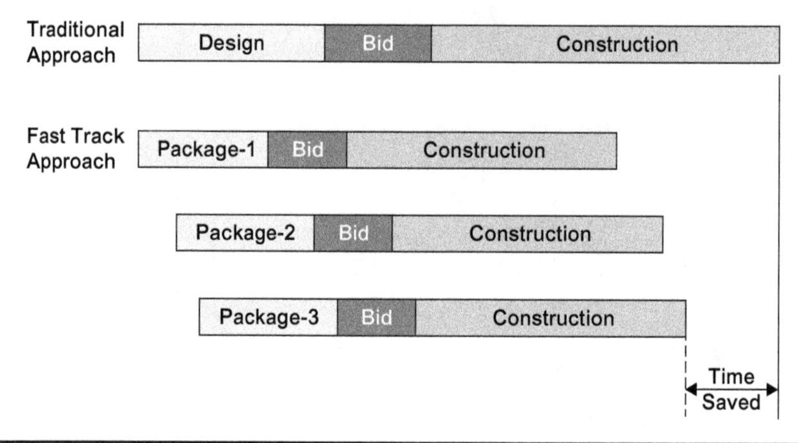

Figure 7.12 Fast tracking a project saves time.

On healthcare projects, the CM-agent will typically develop bid packages with scopes of work delineated and solicit competitive bids from prequalified subcontractors or contractors (Fig. 7.11). The owner is an active participant in the bid opening and awarding process, and has the final say in the selection of contractors and subcontractors working on the project.

CM-agent is often utilized to supplement in-house staff with independent professionals having expertise in project management, scheduling, and cost control. CM-agent is able to achieve time savings by fast tracking construction using phased packages as illustrated in Fig. 7.12. Advantages and disadvantages of CM-agent project delivery method are summarized in Fig. 7.13.

Whether or Not to Fast-Track?

This is a very important decision that must be made by the owner early in the process. The major advantage commonly associated with CM-agency, namely faster delivery, is primarily attributable to fast-tracking. But fast tracking also carries certain inherent risks:

- The owner must irrevocably "commit" to undertake construction work at a point in time when the final cost of construction is still largely unknown (because much of the construction work has yet to be designed or bid). This contributes to an increased risk of cost growth compared to DBB.
- Additional fees are typically charged by design consultants for additional services in preparing multiple trade packages.
- Early design decisions are "cast-in-concrete" and are often extremely difficult, expensive, or impossible to change later on as the design progresses.
- Because the design of various components is often not very far ahead of their construction in the field, incomplete or insufficiently detailed drawings and specifications can lead to numerous changes, rework, and other inefficiencies that can lead to claims for additional costs and delays.
- The owner must be in virtually constant communication with the design consultant and CM, and may often be forced to make critical decisions under extreme time pressures.

Construction Manager–Agent (CM-Agency) Advantages and Disadvantages	
Advantages • CM Agent selected on qualifications rather than low bid • Earlier involvement of CM (constructor) bridges design and construction phases • Furnishes construction expertise to designer • Provides the opportunity for "fast-tracking" or overlapping design and construction phases—faster than traditional design-bid-build system • Augments the agency's own resources to help manage cost, schedule, and quality • Procuring separate design and construction contracts is similar to DBB • Provides an independent point of view regarding constructability, budget, value engineering, and contractor selection (no inherent bias towards design or construction) • Potential to fast track early components of construction prior to completion of design • Reduces the agency's general management and oversight responsibilities	***Disadvantages*** • CM has no contractual responsibility with subcontractors • Agency cedes much of the day-to-day control over the project to the CM, adding a level of bureaucracy in the field • Final price is not established until all packages are bid • No guaranteed maximum price • Owner manages multiple contracts • Cost may be higher with multiple prime contractors • Higher owner administration costs to manage project

Figure 7.13 Advantages and disadvantages of CM-Agent project delivery method.

CM with fast tracking should only be considered when time is of essence and achieving an earlier completion date outweighs all of the disadvantages described above.

The CMAA (Construction Management Association of America) (www.cmaa.org/) defines construction management as the

... process of professional management applied to a construction program from concept to completion for purposes of controlling time, cost, and quality.

The AGC (American General Contractors of America) (www.agc.org) has defined construction management as

... one effective method of satisfying an owner's building needs. It treats the project planning, design, and construction phases as integrated tasks within a construction system. These tasks are assigned to a construction team consisting of the owner, the construction manager and the architect/engineer. Members of the construction team ideally work together from project inception to project completion, with the common objective of best serving the owner's interests. Interactions between construction cost, quality and completion schedule are carefully examined by the team so that a project of maximum value to the owner is realized in the most economic time frame.

The AIA (American Institute of Architects) (www.aiaa.org) defines construction management as

... special management services provided to an owner by an architect or other person or entity possessing requisite training and experience during the design phase and/or construction phase of a project. Such management services may include advice on the time and cost consequences of design, construction decisions, scheduling, cost control, coordination of contract negotiation and awards, timely purchasing of critical material and long lead time items, and coordination of construction activities.

The ASCE (American Society of Civil Engineers) (www.asce.org) defines construction management as

... one effective method of satisfying an owner's construction needs. It treats the project planning, design, and construction Phases as integrated tasks. Tasks are assigned to a project management team consisting of the owner, the construction manager (CM), and the design organization. A prime construction contractor or funding agency, or both, may also be a member of the team. The team works together from the beginning of design to project completion, with the common objective of best serving the owner's interests. Contractual relationships between members of the team are intended to minimize adversary relationships and contribute to greater responsiveness within the management group. Interactions between construction cost, environmental impact, quality, and completion schedule are carefully examined by the team so that a project of maximum value to the owner is realized in the most economic time frame.

Construction Manager/General Contractor (CM at Risk)

With CM at risk, the agency (owner) engages a CM to act as the agency's consultant during the pre-construction phase and as the general contractor (GC) during construction. During the design phase, the CM acts in an advisory role, providing constructability reviews, value engineering suggestions, construction estimates, and other construction-related recommendations. At a mutually agreed upon point during the design process, the CM and the agency will negotiate a guaranteed maximum

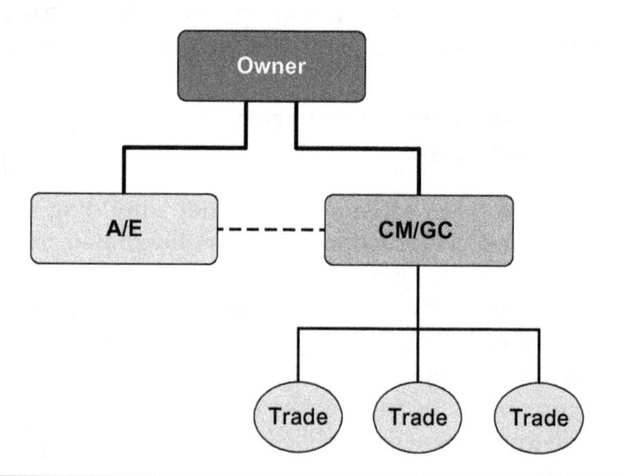

FIGURE 7.14 Organization of construction manager-general contractor (CM at risk) project.

price (GMP). The GMP is typically based on a partially completed design and includes the CM's estimated cost for the remaining design features, general conditions, a CM fee, and a contingency. Figure 7.14 shows the organization of a typical CM/GC project. Advantages and disadvantages of CM at risk project delivery method are summarized in Fig. 7.15.

The contingency can be split into CM and owner components. The CM contingency will cover increased costs due to unavoidable circumstances, for example material escalation or unforeseen underground conditions. The owner contingency would cover cost increases from owner-directed or owner-caused changes. The construction contingency can be handled in different ways under the contract. Unused CM contingency can be returned to the owner, shared by the owner and CM, or, in rare instances, given to the CM.

Owner are increasingly experimenting with sharing the contingency pool with the CM to provide the CM with an incentive to control cost growth associated with change orders to meet the GMP. The agency may elect to remove pricing of some material or work items as part of the GMP if pricing of these items results in an excessively high CM contingency or GMP. For example, in healthcare projects very often due to rapidly evolving medical imaging technology, the owner will purchase and furnish medical equipment for installation by the CM/GC.

After the GMP is established, the CM can begin construction, allowing for the overlap of the design and construction phases to accelerate the schedule. Once construction starts, the CM assumes the role of a GC for the duration of the construction phase. The CM holds the construction contracts and the risk for construction costs exceeding the GMP.

Construction Manager at Risk (CM-GC) Advantages and Disadvantages	
Advantages • Team Concept • CM firm selected on basis of interview/quality rather than low cost • Allows for innovation and constructability recommendations in the design phase, yet the agency still retains significant control over the design • CM holds construction contracts, transferring performance risk to GC • GC puts more investment in cost engineering and constructability review than with CM-Agency • Potential to fast track early components of construction prior to complete design • Reduces agency's general management and oversight responsibilities • Use of a GMP with a fixed fee and opportunity for shared savings provides an incentive for CM to control costs and work within funding limits	***Disadvantages*** • Difficult for owner to evaluate validity of GMP and value of contract • Once construction begins, the CM assumes the role of a general contractor, leading to possible tensions with the owner over project quality, budget, and schedule • Use of a GMP may lead to disputes over the completeness of the design and what constitutes a change to the contract • Agency retains design liability • GMP approach may lead to a large contingency to cover uncertainties and incomplete design elements

FIGURE 7.15 Advantages and disadvantages of CM at risk project delivery method.

Cost Control and Change Order Management

The CM/GC should establish a cost control system that enables the owner to review by trade current costs in relation to final cost projections. This is often referred to as the "open book" approach and is common for healthcare projects. The CM/GC also should establish a change order control system to advise the owner of potential added costs, thereby enabling the owner to make knowledgeable project decisions. The change order control system should be reviewed weekly in the project status meeting. Before submitting change order requests to the owner, the CM/GC should review all specialty contractor and supplier submittals to verify that they, in fact, represent changes that have occurred as a result of latent conditions or project scope changes. Subcontract changes necessary for coordination should be paid for with the CM/GC contingency funds after mutual agreement between the CM/GC and the owner.

Case Study: CM at Risk

On May 9, 2012, the Monroe Carell Jr. Children's Hospital at Vanderbilt celebrated the opening of a $30 million, 33-bed, 30,000-square-foot expansion to the hospital. The opening of the hospital's expansion culminates a year of construction, creating additional acute, neonatal intensive care, and medical-surgical beds. The expansion provides other enhancements such as growth to programs including pediatric bone marrow transplant, cardiac surgical intensive care and congenital heart disease. Since opening in 2004, patient occupancy at Children's Hospital has remained consistently high. The new space positions the hospital with additional capacity to meet the growing demand for the specialty and subspecialty services. A portion of the hospital's expansion was created to accommodate additional premature babies transferred from outlying community hospitals to Children's Hospital's level 3 neonatal intensive care unit, a service providing the region's most sophisticated level of care.

(*Courtesy: Bob Kiger, Vice President of Operations, Balfour Beatty Construction.*)

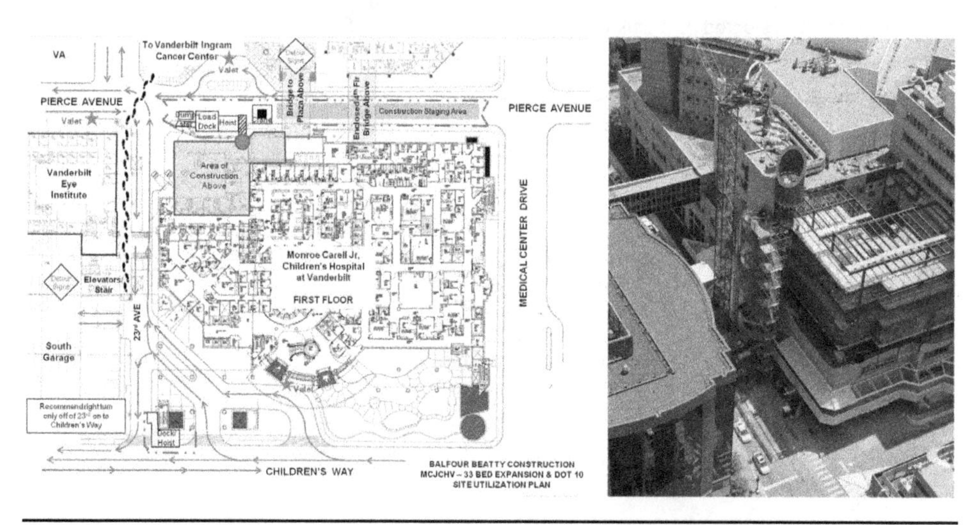

FIGURE 7.16 Thirty-three-bed addition to Monroe Carell Jr. Children's Hospital, Vanderbilt University, Nashville, TN; site traffic plan (*left*) and crane location (*right*).

Construction for the expansion began in March 2011, and was carried out by Balfour Beatty Construction. The 5 story expansion is an extension of the existing building's patient areas on the fourth through eighth floors. The additional 30,000 square feet brings the size of Children's Hospital to nearly 650,000 square feet total. A unique feature for the expansion is a lead-lined room to provide a pioneering radiation therapy process for patients with neuroblastoma, a cancer that develops in nerve tissue. The lead lining, which weighs 27,000 pounds and encompasses the entire room, protects others in the hospital from exposure to high-dose radiation. Figure 7.16 shows the site traffic plan and crane location for the Monroe Carell Jr. Children's Hospital.

Owner: Vanderbilt University

CM at Risk: Balfour Beatty

Budget and schedule

- Project approved by the Board of Trust September 7, 2010; full funding released December 13, 2010
- First release package of the construction documents issued January 20, 2011, for exterior skin, structural steel, and HVAC equipment
- Pierce Ave. closed down to one lane of traffic to make room for the tower crane—barricades were installed March 7, 2011 (Fig. 7.16)
- Total project budget = $25,780,000
- Work scheduled to be completed in May, 2012

CM-facilitated contributions

- Coordinated and facilitated "user group" meetings for each floor
- Facilitated estimating by third party estimating company
- Contributed to site issue resolutions
- Contributed to resolution of options for infection control and disruption mitigation on existing units, including surgical pathology lab and perioperative committee concerns
- Pursued alternate fire zone option, which resulted in lower cost and return of patient rooms ahead of schedule for revenue purposes (rooms taken out of service for "tie-in" work were returned 7 weeks ahead of schedule)

Design-Build

Design-build (DB) is a project delivery system involving a single contract between the project owner and a DB contractor covering both the design and construction of a project. The design-builder performs design, engineering, and construction according to design parameters, performance criteria, and other requirements established by the owner/agency. Figure 7.17 shows the organization of a typical DB project.

Several owners have used an approach called *modified design-build*, also called *low-bid design-build* or *detail-build*, where the agency completes a significant portion of the design before selecting the contractor using a low-bid solicitation or qualified low-bid process. The documents prepared by the agency are sometimes called *bridging documents*, and provide a basis for the bids defining scope and quality, without a full set of construction plans and specs. The design-builder then completes the remainder of the design work and constructs

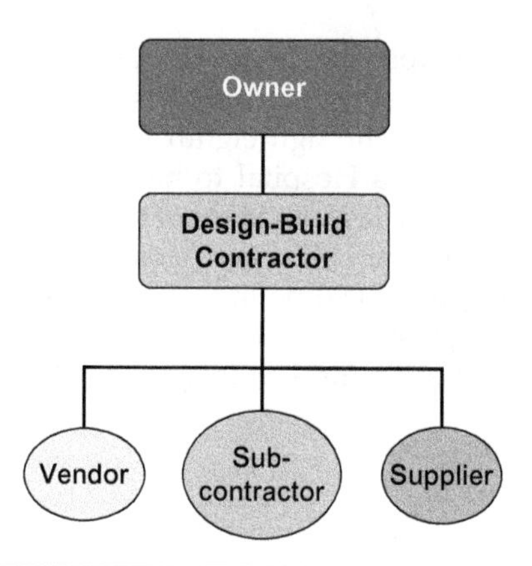

FIGURE 7.17 Organization of a design-build project.

the project under a single contract. Modified DB is primarily used in public projects where the state law prohibits the procurement of construction services using a method other than low bid or before the design is substantially complete, and the agency administers the project using traditional practices and retains greater responsibility for project performance.

The approach recommended by the Design-Build Institute of America (DBIA), is one where the owner completes the conceptual design to a lower level, and then procures the design-builder under a two-step best-value proposal process. This two-step best-value approach allows for an earlier involvement by the design-builder and shifts greater control and responsibility for the design and project performance to the design-builder.

Sometimes, large engineering and construction firms, with the capability of performing all the engineering, design, procurement, and construction themselves, are engaged to execute the total scope of the project. In other cases, teams are formed consisting of an engineering firm and a construction firm joining together (joint venturing) to deliver a "seamless" organization that is responsible to the owner for the project's total execution.

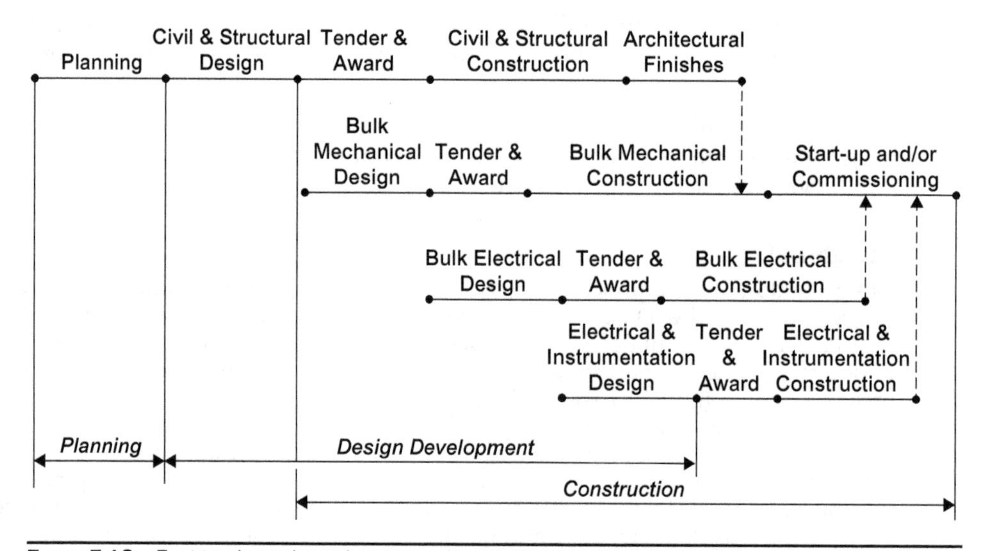

FIGURE 7.18 Fast-track or phased construction approach with design-build.

The use of a single entity to execute projects in this manner has become more commonly known as "fast-track design-build." In the "fast-track design-build" approach, the engineering and design, procurement, and construction processes are overlapped to produce a shorter overall schedule. Figure 7.18 presents a fast-track or phased construction approach using design-build arrangement.

Variations of Design-Build

A design-build contract may also include responsibilities that extend beyond the design and construction phases of a project, shifting more performance risk to the private sector. These include

- *Design-build-maintain.* A single entity designs, builds, and maintains the project works for a specified period of time under a single contract. Payment beyond completion of construction is typically tied to meeting certain prescribed performance-based standards for a period of years.
- *Design-build-operate.* A single entity designs, builds, and operates the project (e.g., a toll road) for a specified period of time under a single contract.
- *Design-build-finance.* Also known as build-operate-transfer (BOT) or build-own-operate-transfer (BOOT) is a form of project financing, in which a private entity receives a concession from the private or public sector to finance, design, construct, and operate a facility as stated in the concession contract. This enables the project proponent to recover its investment, operating, and maintenance expenses in the project.

Forty-four states allow the use of design-build on public works projects. The states that have most actively use design-build include Florida, Michigan, Ohio, and Pennsylvania.

Design-Build (DB) Advantages and Disadvantages	
Advantages	***Disadvantages***
• Single point responsibility for design and construction. • Accelerated project delivery by fast tracking design and construction. • Close coordination between designer and contractor. • Early contractor involvement to enhance constructability of plans. • Cost containment by minimizing owner's exposure to design errors and omissions. • Earlier schedule and cost certainty. • Owner's contract administration and site representative risks and costs are reduced, as the design-build contractor is responsible for all coordination efforts. • Innovation and quality improvements through alternative designs and construction methods suited to the contractor's capabilities. • Flexibility in the selection of design, materials, and construction methods.	• Reduced opportunities for smaller, local construction firms. • Fewer competitors and increased risk may result in higher initial costs. • Elimination of traditional checks and balances. Designer is no longer agency's advocate. Quality may be subordinated by cost or schedule considerations. • Less agency control over final design. • Higher procurement costs • Traditional funding may not support fast tracking construction or may require accelerated cash flow. • Accelerated construction can potentially overextend the workforce.

Figure 7.19 Advantages and disadvantages of the design-build project delivery method.

Advantages and disadvantages of the design-build project delivery method are summarized in Fig. 7.19.

BOT finds extensive application in infrastructure projects and in public-private partnership. In the BOT framework, a third party, for example the public administration, delegates to a private sector entity to design and build infrastructure, and to operate and maintain these facilities for a certain period.

Design-build delivery has been expanded to a public-private partnership concept, where a private entity or developer takes part in financing and leasing a transportation project in return for monetary compensation based on contractual authorization to collect toll revenues, or pursue development rights with the contracting agency. The private entity will be responsible for financing, design, and construction, and often will operate and maintain the roadway or bridge for a specified duration. The public-private contract may give full or partial contracting authority to the private entity.

Design-build and its variants have very rarely been utilized for healthcare projects in the United States. However, they have received wide acceptance in the United Kingdom, Australia, New Zealand, and Canada.

Case Study: Design-Build

Project: Smyth County Community Hospital

Location: Marion, Virginia

Project type: Design-Build (DB)

Project cost: Approx. $47 million

Figure 7.20 Smyth County Community Hospital (SCCH), Marion, Virginia. (*Courtesy: Hal Jones, VDC Director, Skanska USA Building.*)

Project Overview

Smyth County Community Hospital (SCCH), in Marion, Virginia, and as shown in Fig. 7.20, is owned by Mountain States Health Alliance (MSHA), a hospital system based in East Tennessee with 13 hospitals in the region. SCCH utilized a DB procurement method for its new 156,759 square-foot replacement hospital and professional office building (POB). The new hospital, located on a greenfield site, replaced an existing facility that was also in Marion. The new facility included a 13-bay emergency department, 3 operating rooms, 4 ICU rooms, 8 PCU rooms, 14 rehabilitation beds, 18 medical-surgery rooms, dining services with a dining hall, an imaging department (including an MRI suite), oncology suite, administration, and other support services.

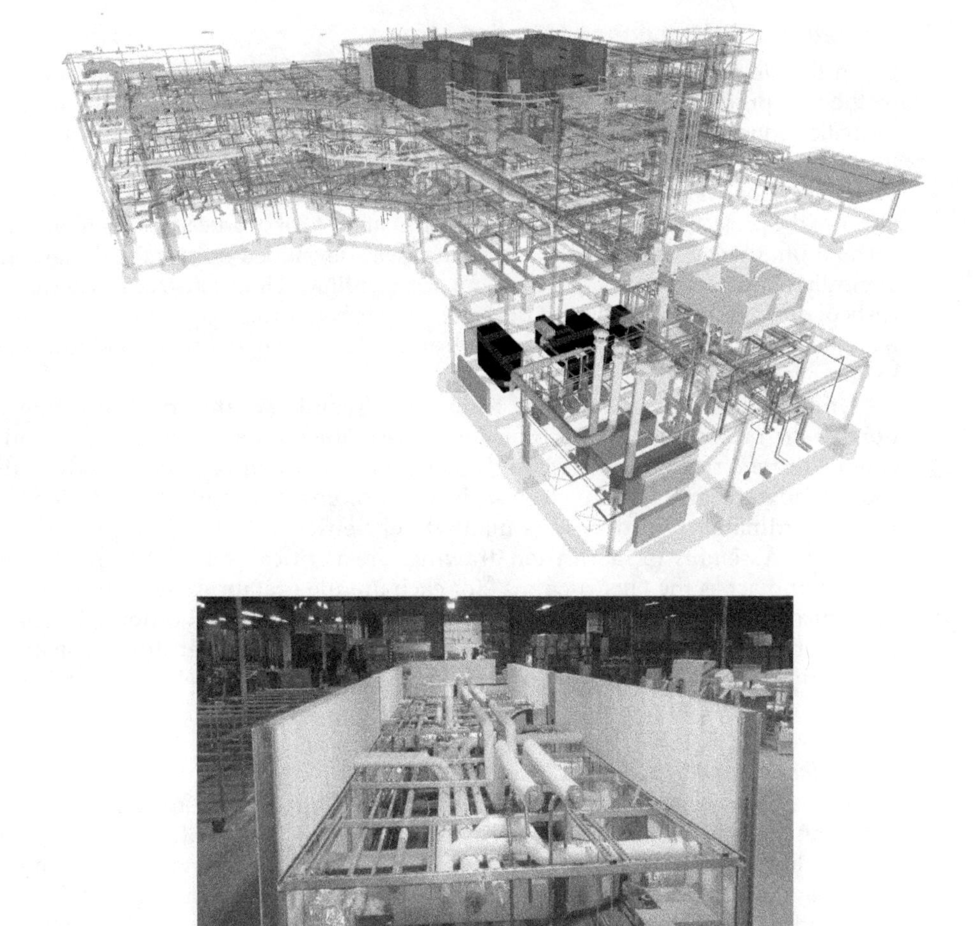

Figure 7.21 Building information model (*top*) and prefabricated unit from the model (*bottom*). (*Courtesy: Hal Jones, VDC Director, Skanska USA Building.*)

The Decision for Design-Build

MSHA was faced with a tight project budget and needed to find an innovative project delivery method aimed at lowering costs and tightening schedule while maintaining or improving quality. They understood that controlling the budget during the design process could significantly reduce overall project costs. Compared to the traditional design-build-bid approach, the DB method requires increased communication and collaboration from day one. The result is a strong collective focus on bringing value to the project from the entire team. Design-build required a collaborative effort that integrated various resources involved in the development of the project and provided incentives to produce a high level of technical performance and consistency with contractual budget and schedule terms.

Utilization of BIM and Prefabrication

Given the DB format of this project with one unified project team, the most effective method for producing high-quality construction documents was the use of building information modeling (BIM). Utilizing BIM from the start of the design process allowed the team to perform clash detection in parallel with design, and to complete more of the design work before construction began, which reduced changes during the construction phase of the project. For example, a fully coordinated set of overhead mechanical, electrical, and plumbing (MEP) rough-in drawings was developed during the design phase. Typically, overhead coordination does not occur until well into construction and is often rushed, due to scheduling. Using BIM allowed for better trade coordination, improved material procurement and quality control, and reduced material waste and rework due to improved coordination.

Utilizing BIM also helped the team move seamlessly into prefabricating portions of the project that would typically be assembled on-site. The project team decided to prefabricate overhead MEP racks, patient room bathroom pods, patient room headwalls, mechanical chases, electrical rooms, and central plant pipe skids. Fully coordinated BIM drawings made it very efficient to produce prefabrication drawings. Assembly or fabrication drawings are a critical step in the prefabrication process and act as the "instructions" for each unique prefabricated unit. It is a common misconception that prefabrication needs to be repetitive in order to be successful. By utilizing BIM, every prefabricated unit can be unique yet still be constructed efficiently. Figure 7.21 shows the building information model and prefabricated unit from the BIM model.

Prefabrication and Expediting the Schedule

Working in a warehouse had many schedule advantages over conventional construction. Craftspeople were able to work in a controlled environment, eliminating typical weather limitations. The entire warehouse was set up like an assembly line and each trade team was able to work through the MEP overhead racks, patient room bathroom pods, and headwalls quickly and efficiently.

This process sped up the schedule significantly. Deep foundations and concrete footings were installed during a very harsh winter. In total, the project lost close to 40 working days due to inclement weather. Since prefabrication allowed additional work to be completed indoors, unaffected by the weather, the project team was able to make up these lost days allowing the project to stay on schedule.

Prefabrication and Safety

One of the most important benefits of the prefabrication process is enhanced safety. With conventional construction of overhead MEP racks, work is completed on 8 to 12 feet ladders. However, because this work was performed in a warehouse and workers were at ground level; not a single ladder had to be used. The project safety was greatly improved and the risk of potential falls, worker fatigue, and eye injury were reduced. Working in a warehouse also minimized the negative effects of weather on worker safety.

Prefabrication and Quality

In addition to safety, working in a controlled environment increases quality on the job. Prefabrication allowed workers to install ductwork, piping, insulation, and electrical conduit at the ground level, allowing for 360 degrees of accessibility, and, thus, higher quality installation. In addition, the increase of accessibility improved quality control inspections. In typical construction circumstances, workers are unable to view the top of a piece of ductwork to make sure the insulation is properly installed, but with prefabricated overhead MEP racks, workers were able to inspect 100 percent of the work.

Summary

Overall, the Smyth County Community Hospital project was a success and a great example of how innovation and collaboration can result in a better project. Utilizing BIM allowed for advanced planning with fewer mistakes made in the field—saving time, improving quality, and reducing material waste. Prefabricating portions of the project also provided benefits. Working in a climate controlled warehouse provided a safer environment for workers—removing the risk of falls and negative impacts from weather conditions. It allowed the team better access to building components compared to traditional construction, resulting in a higher quality product. It also allowed for the construction of interior components of the project to begin early, while the exterior structure was being completed, saving time from the schedule. Because of BIM and prefabrication, the team was able to deliver the completed facility to the client two months ahead of schedule.

(*Courtesy: Hal Jones, VDC Director, Skanska USA Building.*)

Case Study: Design-Build-Finance

Project: St. Michael's Hospital

Location: Toronto, Canada

Project type: Design-Build-Finance (DBF)

Price of contract: To be announced following closing of the contract

(http://www.infrastructureontario.ca/Templates/Projects.aspx?id=2147489001)

St. Michael's Hospital is a teaching hospital and medical center in downtown Toronto, Ontario, Canada, shown in Fig. 7.22. It was established by the Sisters of St. Joseph in 1892, with the founding goal of taking care of the sick and poor of Toronto's inner city. The hospital provides tertiary and quaternary services in cardiovascular surgery,

Figure 7.22 St. Michael's Hospital, Toronto, Canada.

neurosurgery, inner city health, and therapeutic endoscopy. It is one of two level 1 adult trauma centers in Greater Toronto.

Infrastructure Ontario and St. Michael's Hospital issued a request for qualifications (RFQ) to design, build, and finance a new 17-story patient-care tower at the corner of Queen and Victoria Streets and the renovation of approximately 150,000 square feet of existing space.

The new tower will allow St. Michael's Hospital to relocate patient beds from an 85-year-old wing and provide larger space for programs that treat some of the most critically ill patients from across Ontario. These programs include patients from the medical-surgical intensive care unit and the largest adult cystic fibrosis program in North America.

Highlights of the project include

- Five new operating rooms, each large enough to include state-of-the-art medical imaging equipment. These hybrid operating rooms will allow surgeons to perform minimally invasive, image-guided or catheter-based procedures, as well as to undertake open surgery in the same operating room.

- Enlarged, state-of-the-art inpatient facilities for orthopedic surgery, oncology, and coronary care—including the cystic fibrosis program—as well as critical care space for the coronary and medical-surgical units.
- Expansion of the current emergency department, which was originally designed to accommodate 45,000 patient visits a year, but now accommodates over 70,000 a year—a number that continues to grow with the population. This expansion will allow St. Michael's to continue to fulfill its mandate as a regional trauma center.

The request for qualifications is the first step in the process to select a team with both the construction expertise and financial capacity to design, build, and finance the project. Based on a request for qualifications process that began in December, 2012, the following three companies were shortlisted:

- St. Michael's partnership (Bondfield Construction Co. Ltd., NORR Ltd. with Farrow Partnership, and Rocklyn Capital Inc.)
- Integrated team solutions (EllisDon Corp., Kasian Architecture, and EllisDon Capital Inc. with Fengate Capital Inc.)
- PCL partnerships (PCL Constructors Canada Inc., B+H Architects with Silver Thomas Hanley, and TD Securities Inc.)

The companies were invited to respond to a request for proposals, to be issued during summer 2013. Each includes a developer, design and construction firms, and a financial adviser.

"This is a significant step toward renewing our hospital," said Rob Fox, vice president of Planning and Development. "We're on schedule to begin construction next year, and we look forward to working with the successful bidder."

Comparison of Traditional Project Delivery Systems

Achieving a quality healthcare project on time and within the stipulated budget is a goal of all owners. While effective selection of the project participants and proficient execution of the project are important, selecting the optimum method of execution (the project delivery system) is *critical*. Over the past 20 years, many healthcare owners have attempted little or no change in the way they execute capital and/or renovation projects.

The Construction Industry Institute (CII) chartered a research project in 2004 to compare the key attributes for projects—cost, schedule, quality, and safety. Data from 127 building projects was analyzed by CII Research Team 133. The projects covered 37 states and a period of time spanning 7 years (1995–2002). Data was adjusted for time and location using historical indices. Size of the projects ranged from 50,000 square feet to 2.5 million square feet. The unit prices for the projects ranged from $30 to $2000 per square foot.

Design-build had the best performance of the three (DBB, CM, and DB). The median cost growth for DB was 2.17 percent, less than half the result of DBB. DBB had a

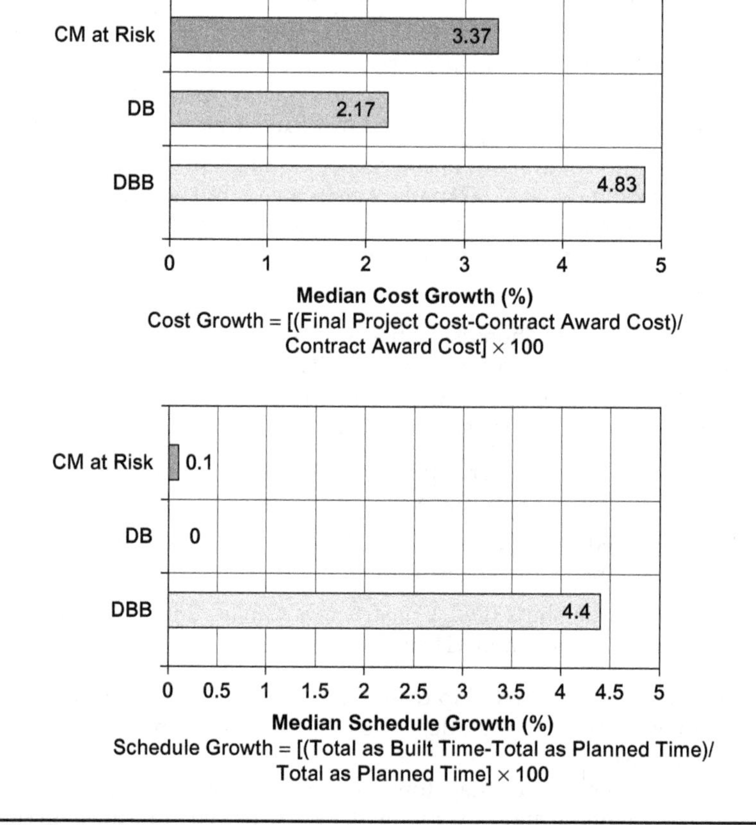

Figure 7.23 Comparison of cost growth (top) and schedule growth (bottom) (CII Research Summary 133-1, 2004).

50 percent likelihood of experiencing cost growth between 2 and 11 percent. When comparing schedule growth, DBB again fared the worst with a median growth of 4.4 percent versus no change for DB and CM at Risk. Figure 7.23 illustrates comparison of cost and schedule growth for the three project delivery systems.

Conclusion

In their article in *Healthcare Design* (November 2011), William R. Seed, VP of Design & Construction, UHS Inc., and Kenneth Lindsey, senior project manager, Southland Industries, suggest that "While other industries streamline their processes to become more productive over time, the building industry has faced a decrease in productivity, especially during the construction phase. This is especially true with healthcare facilities, which are becoming increasingly more expensive to build. In fact, by some calculations, the quantity of work constituting waste on a typical healthcare project may approach 50 percent of the total work performed. These inefficiencies are being absorbed by owners in the form of increasing costs of construction."

Given this situation, owners are finding that picking the right project delivery system, and constituting the right delivery team, is a priority in order to minimize costs and maximize quality for healthcare projects.

References

2012 U.S. Construction Industry Productivity (FMI 2012), http://www.fminet.com/media/pdf/report/FMIProductivitySurvey_2012.pdf

A Guide to the Project Management Body of Knowledge, 3d ed. *PMBOK® Guide,* 2004. *Project Management Institute (PMI),* 14 Campus Boulevard, Newtown Square, PA 19073-3299, www.pmi.org.

Integrated Project Delivery: A Guide, American Institute of Architects, Version 1, 2007.

LEED 2009 for New Construction and Major Renovations, U.S. Green Building Council, Inc. (USGBC®), Washington, DC 20037.

CHAPTER 8

Modern Project Delivery Methods for Healthcare Projects

Project Alliancing

The environment in which construction projects are accomplished today often involves completing complex projects within tight budget and time constraints while meeting stringent quality and safety requirements. In this environment "change" is a defining characteristic and is inevitable. Unfortunately, the traditional project delivery methods do not embrace change, but instead treat it as an anomaly by trying to specify every possible contingency and assign liability in the event change occurs. As projects become more dynamic, this increasingly leads to detrimental adversarial relations as individuals focus on protecting profit and not on collaborating to maximize performance.

In response to traditional contracting limitations, project alliancing, originally developed by British Petroleum (BP), and widely adapted by Australia's public sector to handle high visibility, has been implemented on complex capital projects. Alliance contracting was developed in the early 1990s for high-risk oil and gas projects in the North Sea by BP, in particular the Andrew Drilling Platform project, to create a more collaborative work environment, and to share project risks more evenly among project teams (Raisbeck et al., 2010). In 1994, the Wandoo Project was Australasia's first alliance project. The project was to develop a drilling platform in 175 feet of water, and its owner, Ampolex, chose to develop this field under an alliance contract. Several key management decisions enabled the success of the Wandoo Alliance. For example, Ampolex dedicated $1 million to behavioral workshops, training, and collaborative sessions. All parties agreed to shift from a confrontational approach for pricing to a collaborative "open book" policy. This was tested during construction, when there was a breech in the sea wall, and construction was brought to a halt. A solution was developed in a week and construction resumed.

Project alliancing is a dramatic departure from traditional contracting methods, in that it encourages project participants to work as an integrated team by tying commercial objectives (i.e., profits) of all parties to the actual outcome of the project.

This chapter was written by Sanjiv Gokhale.

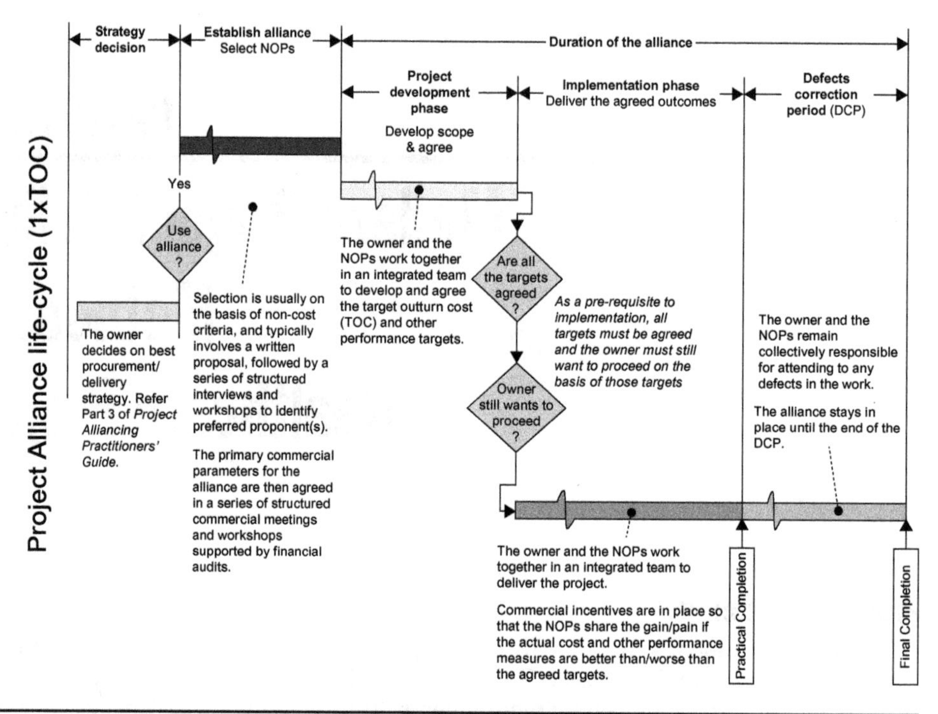

FIGURE 8.1 Project alliance life cycle (*Project Alliancing: A Practitioners' Guide, 2006*).

Under project alliancing, an agency and one or more service providers (constructors, consultants, designers, suppliers, or a combination thereof) collaborate on the delivery of a project. Alliancing uses contractually established financial incentives to encourage superior project performance and cooperation among the alliance participants. A typical life cycle for an alliance project is shown in Fig. 8.1.

Typical characteristics of a project alliance include the following:

- The alliance team members jointly develop and agree to a "target cost," which is then verified by an independent estimator.
- At project completion, the target cost is then compared to the final cost, and the underruns or overruns are shared through pre-agreed ratios among the participants based on their relative contributions to the leadership, performance, outcomes, and overall success of the alliance. In this manner, all participants have a financial stake in the overall project performance.
- Project risk and responsibilities are shared and managed collectively, rather than allocated to specific parties.
- All participants have an equal say in decisions for the project, with decisions made unanimously on a "best-for-project" basis, rather than to further individual interests.
- All participants provide "best-in-class" resources. Full access is provided to the resources, skills, and expertise of all participants.

- The alliance agreement creates a no-fault, no-blame, and no-dispute culture. No legal recourse exists except for the limited cases of willful default and insolvency.
- All transactions are open book.

The use of project alliancing to establish and deliver a project generally entails four phases, with the alliance remaining intact until the end of the final phase. *Project Alliancing: A Practitioners' Guide* (2006), published by the State of Victoria, Australia, describes these phases as follows:

- *Alliance establishment phase.* The owner/agency selects project participants on the basis of non-cost criteria, such as technical expertise and experience, financial and management resources, quality and time record, and willingness to commit to a cooperative relationship with the owner. The agency may either select each of the key participants (e.g., designer, contractor, supplier, etc.) in separate selection processes, or allow the industry to establish its own teams and submit proposals as an integrated team or consortium. Although conducting separate selection processes allows the agency to select the best individual companies, this approach can be time consuming and may not necessarily yield the best overall team. For such reasons, agencies more commonly choose the integrated team approach to alliance participant selection.

  Following participant selection, the agency will conduct a series of meetings and workshops with the selected participants to establish the commercial framework and primary alliance parameters, including the compensation structure, fees for overhead and profit, and the gain/pain share arrangement, which are then formalized in an alliance agreement.
- *Project development phase.* The agency and the selected alliance participants will work together as an integrated team to develop and agree to a target cost and other performance targets (e.g., timely completion, maintenance costs, quality, etc.).
- *Implementation phase.* Once the targets are established and agreed to, the alliance team works together to deliver the project with the objective of achieving or exceeding the agreed-to targets.
- *Defects correction period.* The participants remain collectively responsible for addressing any defects in the work (typically for a period of about 24 months).

Compensation to the non-agency members of the alliance team is typically based on a *three-limb model* (Fig. 8.2) that compensates each participant as follows:

- *Limb 1 fees* consist of all direct project costs and project-specific overhead incurred by the alliance team members. These fees are viewable by all contracting parties using 100 percent open book accounting.
- *Limb 2 fees* consist of corporate overhead and profit. These fees were determined during the alliance establishment phase through a series of financial audits of the participants.
- *Limb 3 fees* are based on a predetermined gain/pain share arrangement that is dependent on how the actual cost (Limb 1 fees) compares to the target cost. Losses are capped at Limb 2 fees; therefore, participants are at least guaranteed to recover all direct costs (Limb 1 fees).

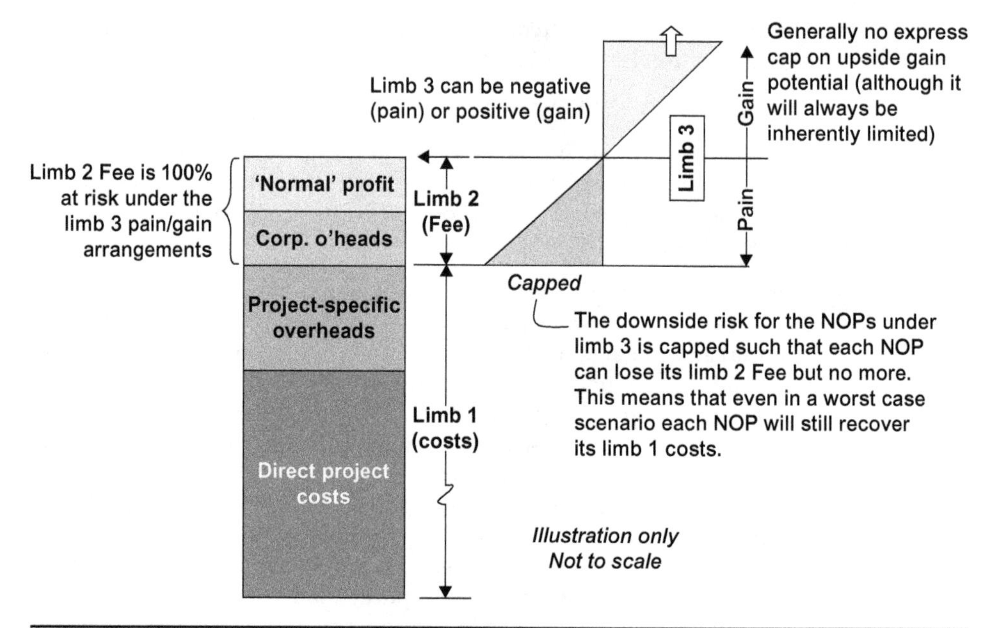

Figure 8.2 Three-limb model for compensation of project participants in an alliancing project (*Project Alliancing: A Practitioners' Guide, 2006*).

Integrated Project Delivery

Integrated project delivery (IPD) has parallels with Koskela's lean construction movement (Koskela, 1992), which has aimed to translate product manufacturing and production methods to construction. Koskela defines these methods based on concepts that originated in the Toyota production system developed in Japan in the 1950s. Specifically, the genealogy of these methods can be related to the production concepts of just in time (JIT) and total quality control (TQC) (Koskela, 1992). The Lean Construction Institute (LCI), which is a coalition of academics, consultants, large software vendors, and contractors from different disciplines based in Southern California, appears to be a key point for the transfer and dissemination of early documents about IPD (Matthews and Howell, 2005). Gregory Howell, a cofounder of the LCI, notes that the name IPD is trademarked in the USA.

Howell notes that IPD was not inspired by alliancing but has much in common with it. Howell succinctly views alliancing "as a form of contract and organizational governance" and lean construction "as the operating system." In this view, IPD can be seen as the combination of alliance governance structures with lean construction operational systems. Both alliancing and IPD are in marked contrast to traditional procurement models, in that traditional project management, contracting, and organizational practices attempt to optimize the whole by optimizing each piece—lump sum contracting connected with centrally developed and managed critical path method (CPM) schedules is the best example. By contrast, alliancing contracting optimizes the project not the pieces.

IPD has clear links to concurrent engineering (CE) theories. CE describes the "method of concurrently designing both the product and its downstream production and support processes" (Kamara, 2007). This approach has parallels in the IPD early stage workshops and the use of collocation in a "big room" environment. As with IPD, central to the idea of CE are two guiding principles: "integration and concurrency." Integration aims to share and transfer information and knowledge "between and within project stages and all of the technologies and tools used in product development process." Concurrency determines "the way tasks are scheduled and the interactions between different actors (people and tools) in the product development process" (Kamara, 2007). As with CE, IPD design processes require early stage briefing, analysis, and consideration of life cycle issues by multidisciplinary teams.

The above comparison points to the obvious similarities between IPD and alliancing. However, there are a number of key points of difference between the IPD model and alliancing. The most notable of these are collocation in a "big room" environment and the mandated use of building information modeling (BIM). An increasingly popular tool for collaboration in IPD projects is the use of physical maps that allow project teams to discuss sequencing as an integrated team in the "big room" environment.

In IPD, BIM is the fundamental platform that enables three-dimensional model integration and data sharing between team members. In the IPD model, BIM serves as the project's base IT infrastructure. It is the digital modeling that drives innovation in the project.

Lean Project Delivery

Another term often used to refer to a form of IPD is lean project delivery system (LPDS), a term developed by the Lean Construction Institute (LCI). Many of the principles attributed to lean project delivery are similar to those attributed to IPD. In fact, in this era of evolving terminology, many refer to IPD as lean project delivery, where the application of lean thinking and lean principles are applied throughout the project.

A common definition of lean design/construction is *the continuous process of eliminating waste, meeting or exceeding all customer requirements, focusing on the entire value stream, and pursuing perfection in the execution of a constructed project.*

Key definition: Waste:

- Overproduction
- Waiting
- Unnecessary transport
- Over-processing
- Excess inventory
- Unnecessary movement
- Defects and rework
- Not using employee talent
- Environment/energy

Followers of IPD treat lean principles along with the resulting efficiencies and elimination of waste as givens. Followers of lean treat collaboration and the use of technologies as givens. In the end, lean and IPD both strive for the same ultimate outcome. They are just two different paths to get to the same place: to a project that has been optimized to maximize the value. Whether the project is optimized by applying lean principles first, then IPD principles, or by applying IPD principles, then lean, does not matter. Early adopters of both have shown that the application of both lean and IPD principles is natural and leads to more successful outcomes.

The ideal application of lean begins during the design with the value stream and project schedule mapped by the team. Production of documents proceeds based on the commitments each party makes to the team. This process develops a sense of camaraderie among the team that should carry through the construction phase of the project. During construction, the project is scheduled throughout as a team from the milestones developed during the pre-construction phase. Each pull-planning session results in a more detailed schedule that clearly and accurately shows all of the activities that must occur prior to or concurrently with the next activity.

The key to the increased efficiency of lean is the measurement of adherence to the project schedule. Each party reports on its ability to meet the schedule commitments made the previous week. If commitments are not met, constraints are identified and removed by the team. The power of peer pressure, built on a foundation of mutual respect and understanding over the course of the project, is a powerful motivating force for team members to meet commitments. Each party is incentivized to be a project leader rather than a project laggard in an effort to move the project forward toward successful completion as defined by the value stream.

Building Information Modeling as a Catalyst

BIM is technology that supports the delivery of projects in a more collaborative and integrative way. Collaborative, integrated teams are using building information models in a collaborative, computable way to achieve better decision making. Collaborative decision-making strategies are, of course, fundamental to the IPD process. Even if, hypothetically, an IPD project may be delivered without using BIM and vice versa, the real benefits will be seen only when BIM methodologies are applied to IPD processes. The consistency of the "information (I)" is the real value that BIM can provide to an IPD process: information integration, reliability, and interoperability are at the heart of the tool. This can only happen when the information model is shared transparently and becomes an integral part of the decision-making process throughout the design, construction, and management of the building.

IPD documents crafted by the American Institute of Architects California Council (AIACC) mandate and promote the full scale implementation of digital technologies. For example, contract E202-2008, the BIM Protocol Exhibit, explicitly encompasses a range of acceptable uses for BIM including: model ownership, responsibilities, and authorized uses covering cost estimating, construction scheduling, documents, shop drawings, and project adaptations (Fig. 8.3). Model ownership is established in early stage workshops and is critical to the success of the project. Participants' capability to take a model to a given level of detail—from "100 level" to "500 level"—is also considered at this stage. The team member with the strongest BIM capability will often be assigned to be the "model owner," regardless of their role on the team or parent organization. Under IPD, if the model is inaccurate all parties share the risk. Other features such as schedule sequencing and cost estimating are tied to the three-dimensional model and continuously updated to reflect the estimated cost of the proposed design. This information allows IPD designers to consider multiple streams of information while crafting a preferred design. On IPD projects, "clash meetings" are held weekly to determine if there are any issues in the placement of building systems. Therefore, under IPD,

Level of detail and model content and authorized uses.	100	200	300	400	500
Design and Coordination (function/form /behaviour)	Non-geometric data or line work, areas, volumes zones, etc.	Generic elements shown in three dimensions - maximum size and purpose	Specific elements Confirmed 3D Object Geometry - dimensions - capacities - connections	Shop drawing/ fabrication - purchase - manufacture - install - specified	As-built - actual
4D Scheduling	total project construction duration phasing of major elements	Time-scaled, ordered appearance of major activities	Time-scaled, ordered appearance of detailed assemblies	Fabrication and assembly detail including construction means and methods	
Cost Estimating	Conceptual cost allowance Example $/sf of floor area, $/hospital bed, $/parking stall, etc.	Estimated cost based on measurement of generic element. E.g., generic interior wall.	Estimated cost based on measurement of specific assembly. E.g., specific wall type.	Committed purchase price of specific assembly at Buyout.	Record costs

Figure 8.3 AIACC Contract Document E202-2008—BIM Protocol (AIACC 2008).

a BIM model is used to update data continuously, so that all project teams are working from the latest version.

BIM can be of great value for all owners, both public and private. In the public arena, most owners are also managers of their buildings, and it is here that BIM adds major value. Most have experienced the loss of major project information between the end of construction and beginning of the management phase; as a result, most owners understand how difficult it is to collect, organize, manage, and store the many different types of information required for long-term facility management. BIM can help the owner in this major task. It can be seen as a repository of major sets of information or be linked to other information perhaps not stored within the model. BIM for facility management is the next big step for a real use of this new technology. At this point, little research exists documenting the benefits of BIM for facility management, but it is a natural step in the building life cycle to capture information at the end of construction and beginning of operations.

Sustainability

Sustainability is based on a simple premise: Everything that we need for our survival and well-being depends, either directly or indirectly, on our natural environment. Sustainability creates and maintains the conditions under which humans and nature can exist in productive harmony, that permit fulfilling the social, economic, and other

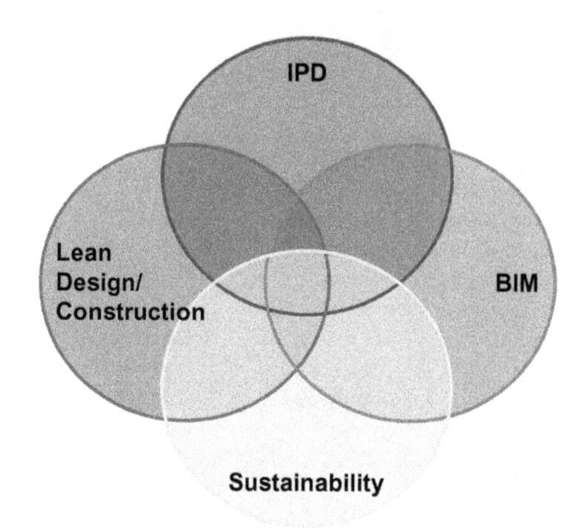

Figure 8.4 Convergence of related trends in modern healthcare design and construction.

requirements of present and future generations. Sustainability is important to making sure that we have, and will continue to have, the water, materials, and resources to protect human health and our environment.

Building owners everywhere, public and private, are thinking about sustainability. Governing bodies, municipalities, and code authorities are also establishing aggressive requirements in terms of energy reduction or sustainability rating system outcomes. The International Code Council (ICC) has published the International Energy Conservation Code (IECC) and the International Green Construction Code (IgCC) to provide minimum regulations for energy conservation and sustainability. The LEED for Healthcare rating system represents a collaboration between the Green Guide for Health Care (GGHC) and the U.S. Green Building Council (USGBC). The GGHC conducted a pilot program that included more than 100 healthcare facilities that informed the development of the LEED for Healthcare rating system included in the USGBC's LEED for Healthcare (2009).

U.S. Energy Information Administration research and other studies show that the construction and operation of buildings are responsible for as much as 48 percent of total U.S. annual energy consumption and 76 percent of annual U.S. electrical consumption, making the built environment the single largest contributor of greenhouse gas emissions. Figure 8.4 shows the convergence of four related trends that are seen as laying the foundation for modern healthcare design and construction.

Prefabrication

Prefabrication, preassembly, modularization, and offsite fabrication (PPMOF) have become more viable with recent advances in design and information technologies. Owners and project teams seeking improved productivity, quality, and safety have recognized that the PPMOF strategy offers a substantial opportunity to improve project performance and to overcome external and internal project challenges.

Definitions

The following definitions are used in the design and construction industry in relation to PPMOF:

- Prefabrication—a manufacturing process, generally taking place at a specialized facility, in which various materials are joined to form a component of a final installation. Prefabricated components often involve the work of a single craft but also can involve multiple trades on building elements, such as a patient room toilet.
- Preassembly—a process by which various materials, prefabricated components, and/or equipment are joined together at a remote location for subsequent installation as a subunit; generally focused on a system.
- Module—a section of a building resulting from a series of remote assembly operations that may include portions of many systems, usually the largest transportable unit or component of a facility.
- Offsite fabrication—the practice of preassembly or fabrication of components both off the site and onsite at a location other than at the final installation location.

In its recent SmartMarket Report, *Prefabrication and Modularization* (May 2011), McGraw-Hill reported that these construction methods were being used on at least some projects by 50 percent of architecture, engineering, and construction (A/E/C) professionals. These numbers are expected to grow as professionals who are not currently using prefabricated/modular elements expect to be doing so on projects.

Prefabrication and modularization boast significant benefits to A/E/C firms such as improved productivity, enhanced competitive advantage, and greater return on investment. And these are not the only drivers to the increasing popularity of prefabrication and modularization. These methods are also gaining traction for their other numerous benefits including

- Increased safety
- Higher degree of quality assurance and quality control
- Accelerated timelines to completion
- Better control of costs
- Waste reduction
- Reduced team traffic on jobsite
- Drastically reduced weather delays
- Factory conditions that allow for more precise cutting and assembly, and for more frequent and more complete component inspections
- Improved ability to address the challenges of structures within a factory controlled environment

While the above advantages hold true for most projects that employ prefabricated and modular construction, healthcare facilities are said to experience additional benefits of employing these methods. In fact, McGraw-Hill states that healthcare is the fastest growing market for prefabricated and modular construction with 49 percent of A/E/C respondents claiming that they are using it on healthcare facilities. The specific benefits of prefabrication and modularization on healthcare construction projects are

- Workers can address unique requirements of infection control and HVAC systems without space limitations.
- Better flexibility when additional space or new technologies are needed as new modular structures can be attached to an existing building more easily and less expensively.
- For healthcare facilities in rural areas, modular construction is even more attractive, as a major challenge in building facilities in less-populated areas is finding skilled workers. Instead of the tradesmen needing to be where the project is, the structure is prepared where the tradesmen are.

Common prefabricated components for medical facilities are patient restrooms, equipment enclosures, operating suites, headwalls, and patient suites. A healthcare system that successfully implemented the use of prefabrication and modularization on the construction of a new 484,000-square-foot facility is Premier Health Partners. Its Miami Valley Hospital in Dayton, Ohio, is believed to be the first major hospital construction project in the United States to employ prefabrication of all major mechanical, electrical, and plumbing components.

By optimizing the processes and maximizing value, owners try to get the most out of their projects, but they must be smart about how they accomplish this. Lean, BIM, prefabrication, and IPD can all be utilized separately, but they are strongest when used together (Fig. 8.5). IPD can be both a collaborative process and a relational contract that

<table>
<tr><th colspan="2">Integrated Project Delivery (IPD) Advantages and Disadvantages</th></tr>
<tr><td><u>Advantages</u>
• Improved ability to manage risks due to the sharing of responsibility and incentive for all participants to proactively mitigate risks
• Earlier involvement of construction and cost planning expertise in the project development phase
• Reduced need for contract administration (i.e., inspection, dispute resolution) allows resources to be focused on achieving project objectives
• Less adversarial system
• Transparent pricing of the project, including contingencies
• Increased efficiency provided by a well-functioning team</td><td><u>Disadvantages</u>
• Absence of direct price competition can lead to overly conservative and easily achievable performance targets
• Absence of legal recourse (with the exception of willful default and insolvency)
• Limited experience in North America
• Participants are exposed to a broader range of risks than on a traditional project
• Participants are liable for the performance of other team members
• Requires high level of involvement from senior management to establish and maintain alliances
• Agency's ability to make unilateral decisions is severely restricted
• Increased procurement costs
• Contractors may be hesitant to enter into a arrangement where risks are shared and selection occurs prior to target pricing</td></tr>
</table>

Figure 8.5 Advantages and disadvantages of IPD project delivery method.

drives different behavior and teamwork. Lean is a mindset and a way of thinking that helps to promote behaviors that inherently help to improve project efficiency and collaboration. BIM is a tool that can be used to practice prefabrication and lean and can be applied to IPD. It is the medium through which these collaborative, efficient behaviors are best employed. Sustainability benefits from all of these factors to provide a more energy-efficient and less wasteful product.

Case Study: Sutter Health Fairfield Medical Office Building, Fairfield, California

Project Description

The project is a three-story, 70,000-square-foot medical office building (MOB) housing primary care medical practices and laboratories, with pediatrics, oncology, rheumatology, and cardiology departments and administrative offices (see Fig. 8.6). The owner, Sutter Health, is one of the largest not-for-profit healthcare providers in northern California. This project is the first built component of a $6.5 billion capital program of which, at the time of this study, several subsequent projects are in advanced stages of design. As such, it gave Sutter the opportunity to test out a new process of collaboratively designing and building facilities in a relatively small project.

Owner: Sutter Regional Medical Foundation (www.sutterhealth.org)

Architect: HGA Architects and Engineers (www.hga.com)

Builder: The Boldt Company (www.theboldtcompany.com)

The Fairfield MOB was the first Sutter Heath project to use a three-way, integrated form of agreement as the basic design and construction contract. Attorney Will Lichtig, whose Sacramento firm has represented Sutter for 50 years, drafted the IFOA used for the Fairfield project. One of the most significant contract provisions has to do with trust: "The Parties recognize that each of their opportunities to succeed on the Project is directly tied to the performance of other Project participants. The Parties shall therefore work together in the spirit of cooperation, collaboration, and mutual respect for the benefit of the Project, and within the limits of their professional expertise and abilities."

FIGURE 8.6 Sutter Health Fairfield Medical Office Building, Fairfield, California (*Integrated Project Delivery: Case Studies, AIA California Council,* 2010).

Early Involvement of Key Participants

The initial project team consisted of Sutter Health (the overall corporate entity), Sutter Regional Medical Foundation (the local Sutter affiliate), HGA, and Boldt. Sutter had issued an RFQ to select an architect in the spring of 2005. HGA interviewed and won the job, in part because of a successful prior relationship with Sutter. Subsequently, Sutter asked HGA to meet with Boldt to see if the firms' cultures aligned. The firms had previously worked together on traditional design-bid-build projects in the Midwest. The principals met and decided that it was good fit and to proceed. The three-way contract called for the core team of owner, architect, and builder to collaboratively select the main design-build subcontractors very early in the design process. Smaller sub-trades were competitively bid with lump sum prices.

Shared Risk/Reward

The integrated form of agreement (IFOA, a "relational" contract) creates a system of shared risk with the goal of reducing overall project risk rather than shifting it between parties. Contingency funds are jointly managed by the project participants, rather than at the owner's discretion alone. The early version of IFOA used for this project allowed for a financial incentive plan but the participants elected not to implement it. "It was all so new," said Bonnie Walker of HGA. "We were still in the mindset of business as usual." Subsequent Sutter IPD projects have used incentives funded by project savings and pooled profits to reward designers and builders for meeting and exceeding agreed project goals. In these projects, most sub-consultants and subcontractors participate in the pool as well.

Multiparty Contract

The IFOA is a three-way contract between the owner, the architect, and the builder. Each party is held accountable to each other as an equal partner. Architect and builder combine their contingencies and are jointly responsible for construction errors and design omissions. All books in regard to the project are open. This contract was the first of its kind to be used by any of the parties and may have been the first such agreement to be used on a construction project in the United States.

Collaborative Decision Making/Control

An integrated project team (IPT) composed of project manager level representatives of Sutter, HGA, Boldt, and the major subcontractors, Rosendin Electric and Southland Industries, met weekly throughout design and construction. The committee was augmented, when appropriate, by representatives of other trade contractors and stakeholders. A higher level core team, consisting of a senior representative each of Sutter, Boldt, and HGA met monthly to resolve issues passed up from the IPT. Any decisions that could not be unanimously agreed to at this level could be referred to an executive level committee with higher level representation from the three partners.

Liability Waivers among Key Participants

The project agreement did not contain a "no-sue" clause. The parties agreed to use alternative dispute resolution: first within the core team, then by agreeing to rely on an expert third party for resolution, and if necessary to mediation. The owner, architect, and builder agreed to indemnify each other and to provide typical insurance, including architects' professional liability insurance, at limits established in the IFOA. The architect's liability for consequential damages was limited to the amount of its fee, and the builder's liability for consequential damages was limited to an amount equal to its fee plus general conditions.

Jointly Developed/Validated Targets

Sutter's internal budget of $19 million was based on a very generic MOB project with little architectural amenity. Boldt's first estimate was $22,250,000. After an intense validation effort, a guaranteed maximum price (GMP) of $19,573,000 was agreed by the three parties. The final construction cost was $19,437,600 which included $836,500 of value-added, owner-initiated scope additions. Benchmarking of comparable medical office buildings was established. A finish date of December 2007 was set. In subsequent Sutter projects, specific metrics called *conditions of satisfaction* have been negotiated for, among other things, improving operations, improving space efficiency, reducing time to build, and reducing consumption of natural resources.

Narrative

Sutter Health, after having had its share of disputatious projects, was looking for a better way to build facilities. It hosted the Sutter Lean Summit in 2004, with help from the Lean Construction Institute. This three-day event set forth a vision for transforming the way Sutter capital projects would be designed and built. Room data sheets and narratives were used to definitively establish detailed requirements. Each room's equipment needs, finishes, utilities, and special requirements were documented. This approach was used to document and preserve decisions made by stakeholders during programming and ensure that the final product met stated needs.

Sutter needed the building delivered in 25 months and that was accomplished despite a three-month delay for reprogramming at the start of the project and with the addition of extra scope. The extensive use of BIM was a new experience for architect, builder, and owner, although the MEP subcontractors had limited prior experience. Live group modeling sessions around a projector were held every other week. Steel structure was modeled along with duct runs, cable trays, plumbing lines, and the sprinkler system. These sessions enabled the IPT team to identify over 400 system clashes that, because they were discovered early, "provided significant cost savings due to increased field productivity, tighter schedule, more prefabricated work, and less redesign," according to Boldt's Jay Harris.

Later, BIM was used with GPS measurement to drop ductwork hangers into the metal decking before concrete was placed. Layout that normally would have taken 2 to 3 weeks was accomplished with greater accuracy in 2 to 3 days. The more accurate hanger placement allowed for much larger sections of shop prefabricated ductwork and less field labor.

The ability of the design team to work directly and interactively with subcontractors was appreciated by both sides, and relieved the general contractor of always having to be the hub of information exchange. For casework, much less detailing effort was needed from the architect—with no loss of design or quality control.

Boldt's project website became the repository of project information and the place where submittals were made and processed electronically. Over 50 percent of the submittals were processed by the architects without paper documentation.

Consideration of change orders was limited to the following categories:

- Owner generated—requested by owner, owner's suppliers, or consultants.
- Unknown conditions—items that could not be anticipated during design or that the builder could not have anticipated during pre-construction.
- Design refinement—added value to the owner. Owner would have paid for work if included in bid documents.
- Construction revision—no added value to the owner. Something had to be added, removed, or reworked once it was installed as a result of design error or omission.
- Governing agency generated—the result of unforeseen agency code changes.
- Interpretations—newly enacted codes or policies being enforced which could not have been anticipated during design or bidding.
- Builder generated—the result of corrective work requiring documentation to record the change, owner accepted nonconforming work, or builder-requested changes.

By the end of the project there were no change orders that had not been initiated by the owner.

"Last planner," "reliable promises," "pull scheduling," end-of-day "huddles," and other lean construction techniques were employed with success. Just-in-time materials management was not used in this project, in part because there were large areas available for staging.

Lessons Learned

Sutter was very pleased with the building and the process. The project was under budget and within schedule. Change orders were virtually eliminated. Lessons learned from this pilot project have been applied to larger and more complex projects Sutter is currently undertaking, including California Pacific Medical Center's $1.7 billion, 555-bed Cathedral Hill Campus in San Francisco and the $320 million Sutter Medical Center in Castro Valley, California.

Subcontractors found that more intense effort is required up front than in negotiated or design-assist projects, but the payback comes later with rework almost completely eliminated. The early commitment inherent in IPD allows them to devote these resources to the pre-construction phase. In future projects, Boldt intends to provide field superintendents with BIM capability in the trailer. In this project, a few of the subcontractors did not want their foremen attending the group scheduling meetings. Boldt now makes this a mandatory requirement. The owner must be kept engaged from earliest design and throughout construction. In this case, during construction the owner's project manager was distracted with another, more troublesome, project and the team felt that this may have slowed decision making. Pre-construction design assist is vital for those trades that have the biggest impact on other systems. Mechanical, electrical, and plumbing/fire protection certainly fall into that category, but Boldt learned that exterior glazing and skin should also be one of the early selected sub-trades that fully engages in early design.

Boldt felt that financial incentives would have been a benefit to this project, with the incentives flowing down to the subcontractor level. All of the considerable project savings in this case went only to the owner. Boldt Group President Dave Kievet thinks the key is the alignment of commercial interests. "By aligning the owner's commercial goals with those of the project team, it is possible to create a win-win situation where any incentive payment becomes an acknowledgement of a job well done and not the driver of it." He believes the way to do that is to put profit in a separate bucket from fee. "One of the lessons learned is that the best way to ensure commercial alignment is to completely separate the cost of the work from the profit. That way, as the team continues to drive down the cost, the partners' actual return as a percentage of revenue goes up." He would apply that thinking to every input from design services to structural steel.

By contrast, Bonnie Walker of HGA is unsure whether the existence of an incentive pool necessarily leads to project-centered behavior. For example, if the architect's fee is a not-to-exceed amount based on a planned number of hours, any savings from hours not used are rolled into the incentive pool with the architect getting a smaller percentage back. "I like having control of our fees," she says, "I believe that a lump-sum fee is a leaner approach. It doesn't take an incentive pool to get us to behave collaboratively."

Participants, when asked if IPD was applicable to all projects, felt that it is ideal for larger-scaled, complex projects and perhaps does not have proportionate value in smaller, simpler projects. This is perhaps more a reflection of the up-front time it takes to establish IPD standards and procedures rather than an issue of scale.

Participants reported a feeling of being respected as equal partners in a collaborative process in which everyone's opinion was valued. In addition to the efficiencies gained from such a process, there was a sense of goodwill, trust, and professional satisfaction.

Case Study: Integrated Project Delivery

Children's Hospital Colorado marked its fifth anniversary on the Anschutz Medical Campus in Aurora, Colorado, by announcing the official completion of a new 10-story East Tower addition completed by the joint venture team of McCarthy and GH Phipps Construction Companies. Figure 8.7 indicates the architects rendering (left) and construction (right) of the project. The tower project reunited the joint venture with design partners H+L Architecture and ZGF Architects LLP, the same design and construction team that completed Children's Hospital in 2007. The new 335,000-square-foot tower was completed under an IPD, tri-party agreement that minimized cost and schedule risk for the hospital.

The new East Tower initially adds 94 beds to Children's, with later expansion available in five floors of shelled space in order to reach a capacity of 500 beds for Children's Hospital. The new tower helps the hospital expand its focus on specialty services, including housing the innovative Colorado Institute for Maternal and Fetal Health, which began seeing patients in September, 2012. The addition also provides space for the hospital's Center for Cancer and Blood Disorders, the Children's Colorado Heart Institute, Orthopedics Institute, Neurosciences Institute, Digestive Health Institute, Breathing Institute, and intensive care.

Figure 8.7 Children's Hospital Colorado (http://www.mccarthy.com/news/2012/12/14/integrated-project-delivery-and-innovative-technology-tools-foster-team-success-on-new-children%E2%80%99s-hospital-colorado-east-tower/).

The Maternal and Fetal Health Institute cares for mothers and newborns throughout a high-risk pregnancy and allows both to remain in the same facility after delivery. The institute has its own entrance to the East Tower, with labor and delivery rooms equipped for full care of mother and child, a maternal operating room, a suite for fetal care and procedures, and infant stabilization rooms. The institute is a partnership involving Children's Hospital, the University of Colorado Hospital, and the University of Colorado School of Medicine.

The new tower is the anchor for the $230 million project, which also includes renovations to the existing hospital that will continue through 2014. In bringing the East Tower on line, Children's Hospital comprises 1.8 million square feet on the Anschutz Medical Campus.

"The reassembly of the original Children's project team created a situation where all three parties could begin right where we left off only two years earlier," said McCarthy project director Doug Mangers.

"By consolidating the contracts through a tri-party IDP contract, we were able to create an integrated project team and harness the strengths of each of the team members," Mangers continued. "The collaboration and focus on a shared set of project goals allowed our team to complete the East Tower on schedule and under the budget, while maximizing the scope that was included in the project. The integrated project delivery method was a success with this team."

The construction team used Bluebeam® document control for collaborative document management and used mobile kiosks in the field to enable all field staff to have access to project information in real time. Unlike the use of only tablet PCs and iPads, the mobile kiosks were available to all jobsite personnel when needed. Rapid changes were tracked and documented as they were developed with the design team, and implemented in the field to minimize rework and improve overall project efficiency. For GH Phipps employees involved in the East Tower project, the opening also was a reminder of five decades of work at Children's Hospital.

"Throughout the process of fast tracking the construction of the new East Tower and continuing into the ongoing renovations at Children's Hospital Colorado, the Phipps/McCarthy team keeps one simple truth in mind: There is nothing more traumatic for a family than having a child in the hospital," said GH Phipps vice president Gary Constant, who has been involved with Children's for 24 years and has had both his sons treated at Children's Hospital.

Since its move to the new hospital in September 2007, Children's says it has experienced a 60 percent increase in the size of its medical staff, a 36 percent inpatient volume increase, and a larger than 50 percent growth in its number of annual outpatient visits.

Case Study: Prefab Gives $1 Billion Hospital Job a Big Schedule Boost

When completed, the University Medical Center (UMC) will be one of the largest hospital campuses of its kind in the country. Facilities include a 560,000-square-foot hospital with 424 beds, a 747,000-square-foot diagnostics and treatment center, a 255,000-square-foot ambulatory care building, and a 546,000-square-foot parking garage for 1400 cars. The facilities are also designed to be resilient during storms. UMC is the replacement for the Medical Center of Louisiana at New Orleans, which closed after sustaining serious flood damage during Hurricane Katrina in 2005.

UMC will be the only level 1 trauma center in southeast Louisiana, so keeping it functional during future disasters will be essential. All mission-critical hospital facilities are located at least 21 feet above base flood elevation. Less critical public and office spaces are placed at ground level. Technology and emergency power can continue to operate for up to a week after a Category 3 storm, according to the project architects— a joint venture of NBBJ, Seattle, and Blitch Knevel Architects, New Orleans.

An extensive prefabrication strategy aims to keep the $1 billion University Medical Center project in New Orleans on target to meet the project's daunting scope and aggressive construction schedule. With more than 2 million square feet of built space, the project for the state of Louisiana is one of the largest healthcare campuses under construction in the United States. Its 31-month construction schedule has prompted construction-manager-at-risk Skanska USA Building, New York City, and joint venture partner MAPP Construction, New Orleans, to pursue prefabrication of mechanical systems, headwalls, and bathroom facilities. Figure 8.8 shows the prefabricated overhead rack (right) and the assembly being brought to the project (left).

Figure 8.8 University Medical Center project, New Orleans, LA (http://texas.construction.com/texas_construction_projects/2013/0617-prefab-gives-1-billion-hospital-job-a-big-schedule-boost.asp).

Major Mechanical

To help keep pace, almost 40 percent of the mechanical systems at the project site are being prefabricated in racks up to 8 feet wide and 20 feet long. The 1100 racks contain HVAC ductwork, utilities, piping, cabling, and relevant mechanicals for each floor. Mike Austin, senior project manager at Skanska, says it's the first time the company has used so much prefabrication on a single project. The company built an off-site fabrication shop and secured a warehouse across town. At its peak, the offsite shop had 120 workers staffing three assembly lines that were constructing up to 90 racks simultaneously. Austin says the shop has framers, plumbers, pipe fitters, electricians, insulators, welders, and HVAC specialists working side by side in a manufacturing-type environment.

Skanska also constructed headwall units and 500 bathroom pods that included rough framing and drywall, with all in-wall services installed. Pod installations were sequenced with the enclosure schedule to maintain the required openings in each room entrance. "We had a schedule so that as each room was ready to receive, we placed the pods and went down the corridors," Austin says.

Controlled Environments

New Orleans-based MEP contractor MCC is responsible for building 700 of the racks for four of the five buildings. Michael Cooper, MCC project executive, says prefabrication allows greater quality control and consistency by having fewer people work on the pods in a controlled environment. Racks in the shop sit just 18 inches off the ground, allowing workers to gain access to the top and sides of work areas that would not be accessible if they were doing it in the field. "You're not dealing with pieces and parts that are 15 feet in the air," Cooper says. "As long as you are planning ahead and doing things in the right order, you don't get any complaints from the field."

Racks are transported into the corridors one at a time, then raised in place by four crew members using duct jacks. While traditional construction might have 20 different trades working simultaneously with ladders down the corridor, this technique requires just a few workers in a clean, wide-open space. "You go down the corridor and see just [a few guys and] one ladder," says Skanska project executive Jim Clemmensen. "There aren't people tripping over each other. It dramatically improves safety."

Austin says because of the heavy use of prefabrication, BIM was employed extensively to ensure quality and avoid field problems. Except for structural and architectural systems, crews remodeled all building systems to meet prefabrication needs. He says the modeling team digitally coordinated 1.6 million square feet of building space in 16 months.

Conclusion

In *A Guide to the Project Management Body of Knowledge (PMBOK® Guide)*, 3d ed., the Project Management Institute (PMI) defines a project as *"a temporary endeavor undertaken to create a unique product or service."* As simple as this definition may seem, there are a few key points that define a project as distinct from ongoing operations. Again, from the *PMBOK® Guide:* "Operations and projects differ primarily in that operations are ongoing and repetitive while projects are temporary and unique." A project can thus be defined in terms of its distinctive characteristics. *Temporary* means that a project has a definite beginning and a definite end. *Unique* means that the product or service is different in some distinguishing way from all similar products or services.

A project delivery method is the comprehensive process of assigning the contractual responsibilities for designing and constructing a project. A delivery method identifies the primary parties taking contractual responsibility for the performance of the work. The delivery method process includes

- Definition of scope and requirements of a project
- Contractual requirements, obligations, and responsibilities of the parties
- Procedures, actions, and sequences of events
- Interrelationships among the participants
- Mechanisms for managing time, cost, safety, and quality
- Forms of agreement and documentation of activity

It is crucial to the success of a project that all participants understand the goals and objectives of the delivery method being used and how all parties are related to each other contractually. The essential elements of any project delivery method are cost, quality, time, and safety. Responsibilities for implementing these elements vary from method to method. A project delivery method is fundamentally a people method, because people remain the most valuable construction resource. The success or failure of any delivery method depends on the performance, trust, and cooperation of the parties.

There are numerous hybrids and variations of the delivery methods discussed in this chapter. It is not the intention of this book to argue in favor of any one delivery method over another. Rather, this book strives to provide readers with a meaningful way to discuss project delivery methods. Most professionals agree that there is no perfect project delivery method. Every project is unique and has its own unique set of challenges. Therefore, industry consensus is that every project should be considered on a case-by-case basis to determine the most appropriate project delivery method.

According to Jennifer Kovacs Silvis, editor-in-chief of *Healthcare Building Ideas* (January 5, 2012), "There was a day when design-bid-build was the obvious choice in terms of how a healthcare construction project would be delivered. However, like any industry, healthcare design and construction is progressing and adapting—and introducing alternatives to what once was the norm." A survey commissioned by the AIA Center for Integrated Practice and sent to nearly 10,000 AIA members underscores the sea change affecting the whole healthcare industry. Characterized by demands for high-performing buildings, BIM, increasingly complex projects, and pressure for innovative business practices, more than half of respondents reported having engaged in some degree of collaborative project delivery, including IPD.

References

A Guide to the Project Management Body of Knowledge, Third Edition (PMBOK Guides), Project Management Institute, 3rd edition (2004).

AIA Document E202–2008, *Building Information Modeling Protocol,* The American Institute of Architects, New York, Washington, DC, www.aia.org.

Integrated Project Delivery: Case Studies, American Institute of Architects California Council, 1303 J Street, Suite 200, Sacramento, CA 95814 (2010).

Integrated Project Delivery Guide, American Institute of Architects California Council (AIACC), 2007, http://info.aia.org/SiteObjects/files/IPD_Guide_2007.pdf.

Kamara, J. M. ed, *Concurrent Engineering in Construction Projects,* published by Taylor and Francis, New York, NY, 2007.

Koskela, L., *Application of the New Production Philosophy to Construction (1992),* CIFE Technical Report #72, Stanford University, September 1992.

Lean Construction Institute, 2300 Wilson Boulevard, Arlington, VA, leanconstruction.org.

Matthews, O. and Howell, G. "Integrated Project Delivery—An Example Of Relational Contracting," *Lean Construction Journal,* Vol. 2, No. 1, pp. 46–61, April 2005.

Project Alliancing: A Practitioners' Guide, published by the Department of Treasury and Finance, Melbourne, State of Victoria, Australia, 2006.

Raisbeck, P, Millie, R, and Maher, A ., *Assessing Integrated Project Delivery: A Comparative Analysis of IPD and Alliance Contracting Procurement.* Proceedings of the 26th Annual ARCOM Conference. Leeds, UK, Association of Researchers in Construction Management, pp. 1019–1028, September 6–8, 2010.

Silvis, J.K., Delivering Healthcare Projects, Healthcare Building Ideas, January 2012.

SmartMarket Report, Prefabrication and Modularization: Increasing Productivity in the Construction Industry, McGraw-Hill Construction, 2011.

U.S. Green Building Council (USGBC), *LEED for Healthcare (2009),* Washington, DC, new.usgbc.org.

CHAPTER 9

Challenges of Additions and Renovations

While new greenfield hospitals and replacement facilities are challenging, with high exposure and typically the largest capital costs, it is much more likely that a construction professional will work on many more additions and renovation projects in his or her career. A 2011 survey by *Health Facility Management Magazine* and the American Society of Healthcare Engineers (ASHE) found that "renovation or expansion accounted for 73% of construction projects at hospitals that responded…." Modern Healthcare's annual Construction and Design Survey, published in March 2013, showed 161 "entire acute care hospitals" completed in 2012 versus 1596 acute care expansions and renovations. Similar anecdotal results can also be obtained from experienced healthcare design and construction professionals, as well as from owners in the industry. One of the authors had a similar experience with a large hospital system. During a 10 year period with unprecedented growth, there were 19 new hospitals built in comparison to over 400 additions and renovations.

Given the strong likelihood of working on many more additions and renovations, it is important for those in the role of managing hospital construction to understand the unique challenges involved in this type of project. In addition to the expected routine issues with construction, such as scheduling manpower and materials, solving technical problems, and managing cost and claims, the manager must also coordinate the work around the hospital's day-to-day operations. One of the most important lessons to learn in managing a hospital addition or renovation is that healthcare must take precedence over construction. While the hospital staff and patients will have to tolerate the construction activities at times, healthcare is always the priority as it can literally be a matter of life or death, if for any reason the caregivers cannot treat a patient in the event of a medical emergency. This concept must be central to the front-end construction planning on all addition and renovation projects, and ideally, should be considered during the design phase as well.

The idea that construction activities must accommodate healthcare has to be communicated to the construction contractor, subcontractors, and workmen from the early planning, negotiations, and bidding throughout the project. At the same time, the owner needs to allow the construction work to proceed in a practical and timely manner, or the cost and schedule could be out of control. There must be a balance of healthcare and construction activities, but with the understanding that work activities can never

This chapter was written by Tom Gormley.

prevent ambulances from reaching the emergency room or cause lights to go out in the operating room (OR) during surgery. As a result, significant front-end planning is required in a collaborative manner between all of the stakeholders. A team attitude should be built with the hospital to aid decision making and to prevent costly complications and delays during the execution of the addition/renovation project. This must involve not only the hospital and the builders, but also the design team, as their decisions can make it easier or more difficult to actually do the work.

This chapter addresses some common types of addition and renovation projects, the role of the contractor or construction manager during the design phase, the additional code requirements for renovations, and contracting strategies to minimize cost and claims. From there, it will explore approaches and tools to be used on additions and renovations, as well as during "shutdowns," protection of critical systems, and interfacing with the facility management staff. Responses to hazardous material and emergency response planning will also be addressed.

Infection control risk assessment (ICRA) and interim life safety measures (ILSM) plans should be done for all work areas in an operational facility. These are addressed in Chap. 12 of this book.

Understanding Hospital Operations

Before engaging in a discussion on the construction of additions and renovations of hospitals, it would be helpful to have a general understanding of the key departments and functions in a typical hospital. Hospitals vary widely in the services they offers, and therefore, in the departments they have. The basic functions that are in most of the roughly 5000 hospitals in the United States are the emergency department, imaging, surgery, intensive care, and patient rooms. They may have higher acuity or specialty services, such as the following:

- Cardiology
- Pediatric intensive care unit
- Neonatal intensive care unit
- Cardiovascular intensive care unit
- Neurology
- Oncology
- Orthopedics
- Obstetrics and gynecology

Most hospitals will offer outpatient services, and some will have specialty treatment units, such as behavioral health services, rehabilitation services, burn units, and physical therapy. Academic medical centers will typically also have research functions and teaching capabilities. Common support units include a dispensary or pharmacy, pathology, lab, and central sterilization while, on the non-medical side, there generally are medical records departments, accounting/business offices, information technology, facilities management, food service, material management, administration, and security. In many hospital systems, some services, such as the business office and material management, have been centralized for a region at an off campus location.

With knowledge about the hospital functions, the construction can be woven in and around the day-to-day operations and/or effectively isolated from the hospital through phasing of the work in areas vacated by the facility. While the "general hospital," as defined in the *Guidelines for the Design and Construction of Health Care Facilities* published by FGI, covers facilities ranging from large academic medical centers with over 1000 beds to rural hospitals that may have as few as 30 beds, there are many similarities in the functions, departmental relationships, and individual spaces, as well as many differences. The Guidelines in 2010 did start a new section for Small Primary Care Hospitals, with 25 beds or fewer, to address the unique features and requirements for these critical access hospitals; however, they are all required to meet the standards of Sec. 2.1 "Common Elements for Hospitals" of the Guidelines. Regardless of a hospital's size, these "common elements" are typical, and the Guidelines provide a good roadmap for exploring the functions and construction considerations on addition and renovation projects.

Site

The site requirements are basic and address access for the community, delivery, and emergency vehicles, transportation including public and private, as well as parking. Security for patients, families, staff, and the public is an important topic, along with availability of utilities ("water, gas, sewer, electricity"). The Internet has become an essential element, as well. Adequate signage and lighting are critical elements for safe access, transportation, and security. During construction on an existing hospital, plans must be made and implemented that keep these functions operational. These steps may include providing temporary entrances for any of the constituents, such as patients, public, staff, service/delivery people, and emergency vehicles. The emergency vehicles may include ambulances trying to reach the emergency entrance, fire trucks coming to respond to an emergency at the hospital, and helicopters transporting victims by air.

As each location is different and the services provided vary, so the site plans for an addition/renovation will be unique to each facility. Some of the common steps include

- Providing temporary signs to redirect traffic as needed
- Identifying additional parking for the construction workers and/or hospital staff, with off-site parking and buses sometimes serving as an option
- Offering valet or assisted parking for patients with mobility problems
- Installing temporary bus shelters for use with public transportation
- Providing covered walkways to allow for safe access through construction areas
- Providing additional security personnel and/or flagmen to coordinate traffic during busy times, such as large concrete pours or steel deliveries
- Installing temporary entrances for the emergency room, front entrance (see Fig. 9.1), loading dock, or other access points
- Adding extra lighting, security cameras, and/or emergency phones in parking areas for additional safety
- Providing temporary utilities during addition or rerouting of services

FIGURE 9.1 Temporary emergency entrance. (*Courtesy of Multivista and Alamance Regional Medical Center.*)

Figure 9.1 demonstrates an approach used for a temporary emergency room entrance. Note additional lighting, security officer, wheel chair storage, signage, and covered entry.

Utilities are another important element to be addressed when planning an addition and/or renovation to an existing hospital. Power is typically provided by an underground service. Most hospitals have redundancy, or two feeders to provide backup in the event one is lost. This should be clarified and both feeders located at the start of the project, so they can be protected in case there is excavation work to be done in the area. Similar steps should be taken for water, gas, and sewer lines, although these typically have only one main line serving a facility, unless it is a large campus. Once these are located, any work near them should be carefully planned with backup options to service the hospital. Internet/phone service typically is also provided through a buried conduit, but on smaller facilities, this may be an overhead wire. Again, any work around these should be carefully planned. The other very critical service that is typically buried is the medical gas lines from the tank "farm" (see Fig. 9.2) or storage area which contains oxygen, nitrous oxide, nitrogen, and possibly others. These require protective walls per the codes due to the potential for explosions. The medical gas lines are often serving critically ill patients, so extensive research, planning, and protection measures should be taken when work is in close proximity.

In all cases, the construction team should notify and coordinate with the hospital's engineering or facility management department. Good communication with this group is very important to minimize problems and maintain positive relationships. The hospital administrative team, chief executive officer, or administrator look to the engineering department to keep the hospital in operation.

Figure 9.2 Tank farm for medical gas behind central energy plant. (*Courtesy of Multivista and Moses Cone Hospital.*)

Finally, the construction team should evaluate any risk for environmental pollution. This could include spilling of fuel from site equipment, uncontrolled water run-off, excavation of unknown hazardous material, or other issues. The potential for environmental damage should be effectively evaluated and mitigation measures implemented.

Nursing Units

Nursing units typically consist of several elements: the patient rooms and toilets; the nursing station that may be centralized and/or have sub-nursing stations closer to the rooms; support rooms, such as the medicine prep room, nourishment room, clean supply, dirty, or soiled utility and/or holding room, equipment storage, staff lounge and toilets; a multipurpose room for meetings and training; a nurse or floor supervisor office; and a documentation or charting area for patient medical records. During construction these rooms must be kept operational, free from debris and contamination, or alternate locations must be provided. There are three commonly used floor plans for hospital patient units—"Race track" floor plan (see Fig. 9.3), T-shaped floor plan (see Fig. 9.4), and Cross-shaped floor plan (see Fig. 9.5). Ideally, a nursing unit could be closed for renovations so it is easier to isolate the construction work from the occupied hospital spaces. In many cases, this is not practical, due to the number of patients or other reasons. The work may need to be phased, so some patient rooms, or a wing, can be closed and separated with fire and dust resistant walls to protect the patients and staff from the construction activities. When possible, it is best to use fan units to put the construction area in negative pressure relative to the occupied space. This will keep dust and contamination away from the hospital operations, and it can be exhausted directly to the exterior. Another option is to use HEPA

Racetrack Layout

PROS:

- Centralized Support Space-accessible from both sides of unit
- Shorter travel distances
- More options for direct view of patients
- Most flexible floor plate

CONS:

- Visibility of end corridors

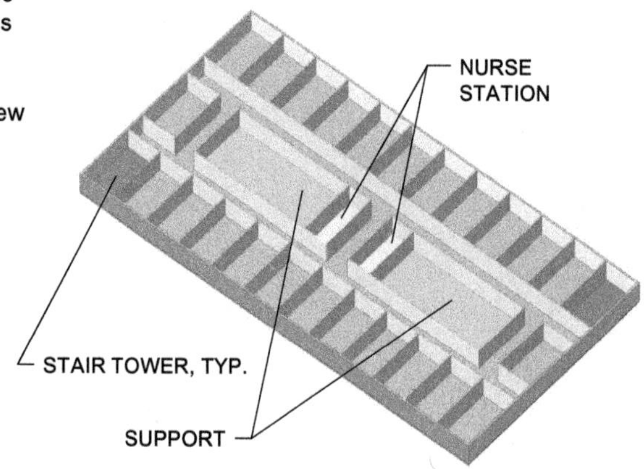

Figure 9.3 "Race track" floor plan. (*Courtesy of Thomas, Miller & Partners, Brentwood, TN.*)

T-shaped Tower Layout

PROS:

- Centralized Support Space
- Simplified Structure
- Best control of visitor traffic

CONS:

- Visibility from Nurse Station
- Travel distance for staff
- Long corridors-increased travel distance for patient care providers

Options:

- Sub-Nurse Station with primary support at each wing to reduce travel distance

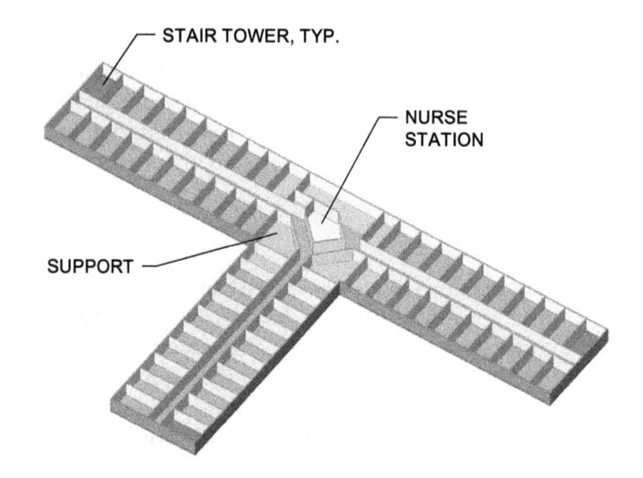

Figure 9.4 T-shaped patient room floor plan. (*Courtesy of Thomas, Miller & Partners, Brentwood, TN.*)

(high efficiency particulate air) filters to catch the dust and mold spores that can be released during demolition and construction. During renovation of a central nursing station and the support areas, it is often necessary to set up temporary space for these functions. Emergency fire exits and fire/smoke walls must be maintained as must mechanical, electrical, and plumbing services to keep the nursing unit fully operational. It is also important to consider the floors above and below the one being renovated, as construction activities can also sometimes impact them as well.

Cross-shaped Layout

PROS:

- Centralized Support Space
- Simplified Structure

CONS:

- Long corridors- increased travel distance for patient care providers
- Multiple stair towers and path to exterior on ground floor. (exit passages)

Options:

- Sub-Nurse Station with primary support at each wing to reduce travel distance

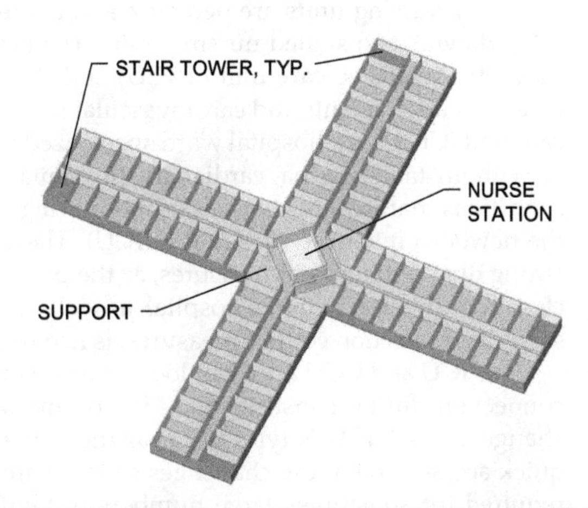

FIGURE 9.5 Cross-shaped patient room floor plan. (*Courtesy of Thomas, Miller & Partners, Brentwood, TN.*)

Another important space that deserves special attention is the airborne infection isolation (AII) room. This unique type of patient room and toilet typically has an anteroom or entrance area for "gowning and storage of clean and soiled materials." All three rooms are kept in controlled negative pressure with a separate exhaust system, as they house patients with airborne infectious diseases, such as tuberculosis. The rooms have air monitoring systems to ensure that negative pressure is maintained, to keep the contaminated air isolated. Prior to doing any construction work in or adjacent to these patient isolation rooms, the engineering department and the floor nursing supervisor should be advised and written approval obtained. The rooms may need to be decontaminated before it is safe for construction workers to enter. If they enter when an infectious patient is there, the worker could be contaminated or the negative pressure relationship upset, allowing the entire unit to be compromised. Extreme care and close coordination with the hospital is required when working in these rooms. Seclusion treatment rooms in psychiatric hospitals also need special consideration, due to the chance of compromising the patient security or safety of the worker or staff.

Specialized Nursing Units

In addition to the medical/surgical nursing units, there are also defined specialized units to serve unique patient populations. These patients may have specific conditions, higher acuity levels, or different age groups. Some of these require other considerations during additions and renovations. The oncology nursing units treat cancer patients who often have weakened immune systems as a result of chemo or radiation therapy. Working in or adjacent to these areas requires the use of negative air pressure zones to prevent contamination by dust or debris from demolition on construction work. The oncology unit has such strict control on possible contaminants that plants, dried flowers, fish tanks, and water features are prohibited by the Guidelines. The hospital construction manager should work closely with the hospital to develop a written plan to show work areas, containment measures, and schedule or phasing.

Other nursing units are pediatric and adolescent, obstetrical, intermediate care or "step down," and skilled nursing units. The next levels of units are critical care (often called the intensive care unit or ICU) and the coronary care unit (CCU), with higher level of acuity patients and cardiovascular surgery patients. A CCU or cardiac intensive care unit (CICU) is a hospital ward specialized in the care of patients after heart surgery or with unstable angina, cardiac dysrhythmia, and (in practice) various other cardiac conditions that require continuous monitoring and treatment. Another high risk unit is the newborn intensive care unit (NICU). These units for higher levels of care require strong dust containment measures, as the patients are sicker and at greater risk. Again, close coordination with the hospital, as well as written plans and graphics showing life safety and infection control measures, is needed to protect the fragile patients.

The ICU and CCU are often located near surgery for quick access. Maintaining this connection during construction is important, along with new signage if routes have changed. The NICU is typically near the delivery rooms or labor/delivery rooms for quick access. One of the challenges of these areas is providing waiting rooms that are required for sometimes large numbers of family and visitors. In some cases, family members may stay at the facility with the patient. During the planning for work on these types of units, consideration should be given to the needs for waiting rooms—access, space, vending, toilets, and emergency exits.

There are two other types of units that have unique patient populations. One is the psychiatric nursing unit, which may contain a wide diversity of patients, most without acute medical conditions. Safety and security for patients, staff, and workers is an important factor. Construction work in these areas may require badging and security system for workers and supervisors. Emergency exits must be maintained for the workers, staff, and patients, but, in some units, the patients are locked-in, so provisions have to be made with supervised keys and controls for the fire alarm system. Some disruptive patients have pulled fire alarms as a nuisance or method to escape, so measures to provide for safe but controlled exiting are needed.

The other unique unit is for bariatric care and, as defined in the Guidelines, "care for the extremely obese patient." Access to waiting rooms, patient rooms, and toilets must be provided for larger wheelchairs. Any temporary facilities for patients must be built larger and stronger than normal standards. For example, the Guidelines require the shower grab bars to be "capable of supporting 1000 pounds." The construction manager should identify these unique spaces and do a risk assessment and mitigation plan, especially if the project involves interior renovations requiring access and temporary patient rooms, toilets, or waiting space.

Diagnostic and Treatment Areas

The primary services or departments in these areas are emergency, surgery, and diagnostic and interventional imaging. Each is unique in the functions and the equipment housed, so there are different items to consider when renovating one of these existing departments. The lines are now being blurred with the new hybrid operating rooms where imaging equipment, such as MRIs, are being introduced into the ORs.

Generally, these three major departments (emergency, surgery, and imaging) are adjacent to one another to make patient flow easier and quicker. In a large medical center in an urban environment, the departments may be on different floors due to the size, utilizing vertical transportation (elevators) for quick access.

Emergency Department

Emergency services are broken down into two phases by the Guidelines. Initial care is to stabilize the patient and "minimize potential for deterioration during transport." Definitive care is defined as "care within a dedicated emergency department... care may range from suturing lacerations to full-scale emergency medical procedures." Some facilities are part of a trauma system. There are four levels, with the fourth being the most severe patients. As with all critical services, access to this on an emergency basis is essential. Construction work on a functional emergency department (ED) must be carefully planned to maintain access for ambulances or walk-ins and their families. Plans should be drawn and displayed in public areas indicating any required temporary entrances, temporarily relocated rooms, rerouted exits, or other changes from the normal flow to accommodate construction work. One of the important areas that should be adjacent to the waiting room is "triage." There, the patients are evaluated for further care. The emergency room typically contains multiple types of rooms for different levels of care, from a trauma room with overhead lights and the look of a small operating room, to exam rooms with just tables and counters with a sink or hand washing station, as defined by the codes. Figure 9.6 shows a typical emergency treatment area.

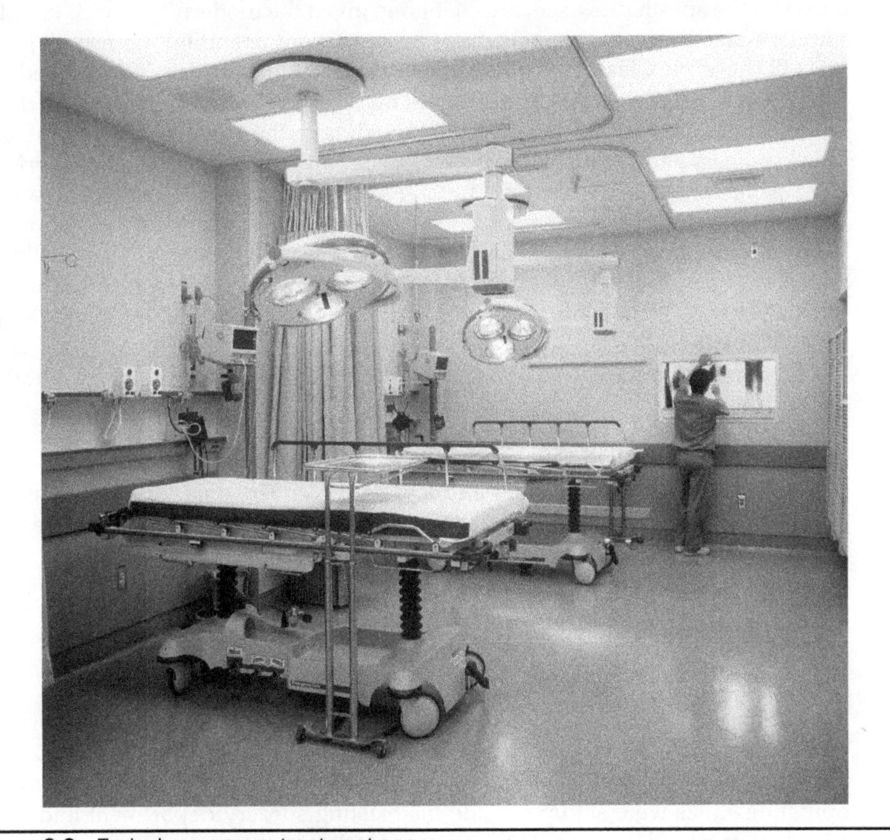

FIGURE 9.6 Typical emergency treatment space.

Temporary entrances made of wood and/or steel with roofs and lights are sometimes required, as the addition is usually attached to the existing ED and the new part is typically built first. Another challenge to renovating the emergency room is maintaining all mechanical and electrical services during the renovations. The ED is even more challenging than other areas, such as surgery, because emergency care is 24 hours per day, 7 days a week. The preference is to relocate portions of the ED to the new space after it is completed, so there can be phased shut downs or closures of areas in the existing ED for the construction work. In some cases, the work has to go on concurrently with the hospital operations, which requires close coordination with the hospital and the subcontractors to maintain heat/air-conditioning, power, water, medical gas, and the other critical mechanical, electrical, and plumbing (MEP) services.

Surgical Services

Surgical procedures are almost always done in an operating room, although some surgery may be done in a trauma room in the ED. The American College of Surgeons (http://www.facs.org/) defines surgical procedure rooms in these levels of care: Class A provides for minor surgical procedures under topical or regional anesthesia without sedation. Class B provides for minor or major surgical procedures with oral, parenteral, or intravenous sedation. Class C provides for major surgical procedures requiring general or regional block anesthesia and support of bodily functions. These classifications, based on the anesthetics used, are changing, given the modern approach to sedation. In the 2014 Guidelines, the classifications for outpatient operating rooms will just include one OR (formerly Class B and C), while Class A will now be labeled a procedure room.

Different classes of operating rooms may be found in a hospital, but most in the United States are Class B or C. When renovating in the surgery department, maintaining cleanliness and the integrity of the "sterile" areas is one of the highest priorities. As always, the first preference would be to shut down part of the area for the construction work. As the surgery department is one of the hospital's busiest areas, and a large contributor to financial performance, closing large portions is generally not allowed. The manager of construction will need to work with the hospital staff to coordinate areas that can closed temporarily or to coordinate doing the work at night after surgery has been completed.

The support areas are also important, because the operating rooms cannot function without them. These areas include

- The prep area, or pre-op, where the patients are prepared for surgery
- The PACU (post anesthesia care unit) or recovery area, where the patient recovers under close nursing supervision before being taken to a patient room or being discharged after an outpatient procedure
- Physician and staff locker and dressing rooms
- Decontamination and sterile processing for the surgical instruments
- Visitor waiting, which typically has seating and vending

The construction planning must provide an approach to safely maintain access and functionality of these spaces. Temporary areas can be provided, if needed, to accommodate the construction. The author was involved in one extreme case, when mobile operating suites were set up outside the existing surgery department and all surgery was done there, while the old operating rooms were gutted and renovated.

Diagnostic Imaging Services

Imaging modalities range from simple mammography rooms to complex magnetic resonance imaging (MRI) suites. If cleanliness is the key for surgery, then properly caring for the equipment is it for imaging. This department is equipment intensive containing high-tech imaging machines, and computer systems worth millions of dollars. The hospital construction manager should coordinate closely with the hospital staff to understand which imaging modalities or types of machines are involved in the work areas. The primary types are

- *Diagnostic x-ray*—which includes tomography and radiography/fluoroscopy rooms
- *MRI*—magnetic resonance imaging, which uses large magnets that must be cooled constantly
- *Ultrasound*—this is a small machine typically set up in exam rooms or may be mobile
- *Cardiac catheterization labs*—where catheters are inserted into the patient to perform cardiovascular diagnostics and procedures
- *Electrophysiology labs*—where studies are done on the electrical signals in the body to diagnose medical conditions
- *Nuclear medicine*—where radioactive chemicals are given to the patient for internal tracking to diagnose gastrointestinal and other conditions
- *Positron emission tomography* (PET) scanners are used to diagnose brain and other cancers, and are often used in conjunction with CT scanners
- *Computerized tomography* (CT)—this equipment uses a series of cameras to take multiple pictures by "slice" or section of the body with a computer used to assemble the "slides" to create three-dimensional images of human organs and systems for diagnostic purposes

Medical Equipment and Information Technology is addressed in more detail in Chap. 11 in this book.

Patient Support Services

There are several essential support spaces that should be considered during any renovation or addition. There are two types

1. Clinical support, such as lab and pharmacy
2. Ancillary support, such as dietary, material management, environmental services/housekeeping

All of these spaces are important to keep the hospital operational, safe, and compliant with the many regulations. Cleanliness and control of dust and contaminants are critical in the lab and pharmacy, as are protecting equipment and keeping mechanical and electrical services functional. Loading dock access can be an issue during construction, as the hospital and, now, the contractors and subcontractors all need to load and unload materials.

Off-campus food preparation can be an option on extensive kitchen renovations, as it is difficult to build in a functioning food service operation. The work can also be phased, and sections renovated a piece at a time, using dust and fire barriers if off-campus preparation is not practical.

There are also administrative areas, business offices, admitting, gift shops, and other softer spaces that can be more easily relocated temporarily, with computer service as one of the critical issue. Coordination with the staff prior to additions and renovations, along with good communication during the work, will produce the best results.

Common Types of Projects

Given the diversity in the 5000 plus hospitals across the United States, and more internationally, there are many different types of additions and renovations. The most common involve adding patient beds or expanding the three primary services—emergency, surgery, and imaging. Often, an expansion of these services will drive the need for other support spaces, such as lab, pharmacy, or dietary. In many cases, the power plant, or some portion of the mechanical and electrical systems, does not have the capacity to serve the newly added or renovated space. This can necessitate adding emergency generators, chillers, or air handling units, as well as upgrading systems, such as fire alarm or medical gas.

Some of the most challenging additions/renovations are the vertical expansions that typically involve adding another floor on top of a fully occupied facility. These provide all of the construction issues related to connecting to an existing or older building, as well as the need to coordinate with the hospital to minimize disruptions of the occupied floor. This is where familiarity with the hospital functions and schedules in the occupied space will help the construction manager develop creative solutions that allow the work to proceed on a reasonable schedule, while protecting patients, visitors, and staff.

Depending on the type of space below the vertical expansion, the issues may vary and may include noise concerns, contamination from dust during demolition, numerous shutdowns, "tie-ins" in for plumbing, medical gas, fire alarm, heating, ventilation, and air-conditioning (HVAC), or other services. When possible, it is best to move people out in sections below the work when it is very disruptive, such as jack hammering concrete slabs, or when it is risky, such as erecting steel. Another important issue that deserves special attention is maintaining a watertight building during the structural tie-in and connections at the building skin. It is best to protect the existing roof and keep it in place until the exterior skin and roof on the vertical expansion are all complete. Waterproof boxes should be built around all of the column connections to allow for access with a method to prevent water intrusion for daily work. Once the addition is dry, then it is safe to remove the existing roof. Typically, at that point, the existing concrete slab will need repairs and patching to prepare it for the finished floors that will be installed.

In some cases, the renovation work is in a department that cannot be relocated or shut down temporarily. These are the most challenging, because the construction work must be done around the hospital activities. In some cases, this may be the only diagnostic equipment in the hospital, so it was mandatory for it to stay accessible. Careful planning with the department head and infection control department is required, so that the work can be done on a room-by-room basis phased over weeks or months. Extensive night work is often required so construction can be done, and the space cleaned before staff arrives the next morning. It can be time consuming, costly, and tedious with a great deal of patience needed from the hospital staff and the construction crews. One important lesson is to always have responsible supervisors at night to monitor and coordinate the construction activities. Quality control is more difficult as well with the work being done in small components. Another challenge is the inspection process by the regulatory agencies. They are not typically available at night, and prefer to inspect the work before it is occupied. Maintaining mechanical and electrical services, such as air-conditioning and medical gas, also takes creative solutions, as these systems are commonly designed in zones. This room-by-room phasing makes it more difficult to shut down a zone, so major upgrades to these systems are complicated. It again requires careful planning, nighttime shutdowns, and coordination to minimize disruptions. In some cases, the construction manager may need to work with the mechanical/electrical engineers to modify the design to better accommodate the phasing. This is where early involvement by the builders for design input can improve results during construction.

Another factor that greatly adds to the complexity of additions and renovations is the presence of contaminants or hazardous material. One of the most common is asbestos that may be found in sprayed-on fireproofing, floor tiles, and pipe insulation. The most difficult is what is known as friable asbestos. This is defined as "any material containing more than 1 percent asbestos" by weight or area "that, when dry, can be crumbled, pulverized, or reduced to powder by hand pressure" (per EPA website, http://www2.epa.gov/lead). Lead-based paint is hazardous material that is sometimes found in older facilities. PCB (polychlorinated biphenyl) is another hazardous or toxic material that may be encountered in transformers, electric meters, and capacitors. A renovation of existing hospital space that contains hazardous materials requires another level of safety precautions, as well as specialty subcontractors with the appropriate training and equipment to execute the remediation. Prior to starting renovations in an existing facility, it is best to test for hazardous material or check with the hospital engineering department for reports from prior testing or projects. The Joint Commission[1] requires hospitals to maintain records of hazardous material and have action plans to address safety while working with these building products. Actual approaches for working in facilities with hazardous materials are discussed further in Chap. 12 of this book.

The most basic project is the pure addition with no renovation of existing space. These are not common, as most projects also include extensive work inside the hospital. Even the simple additions, sometimes referred to as "bump outs," have tie-ins for doors to access the new space, connections to existing mechanical and electrical systems, and interface with the building "skin" systems, usually roof and walls. Careful planning is also required on these projects, but they typically are less costly and less disruptive.

Some projects consist of merely replacing medical equipment. Typically, this includes imaging equipment, such as x-ray machines, MRIs, CTs, and nuclear medicine cameras, although it may sometimes involve sterilizers, operating room lights, or other equipment. As this equipment often has multiple connections to mechanical and electrical systems, the replacement generally involves upgrading or modifying these services to accommodate the new medical or hospital support equipment. In some cases, new electrical service, connections to emergency power, and extra air-conditioning are needed to support the more modern equipment. So what may look like a simple equipment

[1] The Joint Commission (TJC), Oakbrook Terrace, IL, (formerly the Joint Commission on Accreditation of Healthcare Organizations (JCAHO)) is a United States-based nonprofit tax-exempt 501© organization that accredits more than 20,000 healthcare organizations and programs in the United States.

upgrade can sometimes become a complex construction project. In recent years, many hospitals have upgraded their information technology systems to add digital imaging management and storage systems, often called a PACS (picture archiving and communication system), and servers for electronic medical records. These generally require more energy, more power conditioning equipment, and more air-conditioning. In some locations, an equipment upgrade or replacement project can trigger the need for permits from the local building department. A book contributor was involved with a project to replace a cystoscopy table in a procedure room in California. As a result of the plumbing connections, the drawings had to be submitted to OSHPD (Office of State Health Planning and Development), and the approval took several months.

Close communication and coordination with the equipment supplier, such as GE, Siemens, Philips, or others, is very important on these projects. Often the architect and engineer reference the equipment manufacturer's plans as part of the contract documents. These are critical for proper rough-in on the mechanical, electrical, and plumbing (MEP) services needed.

Input during Design

On additions and renovations to hospitals, there are many benefits to having the contractor and subcontractors involved during the design of the project. While this is sometimes not possible with certain delivery methods, such as design-bid-build, this input can help reduce the disruption, cost, and time for "messy" renovation projects. This approach involves developing the phasing plan for construction early in the design process, with the hospital administration and department heads working with the design and construction team to consider the hospital activities that must continue during the renovation work. Based on that analysis, the contractor and subcontractor develop the plans for the construction work including the phasing, temporary relocations, access to the work area, and other factors impacting their activities. With the hospital construction constraints identified, the design and construction team can then identify approaches and options to make the construction less disruptive.

Some examples could include designers having the HVAC zones overlapping with the phasing for the work. In this case, one zone can be shut down for a specific phase of work, as opposed to having multiple HVAC zones covering one work area. This can reduce the number of "shutdowns" of the HVAC system to upgrade ductwork or piping, and reduce the areas of the hospital that are impacted by the "shutdowns." Another example is the addition of valves for water systems to make tie-ins easier. There are typically "risers" or pipes that carry water up to higher floors for distribution to rooms at each level. If valves are added to the top of the risers, then the tie-in for a vertical expansion is greatly simplified. The plumber can simply connect to the riser above the valve and not be forced to drain the riser of water for the tie-in, which impacts all of the hospital floors served by the riser.

A similar approach can be used for the structural system on a hospital designed for vertical expansion. During the design, the engineer can add a "stub-up" or extension of the structural column above the roof level by 1 or 2 feet. Figure 9.7 shows one approach to extending the structural column. This can easily be roofed and flashed to keep it watertight. When it is time for the vertical addition, it is very easy to tie in the structure to go up, without having to remove large sections of roof or demolish the concrete roof slab to reach the top of the existing column. This avoids disruption, noise, and cost for vertical expansion projects.

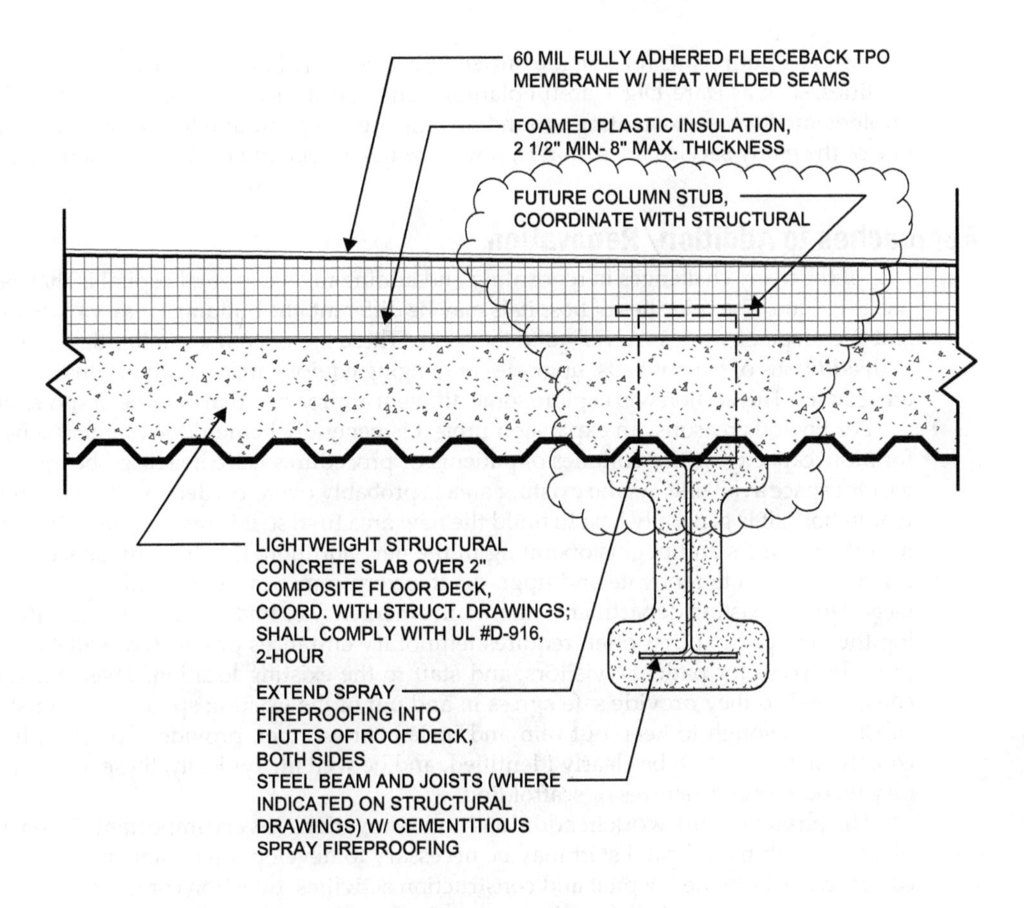

Figure 9.7 Column stub-ups and slab detail for future expansion. (*Courtesy of Thomas, Miller & Partners*).

Another important design approach to ease expansions is to locate the services most likely to expand on exterior walls. The primary services identified earlier—emergency care, surgery, and imaging, should all have pathways to the exterior or be located at the exterior of the hospital. These are the most common areas to be expanded as the hospital grows, due to population growth in the market, consolidation, physician acquisition, and other factors. With these key services on the outside walls, and the corridors laid out for easy extensions to new space, the disruption from additions can be minimized.

Another typical expansion is the addition of patient beds. This is often done with vertical expansions or additional wings. Many times CICU or CCU wings are added adjacent to the operating suite, for quick patient access. New wings are most likely easier than the vertical expansion; however, sometimes there are many underground lines to be relocated. The construction manager should always insist on a thorough review of as-built plans for outside buried lines, and should also use a service to locate the lines, such as Dare-Dig. Careful planning and specific safety procedures should be implemented when excavating around an existing hospital, due to the risk of hitting one of the many service lines, that is, power, water, sewer, medical gas, or natural gas.

Approaches to Addition/Renovation

One of the many challenges in renovating and adding to an existing hospital is that each one is unique. The priorities of hospitals may be different, the buildings may vary widely and the scope of work is never quite the same. These are of course similarities, as well, with additions of patient beds, upgrades of imaging equipment, and expansions of operating suites. This section will explore some different approaches that have been successful.

For obvious reasons, an expansion project is generally being done due to the need for more capacity, as the volumes of patients or procedures have increased beyond the current space available. As the existing area is probably overcrowded, with no room for renovations, it is generally best to build the new area first, so it is ready to use. Once the hospital occupies and begins operating in the new addition, this frees up space in the existing location to renovate and upgrade. It is very common for the addition to be attached to the existing department, such as an emergency or imaging department. Building the new portion first often requires temporary entrances or covered walkways to provide access for patients, visitors, and staff to the existing location. These must be constructed so they provide safe egress in and out of the existing space. They must be substantial enough to keep out rain and inclement weather, provide protection from construction activities, be clearly identified, and be well lit. Typically, these are temporary wood frame structures or scaffolding.

The phasing of the work in additions and renovations is very important. Extensive planning with the hospital staff may be necessary to develop a workable plan that accommodates both the hospital and construction activities. Infection control and facility management are two key functions from the hospital needed in these discussions. Often multiple phases are involved, which require moving temporary entrances and temporary fire/smoke walls to protect the occupied hospital areas. It is best to develop a plan or drawing with color coding showing the phasing, the location of exits, the layout of existing and temporary fire/smoke walls, and how the construction activities will be isolated from the hospital. This plan can also be used to overlay the mechanical, electrical, and plumbing systems, so the phasing of the work can be coordinated. The other use for a graphic demonstration or plan is to communicate with the staff in meetings, or with display boards located in the hospital for the public.

Front-end coordination with the local building department, fire marshal, and other authorities having jurisdiction (AHJ) are essential on complicated additions and renovations. Their input is often needed on the temporary measures that will be required, as their experience on how to safely execute work in an occupied facility can be valuable. Another area where the building officials may need to be involved is with the certificate of occupancy (CO). This official document, and approval to start operating, is commonly issued by the local building department or state hospital agency, confirming that they have inspected the new construction and that it is safe for the intended use. While this is often an extensive process for a new hospital, it can be even further complicated for a multiphase addition and renovation project. The practices for building officials across the country can vary, but it is very common for them to require a temporary certificate of occupancy (TCO) for each phase before it is occupied by the hospital. Obtaining a TCO typically requires interim inspections and sign-offs by the building officials, which must be coordinated by the construction manager. In some cases, there may be multiple TCOs due to several phases, and then a final overall inspection prior to issuance of the full CO. The issuance of permits, CO or TCO, is based on the adopted building code of the jurisdiction, such as the International Building Code (IBC), the International Existing Building Code (IEBC), or a state-specific code, such as the California Title 24 Building Code, the Florida Building Code, the New York Building Code, etc. As mentioned earlier, a meeting with the building officials with jurisdiction prior to start of construction is important in understanding their specific requirements and in developing a trusting relationship that will benefit the project.

In some cases, there is not adequate space at the existing hospital to accommodate the construction. It may be the seemingly simple issue of parking, but the lack of an adequate number of spaces is common around busy medical centers. The building codes typically require one space per patient bed and one space per employee for a routine weekday shift. This requirement, coupled with the parking needed for all of the construction workers that will be coming to the hospital, routinely exceeds the existing available capacity. Solutions include building temporary lots on campus, providing off-campus parking and shuttle buses for construction workers and/or hospital employees, and building new lots or garages prior to the start of the additions and renovations. One creative option, used in an urban area in south Florida, was to rent part of the parking lot at a nearby shopping mall during the weekdays and use shuttle buses for the construction personnel. The mall was not as busy during the weekdays and the construction project had very few people working on the weekend, so it was a practical solution for both parties. Another approach, to minimize the inconvenience of construction for the patients who might have issues walking to parking, is to provide valet service at the front entrance. This promotes positive public relations in the community, and reduces the risk of a patient falling or being struck by a car while walking from a remote parking lot.

In some cases, a hospital department may need to be "gutted" for renovations, and must be relocated for a period of time. One option is to set up a temporary space on campus or to relocate the service off-campus temporarily. Trailers can be temporarily located in parking lots or in landscaped areas to provide space for departments that must be relocated. There are firms that make portable operating rooms that can be used to replace or supplement surgical space. These can be located adjacent to the existing operating suite, to provide for easy access and to keep surgeries underway in spite of extensive renovation work. One creative solution, that worked well on a hospital addition and renovation in west Florida, involved gutting the existing kitchen. After reviewing several options to keep the kitchen open during construction, it was determined that there were few choices that would maintain a clean and safe environment for food preparation, and it would mean a long multiphased process. The hospital staff knew of a high school very close and worked with them to rent their kitchen during the summer, while the students were gone. The staff set up a "cook-chill" process, where the food was prepared, frozen, and sent in carts to the hospital, where it could be rewarmed on the patient floors, so the kitchen at the hospital was not needed. The construction schedule was reworked so the kitchen renovation was done during the summer. The end result was less disruption during food preparation, lower cost by avoiding the multiphased work, and a safer environment for hospital staff.

Shutdown for Tie-Ins

This is a common term heard often in conjunction with additions and renovations. It means turning off, or "shutting down," all or part of one of the many mechanical, electrical, and plumbing systems in a hospital to "tie-in," or connect to, the existing duct, piping, panels, or other device. There are many risks involved with shutting down one of the systems, as some can literally mean life or death for patients, staff, or the public. The planning for a shutdown is very important, so that the hospital staff, both clinical and facility management, have ample early notification, in order to prepare for any potential disruption. These preparations could involve moving patients to other areas, having portable oxygen bottles ready, providing temporary air-conditioning, or a myriad other measures.

The planning for a shutdown can involve a number of steps. Most construction firms and hospitals have forms that must be completed and approved by the appropriate parties. Figure 9.8 shows a sample form from a construction company. It addresses the system to be impacted, the time frame, the responsible parties, a description of the work, and the backup precautionary measures to be taken. A similar form should be required on all additions and renovations to help avoid accidents, to provide legal documentation, and to effectively communicate with the appropriate parties.

One of the first steps in planning a shutdown is to research and clearly confirm the areas that will be impacted. While this may be obvious in some situations, such as the main supply air duct serving one wing of patient rooms, it can be very difficult with electrical circuiting in an old facility. Over the years, during renovations, multiple floors or different areas may have been connected to one electrical panel. Labeling of circuits in a panel may not be present, or may not be correct, so when a circuit is turned off, it may shut off power in areas not anticipated. Of course, this could have disastrous consequences, thus the reason for thorough research regarding the areas served before any shutdowns occur.

Once the scope of work is clearly defined, the areas impacted by the shutdown are confirmed, and the hospital staff (clinical and facility management) provides their input on minimizing disruption, the shutdown form can be completed and the work scheduled. Often shutdowns are done at night while there are fewer staff and patients in the hospital; however, it is a 24/7 facility and there are always some patients and personnel present. In some cases, it may take hours or days to complete the replacement of large air handling units or a chiller. At those times, it may be necessary to install a temporary unit to continue service to the hospital areas impacted. On tie-ins to critical life safety systems, such as the fire alarm, shutting down an area may require a "fire watch." The code (NFPA 101) requires an "approved fire watch" to be implemented if the system is to be down for more than four hours in a 24-hour period or the building must be evacuated. Fire watch and standby personnel are also required by the International Fire Code (IFC) when fire protection systems are impaired, and when certain types of construction, demolition, or "hot work" is taking place. The fire watch will be discussed further in Chap. 12 of this book.

There are several precautionary steps that can help reduce the risk with shutdowns. One is to have more than one part or device that is going to be installed. This is obviously not practical if the part to be installed is a new emergency generator or chiller, but if it is a circuit breaker or a valve, then having a spare part can reduce the risk of having a shutdown, only to discover the part is faulty and everyone's time is wasted, and that more time will need to be scheduled for another shutdown. Another approach that has been successful on tie-ins for water systems is the use of "pipe freezing"

Hospital Shut Down Approval Form

Step 1: Description of Shut Down

System or equipment impacted:

Date Submitted: ____________________ Date Approved: ____________________

Tentative Start Date: ____________ Tentative Finish Date: ____________________

Planned duration: ___________________________________

Step 2: Project Name and Information

Hospital Name: __________________________ **Address:** ____________________

Director of Engineering: ______________________________

Step 3: Contact List

- **For Medical Emergencies call 911**
- **For work-related concerns or questions contact the following:**

Name	Company	Position	Phone #	Email Address:

Step 4: General Description of Work

- **Describe general overview of procedure. Step-by-step details described in STEP 5:**

  Phase #1:

  Phase #2:

- **Systems affected by Work:**

  Phase #1:

  Phase #2:

FIGURE 9.8 Shutdown approval form.

Step 5: Step-By-Step Procedure

Step	Start Date & Time	Action	Action By	Completed Date & time
	Phase 1 Time:	Phase 1	Sub	Phase 1 Time:
1				
2				
3				
4				
	Phase 2 Time:	Phase 2	Sub	Phase 2 Time:
1				
2				
3				
4				

Step 6: Testing Methods

Step	Start Date & Time	Action	Action By	Complete Date & Time
1				
2				
3				

Step 7: Hoisting/Hauling/Storage

1. Identify routing to space:
2. Identify protection of walls/floors:
3. Identify weights of any items over 70 lbs.
4. Identify weight capacity of elevator if appropriate:
5. If building structural system is utilized to support object, identify structural component, capacity of structural component, and method of support:

Step 8: Protection Requirements

1 . T elecommunications Equipment/Computers/Cable/:
2. Protection for Personnel/Tenants:
3. Electrical/Mechanical Equipment:
4. Miscellaneous Equipment:
5. Fire Alarm
6. Fire Sprinkler

Figure 9.8 *(Continued)*

Step 9: Safety Requirements

1. Fire Protection:
2. Fire Watch:
3. Dust Protection:
4. Asbestos Containment:
5. Smoke Containment:
6. Noise Containment:
7. Water Containment:
8. Mold Remediation:

Step 10: Special Tenant Needs/Requirements

This section is to be completed by User Group/Tenant as applicable.

Step 11: Security Requirements

Step 12: Contingency Plans

This is an extremely important section, which must be completed in detail. In the event of an unforeseen problem associated with any aspect of the shut down., a contingency plan is to be developed. Identify potential problems, which may arise. Include safe stop points.

Include back out procedures. Spare parts, materials, etc. shall be identified for contingency plans.

Potential Problems:

Contingency Plans:

Back Out:

Step 13: Shut Down Signature Sheet

1. **If work procedures do not impact the party identified, signature is not required. If you want to be contacted if contingencies are implemented, circle Yes below.**

Figure 9.8 *(Continued)*

2. **Procedure for obtaining signatures shall be as follows:**
   a **Signatures shall be in the order indicated.**
   b **E-mail processing shall be an acceptable method of confirmation of concurrence of MOP.**
   c **Originator of Shut Down shall indicate N/A adjacent to individuals who are not involved and or affected by the MOP.**

Name	Emergency Contact	Emergency Phone #	Date	Signature
(Name of Contact) Fire Protection Contractor	YES NO			
(Name of Contact) HVAC Contractor	YES NO			
(Name of Contact) Electrical Contractor	YES NO			
(Name of Contact) Pneumatic Tube	YES NO			
(Name of Contact) Contrarctor	YES NO			
(Name of Contact) Hospital	YES NO			

Figure 9.8 *(Continued)*

technology. In this scenario, there is the need to connect to an existing water line, but no zone valves are available, so it appears the entire system must be shut down and drained. With "pipe freezing," the water in the pipe near where the valve is to be added is frozen to prevent water from flowing, the pipe is cut and the valve installed. This can be done without the disruption of draining the water for an entire area.

Involving the subcontractors in the planning and preparation for shutdowns is very important. Typically, it is a "subcontractor" that will be doing the work so they will have a better understanding of the time frame required, the risks involved, and possibly can provide creative solutions to minimize the impact. The subcontractors should always be required to have a supervisor there during the work, rather than relying solely on workmen, to effectively manage the effort. The construction manager or general contractor should also be required to have a supervisor monitoring a shutdown and documenting the activities. As with all of the work, pictures should be taken to document the existing conditions, any unforeseen issues, and the completed work. The supervisors should also have an emergency response plan, in case something goes wrong during the execution. The plan should include a notification list with emergency numbers for the hospital engineer, the construction manager, the fire department, and the ambulance service. The plan should also include clear directions on how to return the system to operation if the tie-in cannot be completed successfully. In some cases, staff from the hospital facility management department, the design team, or the building inspection department may need to be present.

Protection of Critical Systems

While working in and around a functioning hospital, it is very important to protect the critical systems. This can be very challenging, with so many systems needed to support good patient care. The priority systems that come to mind are life safety, such as fire alarm, fire sprinkler, and fire/smoke walls; HVAC, which includes hot/cold air, exhaust, and pressure relationships; power, which involves normal and emergency, as well as medical gas and clinical systems, such as physiological monitoring and imaging.

Since every hospital is different and every addition and renovation project has different issues, there are no simple steps to effectively address all situations. The best approach is to assemble an interdisciplinary team, including members from the hospital facility management or engineering, the contractor, the key subcontractors, the design team, and hospital clinicians, to evaluate the critical systems that may be impacted by this particular addition and renovation project. This team can review the specific work to be done, the hospital departments involved, and the critical systems, so that the risks can be identified and mitigation measures implemented. In addition to the steps to control the risks identified, some firms set up contingency dollars in their budget to address potential issues. By doing this, the project not only has the steps defined that may be required, but there are also dollars committed to implement the measures.

A few examples may help clarify this approach. On a vertical bed addition in Florida, the existing rooftop air handling unit had to be relocated to make room for the new space. The units are needed to maintain heating and air-conditioning to the occupied patient rooms below, which was easily defined as a critical issue. With input from the hospital, the designers, and the subcontractors, a plan was developed to keep the existing unit operational until the new bed floor structure was completed. The new air handling unit was installed, tested, and operated to ensure it was ready for use. The ductwork for the connection was fabricated so the tie-in could be made in a 4-hour period. Temporary "area-cooling units" were installed in the occupied floor to maintain proper temperature and humidity during the tie-in period (Fig. 9.9).

The budget was established with monies for these temporary measures, as well as additional manpower during the tie-in, so all of the resources needed would be available. The conversion to the new unit was successfully executed in less than the 4-hour "window," with minimal impact to the facility.

Another example took place on a project in Texas with the addition of two new operating rooms. Since ORs are not typically utilized at night, the best approach there was to shut down and tie-in the medical gas system during the evening. To avoid multiple shutdowns and certifications, the plan developed was to complete the new addition so it was ready for occupancy. The medical gas lines were "stubbed out," or installed ready for connection near the tie-in point to the existing lines. This left a short distance to add the pipe to connect the new lines to the existing lines. To minimize the time required to do the work, access doors were installed in the drywall ceiling to make the tie-in easy. For an OR, this typically involves oxygen, vacuum, medical air, nitrous oxide, and sometimes carbon dioxide and nitrogen. The tie-in was scheduled only after the proper documentation was completed, notices were given to facility management, and the risk assessment was done. The medical gas system required certification after installation by an independent testing firm, so they had to be available for the tie-in. At

FIGURE 9.9 Temporary mobile AC unit.

least two qualified plumbers were there, in case one could not complete the work due to sickness or injury. Supervisory personnel were also available. Using this approach with one tie-in, the third party certification firm was only needed once, to reduce cost by doing the work at night. There was no impact on the OR schedule.

While the need for protection of critical systems is obvious in most cases, there are times when the potential impact of an addition or renovation may not be so apparent. For this reason, it is important that this multidisciplinary risk assessment team look at not only the areas of the hospital involved in the work, but also adjacent areas or different areas that may provide support for the addition or renovation. On many additions, there is work involved in the powerhouse, or central energy plant, that is needed to support the additions or renovations. This may involve adding electrical panels, tie-ins to emergency power, upgrading vacuum or air pumps to provide extra capacity, or a multitude of other issues.

A project example from south Florida required an air intake for a department that was not involved in the addition work. The emergency room expansion project included adding new space, as well as renovating the existing area. The new emergency space was adjacent to the OR, but did not connect in any way. The issue that was overlooked was that the air intake for the surgical space was adjacent to the ED expansion. The intake pulled thousands of cubic feet per minute of outside fresh air (also known as make-up air) into the ORs and support space. After the concrete slab had been poured for the new emergency room addition, the sprayed-on curing compound was being applied. Due to the direction of the wind and the amount of make-up air being pulled in the OR air handling unit, some of the curing compound was pulled into the surgery air. The fumes, while not toxic, did make some staff sick, and led to closing the ORs, and eventually resulted in a physician lawsuit. The problem was very preventable with any of several steps, such as using roll-on compound or protecting the air intake. This unfortunate case highlights the need to be very careful in considering the potential impact on critical hospital systems during additions and renovations.

Contracting Approaches

Given the complexities of adding to and renovating an operational hospital, consideration needs to be given to the contract terms for all members of the design and construction team. Typically, the contractor or subcontractor cannot approach the work as they would on a new hospital or other types of facilities. The contracts must clearly define that the builders work around and give priority to the hospital functions. This is especially true on multiphased renovations, which often involve night work for tie-ins, schedule changes to accommodate emergency procedures, and temporary measures to keep the facility operational. Failure to include these unique requirements in the contract terms can lead to claims for delays and extra costs. A proactive approach is best, so builders understand the needs of the project, and can include the necessary work in their bids and/or budgets. Likewise, the hospital owner must allow for reasonable access to execute the work and realistic time frames for the construction, or they will be faced with significant costs for their projects. There are several steps that can be taken to develop a practical and cost-effective approach.

One of the first steps is to develop a realistic schedule and phasing sequence early in the design phase, working with the hospital staff, the design team, and the builders. This is more difficult in the traditional design-bid-build procurement, as the contractor is not involved until after bidding, but the owner can have a consultant or program manager work with them to develop a schedule that can be included in the bid package. This will help ensure that the bids reflect a realistic schedule and phasing sequence to accommodate the hospital's requirements. With early selection of the construction manager or contractor, the "constructability" input can be provided in the design phase, which is much better. Delivery methods are explored more in Chap. 7 of this book. While the schedule seems to always change in some way during most projects, having it and the phasing as part of the bid documents helps avoid later claims that the builder or subcontractors were not aware of the complexities, or that they may have to provide multiple mobilizations and extend beyond regular work hours.

Another step that is helpful on hospital and other construction projects is a prebid meeting. The schedule and phasing documents can be explained to the contractors and/or subcontractors. The requirements for temporary entrances and services or other unique conditions can be discussed. The need for risk assessments, constructability analysis, interim life safety plans, after hours work, temporary certificates of occupancies, or other processes driven by the renovations can be established. By clearly communicating and documenting the requirements, all bidders can include these measures in their price, reducing the validity of claims for extra charges later during the project.

The additional steps needed on the addition and renovation projects should also be carried forward into the contract documents, so the contractor and/or subcontractors are legally obligated. This can be done in several ways. The owner can have the design team include these requirements in the plans and specifications. Phasing plans, temporary entrances, fire/smoke barriers, and other steps can be shown on the plans or included in the general or supplementary conditions of the project specifications. These requirements should be addressed in actual signed contracts as well. Items such as those mentioned above can be included in the scope and schedule sections, as well as in supplementary conditions.

One of the challenges with citing specific schedule requirements is the difficulty of defining exact dates for a complex multiphased renovation during the contracting phase. The schedule could depend on patient volumes trending upward or downward, so when space is available for renovations is hard to predict. Another example could be vertical additions, where the removal of the existing roof is tied to the weather. Given the number of variables on an addition and renovation project, it is often better to define durations or productivity rates for the different scopes of work rather than setting specific dates. For example, a roofing demolition subcontract could require the subcontractor to demolish the existing roof in a maximum of 10 working days, or to remove 10 squares of roof per day. A contractor for a construction manager could require them to complete the renovation of a certain number of patient rooms per month, or whatever time frame is applicable for that specific project.

Liquidated damages are a contractual measure often used in the construction industry on time-sensitive projects. These are not commonly seen on healthcare additions and renovations, and are not considered a good approach by this writer. There are so many factors and variables that are often out of the owner's and contractor's control that it is difficult to enforce liquidated damages. Incentive payments have been successful on additions and renovations, if they can be effectively tied to returning the hospital space more quickly or minimizing disruptions. A successful example was based on having lump sum general conditions for the duration of the project. The contractor kept any savings resulting from early occupancy, incentivizing them to expedite the project and allowing the hospital to start using the space sooner, increasing revenues. This provided a win-win solution for both the hospital and the contractor. Another example was tying an incentive or bonus for the contractor to minimize the disruption to the hospital. It was effectively used on a very messy renovation that involved replacing all of the galvanized domestic water piping in a facility while it was operational. The hospital staff completed biweekly satisfaction surveys that measured factors such as disruption, noise, and dust with good scores resulting in a bonus. This incentivized the contractor to get the work done without negatively impacting the facility.

There are so many variables with addition and renovation projects that it is impossible to cover all scenarios. The main goal of the contract is to ensure that the contractor and subcontractor are legally required to accommodate the hospital operations while completing the work. It is best to develop strategies where the goals are aligned to promote win-win results, such as the hospital reopening sooner and the contractor getting a bonus. The undesirable converse of this would be that the contractor makes more on a lump sum bid by rushing through the renovation, with no regard to the impact on hospital operations.

Project Completion

Another challenge on multiphased additions and renovations comes at the completion of different areas. On a new hospital, the project is done at one time, so the close-out information and documentation is one step. On an addition and renovation project, there may be many phases completed over months, or even years, so there must be a strategy for executing the important functions that occur at completion. Some of these important "close-out" steps are explored in this section and more information about owner occupancy is provided in Chap. 14 of this book. Many of these are not exclusive to hospitals.

- *Certificate of Substantial completion.* This is an important legal document, establishing the date the project is ready for its "intended use," thus completing the contractor's obligation to the owner. It has several significant implications, including the transfer of responsibility for insurance, utilities, and security, as well as triggering the start of the warranty period. For multiphased expansion projects, there must be an understanding in the contract of how this will be managed. On a project with a few large phases, it may be practical to issue separate certificates for different phases. On a messy renovation project with many small phases, it may be more practical to have one certificate covering all the phases, if they are not spread over an extended time frame. The responsibility for the items mentioned above, such as warranty period and insurance coverage, should be clearly and equitably defined to reduce the risk of disputes.
- *Operation/maintenance manuals and as-built drawings.* This package, which may be paper documents, electronic files, or a combination of both, is provided to the hospital facility management staff to assist them in safely and efficiently running the hospital. Typically, it is turned over to an owner at the completion of a project, hospital or otherwise. In addition to this information, there is often in-service training, provided to the facility managers by the manufacturers, on how to operate and maintain major equipment, such as boilers or chillers. On multiphased expansions, there must be a plan developed by the design and construction team, in conjunction with the owner, regarding effectively managing the process. The owner's facility management staff will need this information to run the hospital. However, the contractor cannot totally complete this process until all of the phases are done. As with the substantial completion form, this can be done by phase, with a comprehensive package prepared at the conclusion of the entire project. The contractor and subcontractors will want to turn over the operation of the different phases as soon as possible, as they are not being paid for this effort. There is no simple solution that covers every project, but it is an important issue that should be addressed in front end planning and in the contract terms. Electronic close-out processes are successfully being done using the BIM model for facility management.
- *Punch lists.* On almost all construction projects, the contract allows the owner, or the architect on their behalf, to prepare a list of minor deficiencies or quality-related problems at the conclusion of the project. The contractor and subcontractors are required to correct these before receiving their final payment. This list of issues, called the "punch list," historically has been done on paper; however, there are now many electronic approaches or systems. While there has

been a push in the construction industry to have "zero punch list" jobs, this is often not realistic. Typically there always seem to be issues that must be addressed at the completion of the project, after the owner moves in, that are not part of the warranty. Again, multiphased expansions adds some complexities, regarding when the "punch list" should be done and where the contractor's responsibilities end. Typically, the contractor would want to complete the punch list at the end of each phase, before the owner moves in. This avoids the contractor being responsible for damages to walls and floors as furniture and equipment are installed in the completed area. This is the most common approach, but, on projects with many small phases, they may require the architect to make numerous trips to the hospital, potentially adding cost. The other issue, from the owner's perspective, is that the approach does not provide a comprehensive final inspection for all of the renovated areas.

Interface with Hospital Facility Management

One important step for a construction manager, contractor, or subcontractor is to develop a good working relationship with the facility management staff. During an addition and renovation project, the facility managers will have an integral role in coordinating work with the contractor, arranging access to areas of the hospital, and scheduling shutdowns for tie-ins. From the owner's perspective, it is important for the role of their facility managers to be clearly defined, so they can continue to operate the hospital in a safe and effective manner during the construction project. In addition to operating the physical plant during construction, the facility manager also should help make sure the new work is done so that it can be operated efficiently. The design team and the commissioning agent, which have been discussed further in Chap. 13 of this book, should consider the operation of the facility, but the facility manager should also be involved. Issues to be addressed include access to valves or equipment for maintenance, use of different manufacturer's products that may increase the number of spare parts required, and energy efficiency that could drive up the hospital's operating costs.

The facility managers will also need to be involved with the front end planning regarding phasing of the project. Often their knowledge of the existing facility is invaluable on an expansion project. The facility managers should know the issues with the existing facility that may impact the construction, and have as-built plans to show the locations of mechanical, electrical, and plumbing systems. The facility management department, also sometimes known as engineering, would be the point of contact for the statement of conditions required by the Joint Commission, as well as any surveys during the construction project. Positive relationships and good coordination with the department can greatly contribute to the success of a project.

Conclusion

Additions and renovations on acute care facilities typically cost less than entire hospitals. In Modern Healthcare's March 2013 Construction and Design Survey, which shows projects completed in 2012, the cost of renovations averaged $4.6 million, compared to entire hospitals, which averaged $68.9 million. While the capital expended may be much greater on new facilities, the expansion project can often be more complex to execute. In addition to the typical construction issues and the unique systems involved with building a new hospital, addition and renovation projects add the peripherals of patients, staff, and visitors to the challenges. The risk is greatly increased as every discussion regarding typical construction problems, such as schedule, budget, and quality, must also now be tempered with the potential impact on hospital operations. The construction manager must coordinate with department heads and clinicians on a daily basis to maintain a safe and functional healthcare environment, as opposed to occasional tours by staff at a new hospital. The keys to success are taking time to learn and understand the hospital functions impacted, taking time to plan and communicate before the work is done, and proactively managing the many people typically involved in expanding an operational hospital.

References

"Modern Healthcare's 34th annual Construction & Design Survey", *Modern Healthcare,* March 2013.

Carpenter, D., "2011 Hospital Building Report: Shifting Priorities," *Health Facility Management Magazine,* February 2011.

CHAPTER 10

Mechanical and Electrical Systems in Hospitals

Why devote an entire chapter to mechanical and electrical systems? If designed and installed correctly, they should be neither seen nor noticed. Here are three reasons why construction managers should pay attention to these systems.

1. Between 35 and 45 percent of the total construction cost of a new hospital is for the mechanical, plumbing, and electrical systems. The percentage is higher in a renovation project and, depending on the age and condition of the existing facility, there may be significant additional expenditures required that are not directly associated with the project area. For example, if a 1970s era patient wing is to be renovated as an ICU, there may not be sufficient air-conditioning or electrical power available in the project area; it may be necessary to add an air handling unit or to bring new electrical power into the area from some distance away.
2. In constructing a hospital, 30 to 50 percent of formal RFIs (requests for information) pertain to these systems. Additionally, coordination of mechanical, plumbing, and electrical trades requires a day-to-day planning effort to make all of the components fit in tight ceiling spaces.
3. Mechanical and electrical systems can occupy up to 10 percent of the building floor area.

Facility managers take a great interest in the building mechanical and electrical systems because of the impact of energy costs. If a nonprofit hospital reduces their energy consumption by $1, it has the equivalent benefit to the hospital's bottom line of an additional $15 to $20 in patient care revenue (www.energystar.gov).

Special Code Requirements and Impact on Mechanical Systems

As described more fully in Chap. 2, hospital construction is subject to both general building codes and to specialized codes that focus on hospital construction. The building codes govern such issues as locations of fire dampers in ductwork that penetrates rated walls, special requirements for high-rise buildings, etc. Specialized codes that greatly affect design of mechanical systems in hospitals are the energy code sections of the applicable building codes and the Facilities Guidelines Institute *Guidelines for Design*

This chapter was written by Rick Wood.

and Construction of Health Care Facilities, adopted and enforced by most states. Some states have their own set of regulations that are similar to the FGI criteria.

The energy codes such, as the International Energy Conservation Code (IECC) and ASHRAE/IES 90.1, list minimum efficiencies of equipment performance and minimum material standards, such as the insulating values of pipe and duct insulation, but can also affect system layout and control. For example, the energy codes list the maximum allowable fan horsepower for each HVAC system. Fan horsepower depends on the amount of pressure the fan must develop to move the air, which is dependent on the length of ductwork. If a fan is located a long distance from the area served, the horsepower required may exceed the code-allowed values. The energy codes also contain requirements for part load control and energy recovery, both of which affect HVAC system layout and system cost.

The FGI *Guidelines*, or equivalents, state criteria, list various hospital spaces (operating rooms or ORs, patient rooms, emergency waiting rooms, etc.) and associated HVAC requirements for that space (air change rate, air pressure relationship, filtration level, temperature/humidity range). There are many different types of air-conditioning systems but, since not all of them can meet the criteria, some are not applicable to hospital construction.

A central air handling unit (see Fig. 10.1), which conditions air to close tolerances of temperature and humidity, and then distributes it via extensive ductwork to many spaces, can be configured to meet the filtration, air change, temperature, and humidity requirements for any hospital space. Unless located on a roof, these units also take up the most building space, and will have the highest first cost of available options.

Smaller units located in the rooms, such as fan coil units and certain types of heat pump units, can be configured to meet the FGI requirements for some spaces (patient rooms, exam rooms, some imaging rooms), but usually cannot meet the requirements of other spaces (ORs, ICUs, labs), except with custom equipment. These units are often located above the ceiling of the space served, which reduces the building floor space required and the amount of ductwork (and conflict with other trades) above the ceiling. Installation costs are usually less for these types of systems.

Hospitals have areas that require special mechanical systems to provide the desired environment. Some areas have unusual temperature and humidity requirements. Other areas have equipment with special exhaust needs.

For most spaces in a hospital, the conventional space indoor conditions, temperatures in the low 70s and a corresponding relative humidity in the range of 30 to 60 percent, are acceptable. However, operating rooms are typically maintained at lower temperatures, from 62 to 68°F, for both comfort and procedural concerns. Surgeons who are fully gowned, and sometimes in full bodysuits, prefer a cooler atmosphere to maintain their comfort and concentration. When room temperatures below 68°F are desired, additional HVAC equipment is needed to maintain an acceptable relative humidity (see Fig. 10.2). This equipment could be a glycol cooling system or a desiccant system, both of which serve to wring moisture out of the air stream. At the other end of the spectrum, pediatric and infant surgery may require that an operating room be maintained at 78°F.

Some areas of the decontamination department should be kept between 60 and 65°F (with corresponding relative humidity between 30 and 60 percent) according to ANSI/AAMI Standard ST 35. (AAMI is the Association for the Advancement of Medical Instrumentation.) Maintaining this condition requires some type of additional air dehumidification, as does maintaining a low temperature OR.

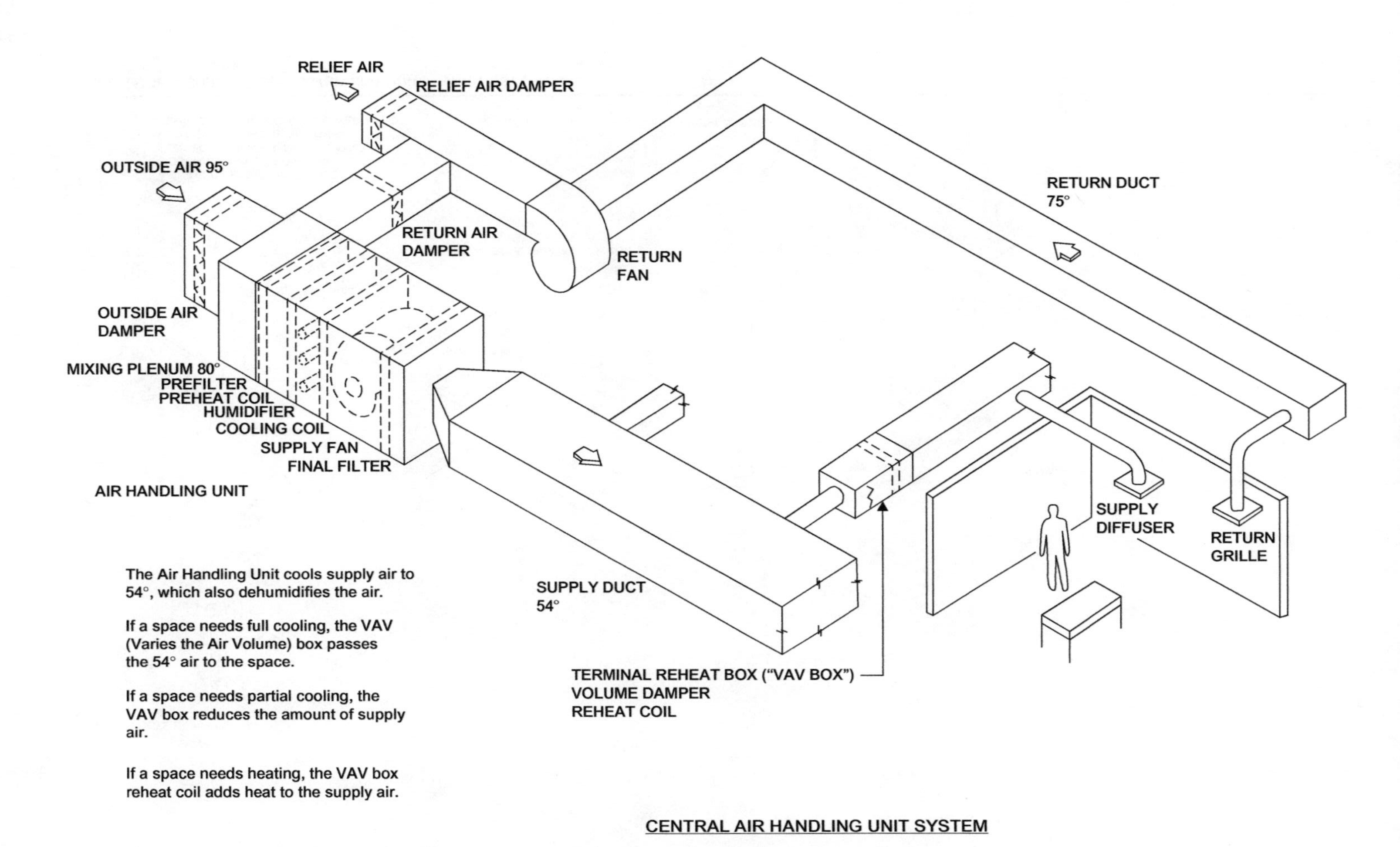

FIGURE 10.1 Central air handling unit system.

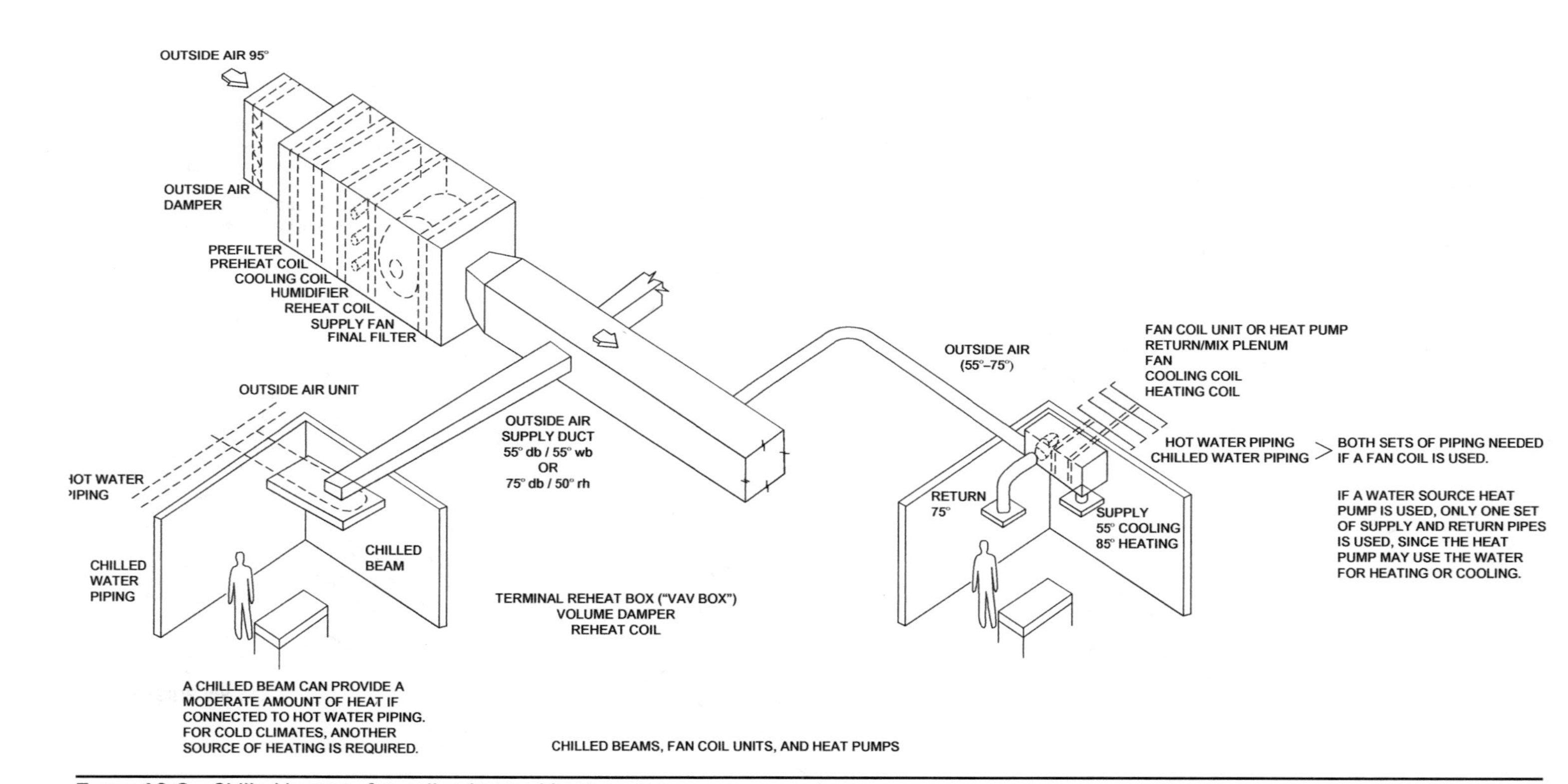

FIGURE 10.2 Chilled beams, fan coil units, and heat pumps.

Hospitals have several areas with equipment that requires special exhaust. The kitchen area will typically have a grease hood and dishwasher hood. NFPA 96 details the installation instructions for the grease ductwork. The laboratory and pharmacy departments will have hoods that require exhaust systems. Grease hoods and lab/pharmacy hoods should not have dampers in the ductwork, so the duct must be protected in fire-rated enclosures from the hood to the exhaust fan if it passes through a rated wall.

If sterile compounding is to be done in the pharmacy area, additional equipment to comply with the clean environment requirements of USP 797 is needed. This is accomplished by using ceiling-mounted fan-filter units that recirculate room air through HEPA filters to maintain 30 to 40 air changes per hour.

Special Code Requirements and Impact on Electrical Systems

Similar to mechanical systems, general building codes will govern some aspects of electrical system and design, while specialized codes will focus on restrictions on energy usage, and on providing the necessary building environment for patient care.

The applicable energy code will affect electrical design primarily in lighting, by use of maximum lighting energy densities for various spaces. This can be met by a combination of quantity and quality (i.e., energy efficiency) of the lighting fixtures selected.

The FGI *Guidelines* have criteria for special lighting in patient rooms and for locations of major components, such as panelboards and switchboards.

NFPA 99 *Health Care Facilities* gives some specific instructions for devices and testing of equipment used in patient care.

NFPA 72 *Fire Alarm Systems* is a comprehensive guide to the specific requirements for fire alarm systems.

By far, the most significant impact on electrical systems in hospitals is NFPA 70 *National Electrical Code*, in particular Chap. 5 and Sec. 517. This chapter details the requirements for the electrical distribution system, including standby power generation for essential electrical loads, and the separation of the essential electrical system into three independent main branches (life safety, critical and equipment).

Medical Gas Systems

One unique aspect of hospitals is the extensive medical gas system. Oxygen, medical air, medical vacuum, nitrous oxide and, sometimes, nitrogen are piped throughout much of the hospital, to be used in the treatment and recovery of patients.

NFPA 99 *Health Care Facilities* and the FGI *Guidelines* provide the requirements for these systems, from piping material requirements to the quantity of outlets required in various spaces (ORs, patient bed, nurseries, etc.). All medical gas piping is copper with brazed joints. A special requirement of medical gas pipe joining is the requirement, found in NFPA 99, that the installing contractor flow nitrogen through the pipe while brazing, to keep pipe temperatures low and avoid copper oxide formation on the inside of the pipe joint. There are also criteria for cleaning and degreasing the pipe before and after installation. Medical gas systems are also governed by the International Fire Code (IFC). IFC Chap. 53 includes general provisions for medical gas systems, such as interior supply locations, fire resistance rated separations, and other general provisions, and also makes reference to NFPA 99.

After medical gas piping is installed, it is pressure tested. Upon successful pressure test, but before it is placed in service, the completed system is tested and certified by an independent agency for gas purity and to ensure that the pipes are not "cross connected" to other piping systems.

Oxygen is usually supplied from a remotely located tank with distribution piping into the building. As these tanks are very specialized equipment, they are most often rented from an oxygen supplier.

Medical air is generated by a medical air compressor, usually located in the central utilities plant or main mechanical room of a hospital, then piped to the points of use. A medical air compressor package usually consists of several small compressors in parallel, to handle a wide swing in load, and to provide redundancy, in case one unit fails.

Likewise, a medical vacuum pump located in the central plant or main mechanical room will produce the vacuum in a distributed piping system. As with the medical air compressor, the vacuum pump package consists of several small units manifolded in parallel, for load swings and redundancy.

Nitrous oxide, used for anesthesia, is supplied in cylinders by a medical gas supplier. Typically, the cylinders are grouped and located near a convenient location for change out by the supplier. A particular note for nitrous oxide is that the cylinders must be stored above 25°F; thus, they need to be in some type of heated enclosure.

Nitrogen is a service gas, used in some surgical procedures to power the surgical tools, or as a "brake" when adjusting equipment booms or articulating arms in ORs or imaging equipment. Like nitrous oxide, it is supplied in bottles by a supplier, but does not need to be stored at a minimum temperature.

Zone valves are used to identify areas served and allow isolation of parts of the system for maintenance, without taking down the entire system. Line pressure alarms monitor system pressure, and provide signals when pressures fall below operational minimums.

Plumbing Systems

The FGI *Guidelines* and the plumbing section of the applicable building code (such as the International Plumbing Code) are the basis of design and installation for plumbing systems in hospitals.

Dual sources of water supply are required for hospitals in the IPC. This may be accomplished by connecting to the city water distribution system in two places, or by having a tap available for a temporary water truck connection. The two sources may be combined into a single pipe entrance into the building.

Hospital mechanical and plumbing systems usually require some type of treatment of city water, because the process of heating water and boiling steam will leave scale deposits on the inside of piping. To minimize this scale, hospitals use "softeners" to reduce the "hardness" (the mineral content of water, primarily calcium and magnesium). A city water analysis is used to specify the softeners.

City water pressure is also a factor in hospital water systems. Hospitals typically use flush valve toilets in patient rooms, which require higher water pressure to operate. If the available water pressure is not sufficient to operate fixtures on the top floor, booster pumps are needed.

Waste, vent, and rainwater systems are typical of those found in most commercial buildings. One application unique to hospitals is that PVC piping is sometimes used for underground waste piping. Hospitals usually have some sort of steam system, and may have locations where hot condensate is discharged to the waste system. Since PVC has an operational temperature limit of 140°F, hot condensate must be cooled before discharge. PVC gives off toxic gases before it ignites. It is not recommended for use above grade in hospitals.

Connections of the waste system to the city sewer system must be considered in design. Most cities use separate sanitary waste and storm sewer systems, so the hospital waste system needs to be designed to connect properly.

The domestic hot water system in hospitals is more extensive than in other occupancies. Sizing and selection of water heaters will include redundancy, in case a unit fails. The current standard is to use semi-instantaneous water heaters that provide hot water supply on demand, as opposed to the older method of using large hot water storage tanks. Large storage tanks were susceptible to Legionella growth, and it is a waste of energy to keep a large volume of standing water hot. Historically, there have been two separate hot water systems, one for domestic use at 120°F supply water, and another for kitchen use at 140°F supply water, but recent FGI *Guideline* changes have reduced the kitchen supply temperature requirement to 120°F, so the projected trend is to simplify with one combined system. The hot water recirculation system is designed to reduce waiting times for hot water delivery at a lavatory, and to eliminate "dead legs" (piping sections with no flow) in hot water piping.

Plumbing fixtures in hospitals have some unique requirements. "Institutional" grade fixtures are specified, because they are typically stronger and able to withstand more usage. The FGI *Guidelines* have specific requirements for different fixtures in particular rooms. Some fixtures may need "hands-free" operation, using wrist blades or knee, foot pedal, or electric eye operators. Patient rooms need to have bedpan flushing devices, which means that patient room toilets must use flush valves, instead of tank type fixtures.

Fire Suppression

Almost all new and renovation hospital projects have automatic sprinkler systems for fire suppression. The design and installation of sprinkler systems is governed by NFPA 13.

The sprinkler system water entrance into a hospital must be separate from the domestic water entrance, and is subject to various constraints.

Sprinkler systems must be maintained at a minimum pressure to provide adequate coverage of the most remote area. A hydraulic calculation is done to determine this minimum pressure, and takes into account pressure losses due to flow through the piping components. If the incoming city water pressure is not sufficient, a fire pump may be necessary.

In a hospital, there are several areas where water discharge from a sprinkler system would be hazardous and/or impose great damage. For those areas, alternative means of fire suppression are used. The grease hoods in hospital kitchens use special self-contained systems that comply with ANSI/UL 300, in which a special wet chemical is used to extinguish, and prevent reignition of, a hot grease fire. Fire suppression in data centers can be accomplished by the use of clean agent or inert gas systems; water-based systems are required to be used if the rest of the hospital has a wet sprinkler system, but some additional controls can be used to delay the activation of the water sprinklers, in hopes of containing and extinguishing the fire before it becomes necessary to activate the water sprinkler. The equipment in imaging areas presents a similar challenge for fire suppression, in trying to minimize damage to equipment, and reduce the possibility of electrocution from water discharge. A cleaning agent and delay controls on the water sprinkler system are used in MRI rooms. Delay controls on the water system are used in CT scan and radiology rooms.

Sprinkler piping that covers canopies and other areas, where the piping is exposed to cold outdoor temperatures, uses a "dry pipe" system, in which the water is held back by a valve indoors and released into the pipe upon activation of a sprinkler head.

Mechanical and Electrical Issues in Construction

There are over 20 separate specification sections in a new hospital construction project, with most of them requiring specialized subcontractors. Some will come and go quickly; others will be there most of the duration of the construction, and often will be working closely together. Communication, coordination, and planning can enable the process to move forward; lack of it can bring the process to a halt.

"I Was Here First"

The sprinkler subcontractor, sheet metal subcontractor, piping subcontractor, and electrical subcontractor all need to install their work in the same space above the ceiling. Mechanical, plumbing, and electrical plans show the design intent. They are diagrammatic, and show the relative arrangement of components, but are not a scaled representation for fabrication of the systems. The plans have been prepared with a general understanding of the building dimensions, but do not account for the specific means of installation of all components and equipment. Miscellaneous steel used for bracing and framing soffits, hangers used to support ductwork, and pipe and other such pieces necessary for construction are not individually shown. The space above a ceiling is always going to be tight; if there were an abundance of room, the owner and architect would ask (rightly), "why are we paying for unused space?" Coordination among the trades is the responsibility of the construction manager. This means not only *where* to install, but *when* to install. If the sprinkler contractor starts first and runs sprinkler piping down the middle of the ceiling space, the ductwork will not fit, and the waste piping will not slope.

"The Pipes Have to Slope"

Waste and condensate piping depend on gravity to work. Waste piping must slope at one quarter of an inch per foot of pipe length per code. If it hits a duct, it cannot rise up. Likewise, steam and condensate piping must slope.

"How Will I Turn That Valve?"

Installed equipment must be maintainable by the owner after the project is finished and occupied. A duct access door is useless if it can't open because a pipe runs adjacent to it. A valve that is hidden on top of a 36 inch wide duct will never be found.

"Who Let the Air In?"

Cold pipes and ducts in an unsealed building will cause problems. When warm, moist air is cooled or comes in contact with a cold surface, the air will cool and, if it cools enough, the moisture in the air will condense, and water droplets will form on the cold surface. If the cold water pipes or supply duct are operable (i.e., the surface temps will be 60°F or lower), but not completely insulated, and the building is not sealed (or the

A/C unit is bringing in outside air for construction ventilation), warm outside air will infiltrate into the construction area, and cause "sweating" on the pipe or duct. This moisture collection will ruin insulation and ceilings.

"Where Is the Power (or Water or Steam) for This Equipment?"

Owner furnished equipment (OFE), such as operating room booms, steam sterilizers, ice makers, etc., need the correct building service to work. The design team uses information provided by the OFE consultant to show electrical power, telecom, and data connections, hot and cold water, and steam and condensate to equipment. If the information did not come during design, or if the information was incomplete, or if the purchased equipment is different from the planned equipment, the required building service may not be available. This results in a hurried, expensive effort to bring in the necessary service. If the original intent was to have a sterilizer that uses 30 psi steam and the delivered unit uses 50 psi steam, a new steam line needs to be routed. If the original intent was to have a warming cabinet that uses 120 volt power, and the cabinet on site needs 208 volt power, a new electrical circuit must be routed. A similar situation often exists in coordinating electrical service to door hardware. If it is left until the construction phase to determine exact requirements of which doors get electric operators, and the power requirements of those doors, there will be a lot of last minute design and construction effort to make the equipment work.

"You Can't Wire a Hospital Like an Office Building"

Read, mark, learn, and inwardly digest NEC 517. The three branches of essential power (life safety, critical, equipment) must be kept separate from the source (usually the transfer switch located near the generator) to the end use device. A large junction box cannot contain circuits from multiple branches. A fan can't be powered from a critical panel. Receptacles can't be powered from a life safety panel. Paying attention the first time saves significant rework time and expense later.

The "Standard" Way of System Design and Construction

As soon as a "standard" way of doing something is identified, there are exceptions noted. However, for the past 30 years, it is safe to say that there has been a prevalent process.

The design team would be led by an architect, who would be the prime design professional contracted to the owner. Civil, structural, mechanical, and electrical engineers worked for the architect, and designed their respective systems. The plans would be finished, put out to bid among general contractors, and the successful bidder would be awarded the project. Advancements in building materials, building methods, and mechanical and electrical equipment brought about more options and choices, with a corresponding increase in range of costs. To help guide the owner and design team through the more extensive field of choices, "general contracting" evolved into "construction management." As opposed to a general contractor, who first became involved when plans were complete, the construction manager became involved earlier in the process, by providing preliminary budgeting and estimating as plans were being developed. The end result would be a guaranteed maximum price (GMP), based on some level of partially completed documents. The GMP would be adjusted if the final documents contained more "scope" than what was shown or stipulated on the GMP documents.

To comply with the stringent codes, most HVAC systems use central air handling units for the medically intensive areas (surgery, imaging, radiology, lab, intensive care, emergency), and either central air handling units or fan coil units in patient rooms and administrative areas. Electrical systems were designed to code. The emergency generator, to contain costs, would not have much capacity beyond code required loads.

While energy costs and standby power availability were considered and discussed, for the most part, the priority was the getting an operational, code compliant building at the lowest first cost.

New Priorities

"The times, they are a-changing" in healthcare construction.

When hospitals evaluate their overall patient quality of care, they now consider how the building environment (lighting, temperature, humidity, air quality) affect the patient's outcome. They also look at how the building systems affect staff performance. (Do the building systems cause any distractions to the staff in performing their jobs?)

In an attempt to develop national metrics of performance, many hospitals now focus on their Energy Star rating. Energy Star is a program developed by the EPA to promote energy efficiency in many areas, from consumer products (refrigerators, air-conditioners, TVs, etc.) to new buildings. The program has a database of energy consumption in hospitals across the nation, large and small, and the Energy Star rating is a percentile rank of a facility's energy consumption compared to similar hospitals. Another national program that promotes energy efficiency is Leadership in Energy and Environmental Design (LEED), administered by the U.S. Green Building Council. LEED is a third party certification process that awards levels of achievement based upon meeting or exceeding certain criteria. The LEED program is very visible to the public; hospitals understand their leadership roles in the community and, as a result, more hospital owners are pursuing LEED certification for their construction projects. Another national program that establishes minimum baseline energy efficiency and green building practices is that of the International Code Council. They publish an energy code (International Energy Conservation Code or IECC) that has been adopted in most jurisdictions, and also publish a green code (International Green Construction Code or IgCC). The IECC references ASHRAE/IES 90.1, while the IgCC references ASHRAE/USGBC/IES 189.1.

Building systems have become more complex as buildings get bigger, yet the size of maintenance staffs is in a downward trend. Hospital revenues haven't kept pace with expenditures. Fewer people are being asked to do more.

The result is that the new goal in hospital construction is to increase *value* rather than focus solely on reducing cost. More hospital decision makers now ask, "If I spend this extra amount here, will it save me more there?" and "Have you [design/construction team] considered this option?"

New Methods of Project Delivery

Conventional and new methods of project delivery are discussed in depth in Chaps. 7 and 8 of this book. These new methods can result in new approaches, and new paradigms, for hospital building systems.

As mentioned earlier, the emergence of the "construction manager" enabled the design team to interact with a contractor during design with preliminary budgeting, as plans were being developed. This process evolved to the point that CMs were able to give meaningful input very early in the project planning, and now can be an integral member of the design team.

Early involvement by a construction manager and the goal of total project value have led to a recent development in project delivery called integrated project delivery (IPD). IPD is covered more extensively in Chap. 7, and can be summarized as a process in which all involved parties (owner, architect, various engineers, construction manager, and specialty contractors) become members of one team, focused on an end goal of the "best" project for the available funds. "Best" project means more than "least first cost." The "best" project meets the specific goals of each particular owner, and will likely vary from project to project. Some owners may place more priority on future operating costs, or duration of construction, or visibility in the community. Each member of the IPD team contributes their expertise toward meeting the overall project goals and priorities. One interesting aspect of IPD is that contractor involvement will start earlier in an IPD project, but design may last longer—some parts of the building may start construction while other parts are still in design. An example of how IPD works is given in the Case Study: Methodist Olive Branch Hospital at the end of this chapter.

One important aspect of IPD is the life cycle cost analysis, which looks at the projected energy consumption, equipment service life and replacement costs, maintenance effort required, and the cost of financing over the projected life of the hospital facility. In order to perform an informative and useful Life Cycle Cost Analysis (LCCA), a wide spectrum of information is needed. If available, much of this information should come from historical data, such as equipment service life and expected maintenance costs.

Another new tool of design and construction is building information modeling (BIM). This refers to the use of an all-inclusive three-dimensional model of the facility. Structural beams, pipe and duct routes, ceilings, walls, doors, etc. are input into a shared model, from which each trade designer/constructor can see how their respective equipment is affected by other trades. The development of the model starts, of course, with the architect, and soon grows as other disciplines add information (beam sizes, major duct routes, etc.). As the design of the project becomes more defined, the model reflects this greater detail. During this time, the purpose of the model begins to change from a means to visualize the concept to a method of coordination and determination of material quantities and fabrication. Responsibility for development of the model shifts with the change in purpose of the model—the designers have the largest input in conceptualizing the project, but give way to the contractors as the model becomes a tool for coordination and material quantities. This period of transition will vary, depending on project contractual arrangements (CM at Risk, bid, IPD, etc.), and will usually start earlier, but last longer, in an IPD project, as contractor involvement begins earlier, but design may continue longer in that type of project.

New Mechanical Trends

Supply rises to meet demand, and the demand for improvements in energy consumption, while meeting increasing design and construction standards, has resulted in the availability of new methods of design and equipment options.

Reducing Reheat Energy

In the "standard way of doing things," a central air handling unit would take return air and fresh outside air, blend them together, heat or cool/dehumidify as necessary, then distribute this air to all spaces. All rooms would get the same temperature air, whether or not the heat load in that space required the supply air to be that cold. If the space did not need the full cooling available, the terminal reheat box would first reduce the amount of supply air then, if the available cooling from the supply air was still greater than the heat load in the room, the box would "reheat" the air, by use of a heating coil in the box. As an example, the design heat load for a space on an August afternoon might require 300 cubic feet per minute (cfm) of supply air at 50°F but, on an April morning, the space might need just 200 cfm at 58°F. The box would reduce the supply air volume to 200 cfm, then "reheat" this air from 50 to 58°F as it is introduced into the room. This concession to comfort comes at cost of energy to reheat air after consuming energy to cool it. Reducing reheat reduces energy consumption, and has been a major reason for some new designs.

One method of reducing reheat energy is to separate the treatment of outside air from the return air. A central air handling unit takes all the air (return and outside air) and cools/dehumidifies it to a point that wrings out the moisture in the outside air. This is somewhat wasteful, as often only the outside air needs to be cooled/dehumidified to a lower temperature. By separating the outside air stream and using a dedicated outside air system (DOAS), sized to handle only the outside air quantity, a smaller quantity of air is cooled to the point of dehumidification. Return air (which does not need as much dehumidification and, thus, does not need to be cooled to a lower temperature) may only need to be cooled to 60°F. The net result is that less air is cooled to a lower temperature, and the supply air to a space is introduced at a higher temperature. If the heat load in the space is less than design, then the air is reheated from a higher temperature (i.e., supply air may be reheated from 60°F instead of from 50°F).

The use of a DOAS then opens up other design schemes that are more energy efficient. One principle of energy systems is that transferring energy via water is more efficient than transferring energy via air. A central air handling unit conditions the air and distributes it via ductwork over a large area; the energy transfer from the point of cooling/dehumidification is by the air. If the point of cooling/dehumidification was located closer to the space, then less energy transfer by air would be required. Locating the cooling equipment closer to the space can be accomplished several ways.

Water Source Heat Pumps

Recent developments in water source heat pumps (see Fig. 10.3) make it feasible to use this equipment for hospital uses, even in operating rooms. New models of heat pumps can provide the necessary filtration. By locating small heat pumps near the spaces and using water, not air, to transfer energy to and from the room, the air ducts are shorter, smaller, and require less fan energy. A DOAS system is used to supply the required amounts of outside air, while the heat pump provides only the heating or cooling required for that space, thus eliminating reheat energy. A heat pump is a compact refrigeration cycle, so no central chiller is used. The heat pump uses a circulating water loop from which to draw heat (in winter) or reject heat (in summer). A boiler and cooling tower add or reject heat to maintain the circulating water system between 60 and 90°F.

A variation of water source heat pumps is the use of the ground as a heat source (in winter) and a heat sink (in summer) (see Fig. 10.4). Inside the building, the system is the

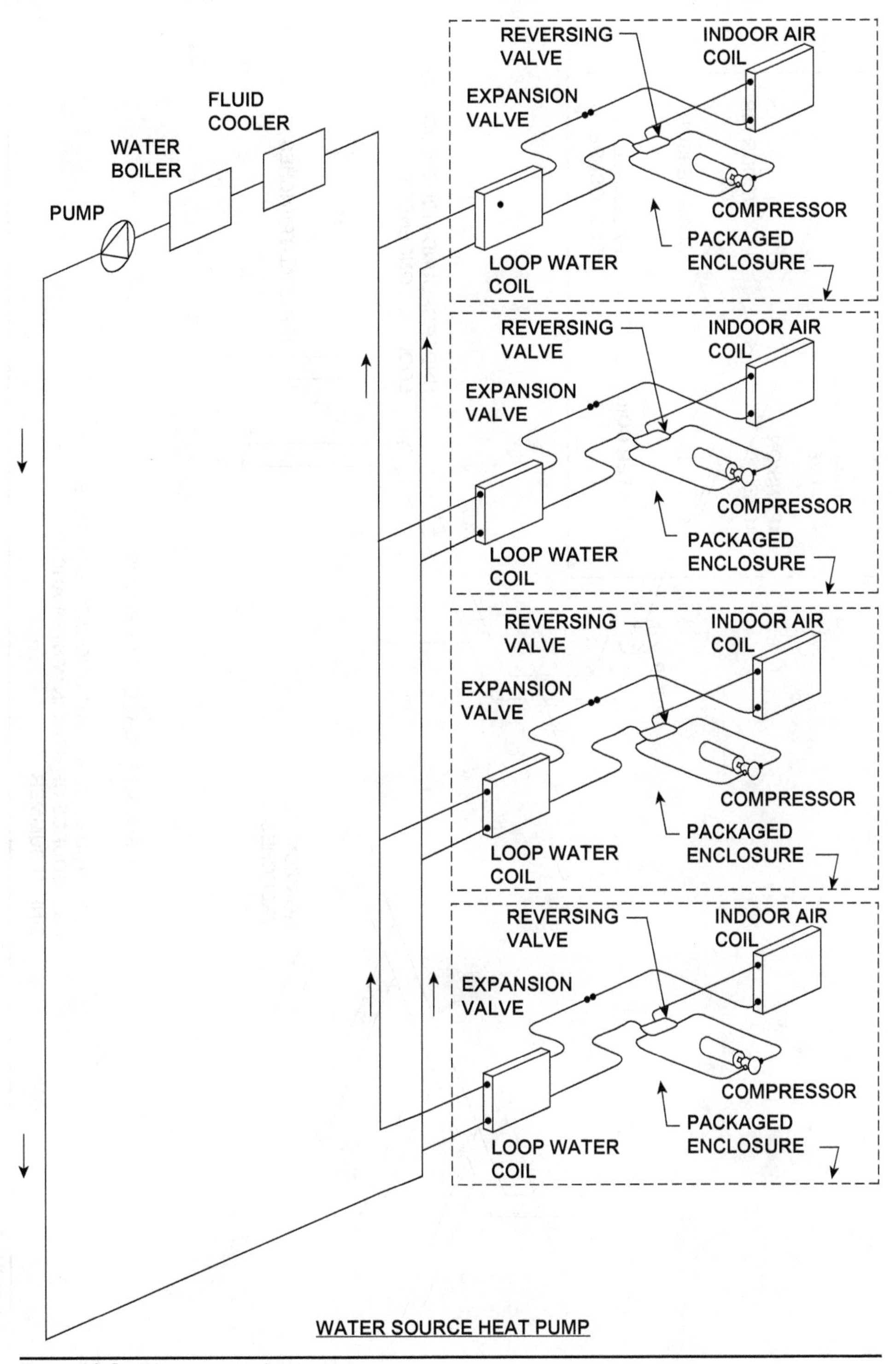

FIGURE 10.3 Water source heat pump.

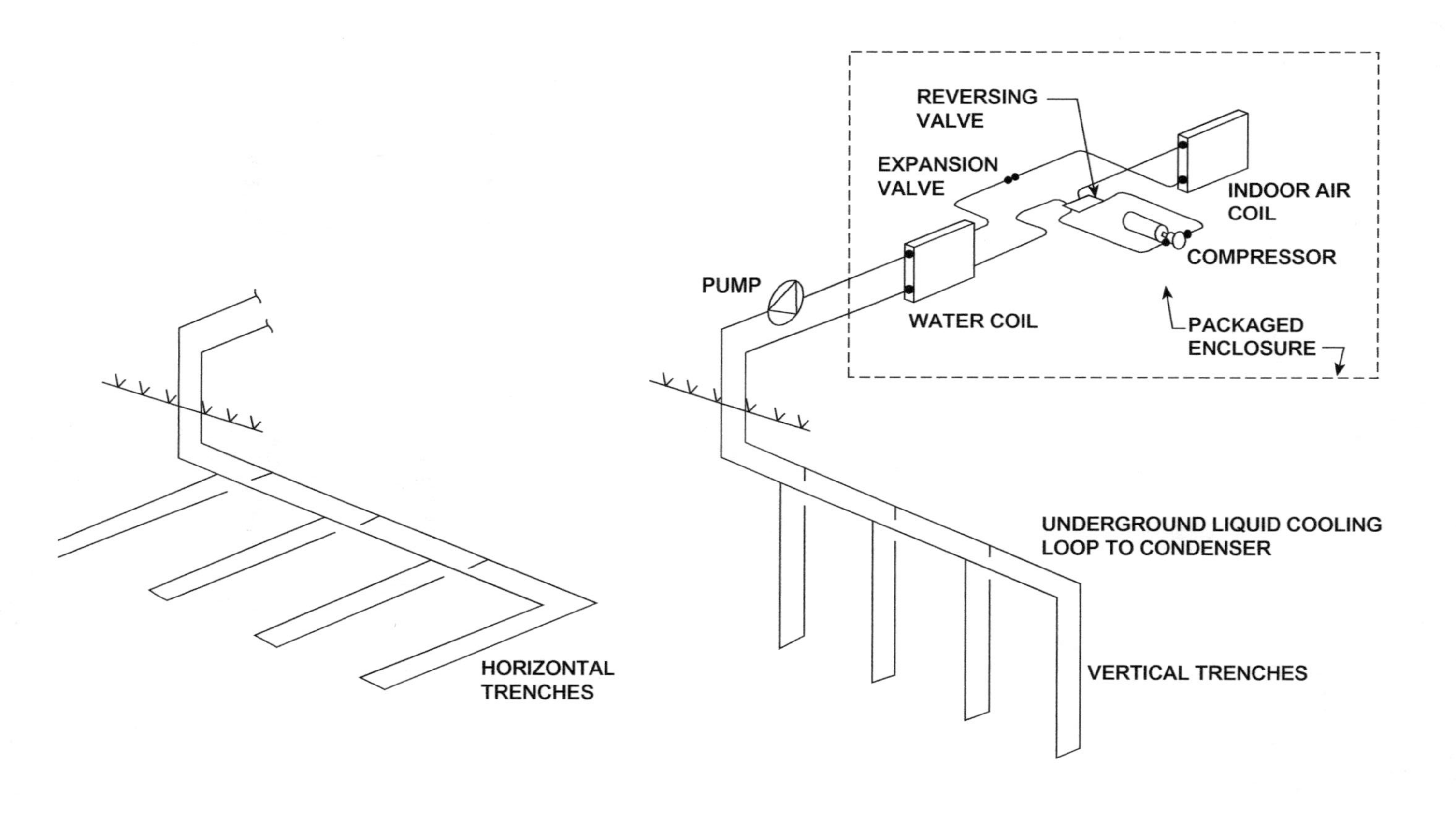

FIGURE 10.4 Ground source heat pumps.

same—water source heat pumps are connected to a circulating water loop, and either draw heat from the loop or reject heat to the loop. Instead of a boiler and cooling tower to add or reject heat from the circulating water loop, the relatively constant temperature of the earth is used. The energy savings are obvious, as boiler and cooling tower energy is eliminated or greatly reduced.

Chilled Beams and Perimeter Heating

Another means of using water to locate the cooling source closer to the space, and reducing energy transfer via air, is to use chilled beams. These units are chilled water coils, with air inlets and outlets, that are located in the room, typically flush with the ceiling. A small amount of outside air from a DOAS is supplied to the chilled beam—the supply air enters the beam and induces a flow of room air with it. This air mixture then passes over a chilled water coil and into the space. The reduction in air fan energy (less air quantity, less ductwork distance) and the elimination of reheat provide the energy savings.

Using water to transfer energy is also more efficient for heating. Perimeter wall heating systems save fan energy, and provide the source of heat right at the greatest need, instead of trying to ensure that supply air sufficiently covers the room with warmth. These types of systems, however, add installation cost, as they are additional equipment. Their use is best justified in cold climates, with a winter design temperature near zero or below, and long heating seasons.

Separate Steam and Hot Water Boilers

A current "paradigm shift" in hospital heating systems is the trend of using separate hot water boilers and small steam boilers. The "standard" design in the past typically used large steam boilers to provide steam for process needs like sterilization and humidification, and also used the steam to heat domestic hot water and heating hot water through steam-to-water heat exchangers. Heating and domestic hot water demand accounted for approximately 70 percent of a typical steam boiler system. The installed cost and operating costs of steam systems are significant, when the costs of a deaerator (used to remove oxygen from water going into the boiler), steam trap maintenance, and continuous chemical treatment are included. If the sterilization and humidification needs are met with a much smaller boiler, the ancillary equipment (deaerator, steam traps, chemical treatment) is also greatly reduced. The heating hot water and domestic hot water needs are met with separate hot water boilers, which do not require deaerators or steam traps, and have minimal chemical treatment requirements.

Variable Refrigerant Flow

Another new development that "decentralizes" the source of cooling is the variable refrigerant flow system. In the past, it was considered difficult and inaccurate to modulate the flow of a refrigerant in response to partial load. Direct-expansion (DX) systems using refrigerant to cool air (instead of using chilled water) were small units that were either on or off, or had a very few (3 or 4) discrete steps of unloading. Thus, precise constant space temperature control was difficult to achieve. The recent development of "variable refrigerant flow" (VRF), first extensively used in Japan, has changed that way of thinking. Now one condensing unit may be connected to seven or eight DX fan coil units in an area. The controls on the unit are able to modulate the refrigerant flow to each space fan coil unit, in response to heat load. Unlike the advances in water source heat pumps, VRF units cannot provide high levels of filtration and are applicable only to patient rooms or administrative areas.

Chilled Water Systems

The advances in digital control have enabled much more precise control of chilled water systems, which has had a corresponding result in energy efficiency. In the past, it was typical to have two loops of circulating water, a "primary" loop in the chiller plant (the purpose of which was to maintain constant flow through the chillers), and a "secondary" loop that distributed water to the air handling units and modulated in proportion to cooling demand. The current standard of control is variable primary flow, in which chilled water flow through the chillers is modulated in response to building cooling load, and no secondary loop is required (which eliminates the secondary chilled water pumps). A recent control strategy carries this concept even further. If the only chilled water pumps are at the chiller plant, then the pumps must be sized to provide flow to the farthest air handling unit. This is wasteful, as most of the units will not require this much pumping power, yet the pumps must operate at this pressure. If a "booster" pump is located downstream, and is sized for those units farther away, it can provide the power needed to reach those units, and the main pumps do not have to maintain a pressure higher than needed for the units closer to the chiller plant.

Heat recovery chillers are becoming more prevalent as a means to recover and use heat otherwise wasted. They use the heat normally rejected in a cooling tower to preheat domestic hot water or to provide heating hot water. These chillers are smaller (usually 30 to 300 tons of cooling capacity). Heat recovery can be accomplished with modifications to larger chillers through the use of heat exchangers.

Airside Energy Recovery

The single biggest cooling load in a hospital is the treatment of fresh outside air to make it suitable for use inside (cooling and dehumidification in warm weather, heating in cool weather). To keep the air balanced in a facility, the same amount of air is exhausted (from toilets, soiled linen rooms, laboratories, etc.). This exhaust air is 70 to 75°F and already dehumidified—it contains heat that could be used in the winter, and can be used to cool 90°F air in the summer, so various methods of heat recovery are becoming more prevalent with the increased emphasis on energy consumption.

Air-to-air energy recovery heat exchangers do just what their name implies—the outside air and exhaust air streams pass through adjacent metal paths that allow heat to transfer between them.

A more complex method of recovering heat involves the use of rotating energy wheels. The exhaust and outside air streams are located adjacent to each other, and a wheel rotates through both paths to transfer the energy between the air streams. The fill material of the wheel may only be sufficient to transfer "sensible" heat (a change of temperature), or it may be a desiccant material, which has the capacity to transfer "sensible" as well as "latent" heat. (Latent heat is the energy associated with a change in phase, such as from gas to liquid.) Desiccant wheels may be classified according to their use. They can be used for "total energy" transfer, or they can be used to lower the dew point of the supply air (a conventional cooling coil not only cools the air, lowering the temperature, but dehumidifies air, by lowering its dew point). Often, the indoor conditions for an operating room are 64°F or below. To maintain an adequate relative humidity at this lower temperature, the air must be cooled and dehumidified below what is normally obtained using standard chilled water temperatures. Glycol can be used and cooled below temperatures for chilled water, or desiccant wheels can be used to achieve the lower supply air dew point.

Sound Attenuation

The use of sound attenuation in hospital mechanical systems is increasing in response to the goal of improving the overall patient experience. Sound in a mechanical system is generated by fans, air movement through metal ducts, and air movement through the diffusers and grilles in a space. Generally, noise generated by fans and "duct rumbling" is in the lower and mid frequency ranges, and noise from outlets is in the mid to high frequency ranges. Typical methods of attenuation include special attenuators located downstream of fans and terminal boxes.

Lessons Learned from Commissioning

Mechanical systems today are much more complex than 20 or 30 years ago. Digital control systems enable more complicated sequences of operation to be implemented. Technological advances in equipment provide better performance, but require more knowledgeable operation and maintenance. In response to increased complexity of systems, and to assure the owner that these systems are operating as intended, the activity of commissioning became a necessary activity in the design and construction of a project. Commissioning is a detailed and well-documented start-up of the systems, but the lessons of commissioning can be incorporated early in the design phase. As commissioning agents have gained in-depth knowledge of system operation, their knowledge has led to implementation of such control strategies as static pressure resets on fans, temperature set-point resets on air and water, and variable control of outside air quantities, all of which are intended to reduce energy consumption as close as possible to the real-time building heating or cooling load.

New Electrical Trends

Attention to energy efficiency and awareness for the total patient experience have led to new paradigms in lighting and power systems.

The energy codes have become more stringent in terms of lighting energy, with specific limits for watts per square foot of installed lighting. Fixtures are selected and placed to obtain the maximum lighting benefits for the power consumed. It is now standard practice to use occupancy sensors to turn off lights when a space is unoccupied; sensors can be placed in each room or "sweeps" can be used to reduce lighting in a larger area.

For many years, the size of the hospital's emergency generator was determined mainly by the code-required emergency services—life safety egress lighting, critical branch receptacles used in patient care, and equipment branch power to keep ventilation fans and heating pumps operating. It is not uncommon now for facility administrators to want cooling systems (chillers and associated pumps) available if the utility power is out, as well as lighting and power availability in noncritical areas for patient and staff comfort and convenience. The desired available time period for standby power has also increased above the code-required 96 hours, as some facilities may want a week or more of emergency power available, to be prepared for extended outages due to natural disasters.

One impact of this increased load on the emergency generator system is the increased motor load on generators, which can increase the harmonic distortion of the electrical system. As a result, closer attention must be allowed in specifications of the electric generator.

Uninterruptible power systems (UPs) are used to maintain power to electronically sensitive equipment in the 10 seconds between the loss of utility power and the start up of the emergency generators. Medical equipment (MRI, linear accelerators, CT scans) and larger data systems often have a need for UPs. Improvements in UPS technology have resulted in the placement of smaller systems place near the point of use, instead of the older, larger, centrally located UPS.

Facility Energy Supply and Demand Options

The new means and methods described above mostly deal with systems inside the building and the systems directly affecting the occupied spaces. There have also been new developments in the energy supply systems for facilities.

Many electric utilities have rate structures in which the price of electricity varied with time of day and/or the season, in response to their relative costs of supplying electricity. When the demand on their system grid is low, they can run their large base load generating stations (usually coal or nuclear), and meet all demand. These generating facilities are usually their most efficient stations, based on fuel used or the economy of scale of operating a large plant. As the demand on their grid increases, the utility may need to start other generating plants, which may not be as efficient for several reasons—fuel used, older, less efficient equipment, smaller scale plants, etc. If demand exceeds their generation supply capacity, they must buy power from other utilities at the current market cost, which is the most expensive time to buy it. Therefore, utilities pass on their variable costs in the form of rate schedules, based on time of day or season. Hospitals can sometimes implement strategies to take advantage of this variation in electric costs, by shifting their power draw from daytime hours to night, or by using their emergency generators to generate part of their electric needs, and reduce the electricity bought from the utility.

One method of "shifting" the electric load is thermal storage. When applied to hospitals, it typically involves the use of ice tanks and chillers that are capable of the low temperatures necessary to freeze water. The ice chiller runs at night, builds up ice, then the chilled water system uses the ice to provide cooling during the day. Thus, chiller electric energy is "shifted" from expensive daytime hours to less expensive times at night. A unique feature about thermal storage at a hospital is that the low temperatures available from an ice system can be used to maintain lower space temperatures in operating rooms (some surgeons prefer that the OR be kept at 64°F or lower, which is difficult or impossible to achieve using conventional chilled water ranges).

Since most hospitals have emergency generators to supply electricity in the event of power loss from the utility, those generators can sometimes be used to supply power to the facility during times of expensive utility power. This practice has several names—peak shaving (reducing the peak electric draw), load limiting, load shedding, or demand side management, but the activity is the same, running the generators. There are variations as to whether the generators run in parallel with the utility power or whether discrete quantities of the electrical load are taken off the utility power, and transferred to the generator. One caveat of running the generators in this way is that there are limits imposed by EPA on maximum allowable hours of runtime, when not being used for utility power outages. If the engine runs more than 100 hours/year for maintenance, testing, and other reasons, such as peak shaving, etc., the engine will be subject to more stringent emissions standards.

Operating an engine-driven electric generator produces significant amounts of heat, which is typically rejected through the radiator and exhaust systems. When this heat is captured and used, it is called *cogeneration* (the cogenerating of electrical and thermal energy from a common fuel source). The exhaust system heat is hot enough to generate steam that can be used in sterilizers or for heating. The heat recovered from the radiator system is low-grade heat (less than 250°F), and is useful for supplying heat to the heating hot water or domestic hot water systems. Cogeneration requires additional equipment, in the form of heat recovery boilers and associated pumps and controls, so there must be a payback in the form of reduced utility energy consumption to justify the added equipment cost. The system must be sized to match the simultaneous heating and electric needs. The economics of cogeneration must be proven by analysis of the heating and electrical loads for a particular hospital, using the local utility rates. The most accurate way is to use daily electric and thermal load profiles (preferably for each month), which will give a picture and quantify benefits of the cogeneration system.

It is important to note that most standby generators are diesel engines. However, using diesel fuel to generate electricity is relatively expensive, and most engines used for cogeneration are natural gas-fired or are bi-fuel (diesel and natural gas). Gas and bi-fuel engines are more expensive than diesel engines, so this premium must be included in the economic analysis.

Case Study: Methodist Olive Branch Hospital

MOBH is a new community hospital in Olive Branch, MS. The hospital is designed for an ultimate build—out of 220,000 square feet and 100 beds, although the initial occupancy will be 60 beds. Design started in July 2011, early construction site work began March 2012, final construction documents were issued May 2012, and construction completion is scheduled for July 2013. This project is an application of new developments in both project delivery and mechanical system layout.

New ways of thinking began at the outset of the planning stage. The owner issued invitations for teams to submit proposals for a new hospital based on general program requirements (number of beds, department programs typical of a community hospital, approximate floor area, and a target budget), and requested that teams design and construct the project using the IPD process. When challenged by the successful team that, to get the most value from the IPD process, they should be willing to think outside the box themselves, the owner agreed to broaden their vision by focusing on overall facility costs, not just first costs, and by being willing to look at nonstandard approaches to system design. The result is a hospital that uses HVAC systems not typically considered for hospital use, but that provide the lowest cumulative costs over a 7-year time period.

Collaboration between designers and contractors began at the initial planning stage. In developing a schematic plan, a "brainstorming" session was held to identify all ideas—no option was dismissed at that point. Those ideas were discussed, examined, vetted, and researched to come up with five possible HVAC systems before the schematic phase was very far along. Contractors on the team priced the different systems, and considered not only the mechanical equipment costs but also associated general and electrical costs. The cost estimates were much more detailed than typical schematic phase pricing. Would one option require less building floor space or more floor to floor to heights? Would another option have more or less electrical distribution associated with it? As the five options were being priced, Turner Construction, the construction manager, developed operating and maintenance costs for each option, and SSRCx, the energy consultant, projected annual energy costs. At the end of the schematic phase, the five options had been narrowed to two—ground source heat pumps or water source heat pumps. This intensive examination of system options meant that the schematic phase was longer than typical, lasting almost 6 months.

Once into design development with the two options, communication and coordination continued to refine the options. A team attitude meant that the project was not "mine" or "yours" but "ours." As an example, the screen walls around the rooftop equipment were priced at $1 million. Led by the owner, the team decided the $1 million could be better spent elsewhere, perhaps in finish upgrades or specialty lighting.

To produce the final construction documents, the plan was to have the mechanical contractor draw the construction document ductwork and piping plans. The idea was that, since mechanical contractors usually prepare their own drawings in a CADD system that automates fabrication, the time spent by the mechanical designer to prepare construction documents could be avoided. This did not work as planned. One problem was that the architect, the mechanical designers, and the contractors all used different CADD platforms. Trading files and trying to see each other's work proved difficult. Additionally, the designers were drawing at ⅛-inch scale and the contractors were drawing at ¼-inch scale, another impediment to transferring and coordinating files. The lesson learned was that, going forward, the designers should prepare traditional two-dimensional design development plans to a level suitable for permitting, show more design intent, but not every coordinated detail, and then let the mechanical contractor start with those plans to develop their drawings, adding necessary detail for coordination and fabrication.

In comparing the experience of this IPD project to a traditional delivery method using separate design and construction teams, the team members offered several insights. Because it is a true team effort, selection of discipline contractors must be made not on low price, but on the total value they bring to the effort. This value is expressed in a collaborative attitude and the ability to work in unconventional ways, expertise and the willingness to share it, and commitment to the team goal of the best project for the owner. Designers found they did more work in schematic design phase than in a traditional project delivery. In the construction phase, team members report that there is the same amount of effort in construction administration, but a different type of effort. There are fewer RFIs (requests for information), but more informal discussions. Designers and contractors still evaluate new ideas and options as opportunities arise in the course of construction.

Five months before occupancy, the team was optimistic that the owner will be satisfied that the team met the project goal of getting the most value for the money.

Conclusion

With healthcare decision makers being asked to do more with less, hospital construction projects begin with a question: What is the best value available? The answer is derived from identifying and evaluating the options, choosing the option that best meets the specific goals of the project, and planning the process to meet those goals. The "standard way of doing things" may not be the best way. Value comes from being flexible and open to innovative ways of thinking.

The construction manager is in a position to guide the owner in a direction to obtain the most value for construction dollars spent. Since mechanical and electrical systems are a major portion of the project cost, a working knowledge of these systems is necessary to determine the right direction.

CHAPTER 11

Medical Technology and Information Systems

Perhaps no area in healthcare has experienced greater change, nor had a greater impact on construction in recent years, than technology.

Technology is advancing at mind-numbing speed. We see the advancement in our everyday lives. *Smartphones* have replaced traditional cell phones, as cell phones are rendering wired "landlines" obsolete. Neither our homes nor our businesses can operate without computers. Laptop computers have all but replaced desktop PCs. Music players like the iPod and others have replaced CD players that replaced traditional turn tables only a few years earlier. Now, online cloud services, coupled with ubiquitous availability of wireless broadband connectivity and smartphone music "apps," are eliminating the need for music players. High-definition flat panel displays have replaced cathode ray picture tubes, offering up to 20 times the resolution, while requiring a fraction of the electric power, and producing a fraction of the heat load. Tablet computers, like the Microsoft Surface and Apple iPad, are now replacing laptops for many applications, while opening opportunities for new applications through dramatically improved portability.

Increasingly, services are available online. Online banking has made check writing all but obsolete. Global positioning systems (GPSs) have replaced printed roadmaps, and now smartphone apps are starting to replace dedicated GPS units. In the past, we carried blueprints while walking the construction sites—now we carry a tablet computer that accesses the full range of construction documents through wireless connectivity.

The Internet has become our primary route for communication of all types, and certainly our primary source for information. Although not yet accepted as a valid source in academia, Wikipedia, still a free service supported by voluntary user donations, has all but made hard copy encyclopedias obsolete for day-to-day casual research. One study claimed that any print media more than 90 days old could no longer be considered accurate. We routinely share information around the world—print media of all kinds are rapidly disappearing—smart walls and surface computing threaten to replace bulletin and chalk boards in schools, rendering pen and paper nearly obsolete.

This chapter was written by Terry Miller.

Why Discuss Technology as Part of Hospital Construction?

What are the implications of this amazing rate of technology development for healthcare planning, design, and construction?

First, it is important to understand what "technology" is. Technology in healthcare falls into two broad categories: building systems and medical technology.

Building systems primarily include fire alarm and building automation/energy management systems, which are planned and designed as parts of the electrical and mechanical disciplines. Energy management systems are frequently integrated with utility companies, to track power usage and utility rates, and tied to other building systems, to automatically close blinds or darken active glass panels on sunny days, or to open louvers and lighten glass during cloudy weather.

Fire alarm systems are integrated with security and energy management systems to capture the picture of someone activating a fire alarm, while closing doors and shutting down nearby fan units in response to the potential fire danger. Similarly, security systems are integrated with building systems to stop elevators with doors open during infant security alarms, or to redirect cameras to capture the picture of an intruder attempting to enter the pharmacy or medication room after hours.

Pneumatic tube systems are integrated with security systems limiting receipt of carriers with sensitive materials, such as lab samples or medications, to only the correct staff, identified by presenting an authorized security card. For many years, nurse call systems have been integrated with televisions, electric blinds, and low-voltage light controls, providing patients easy control of those devices from the handheld pillow speaker or controls built into the bed side-rail.

In the future, we will undoubtedly see increased deployment of smart building elements, like active glass canopies and skin systems to control heat transmission and absorption to and from the outside. All of these technologies will require network connectivity for interface between systems, and connectivity through the Internet for tracking energy and water cost and usage, for remote maintenance, or even for ordering spare parts, as part of the facility management system.

A significantly larger and faster-growing segment of technology is what we call *medical technology*. One could easily find many definitions for the term "medical technology"—probably no two alike. For the purposes of this chapter, we will define medical technology as follows:

Medical Technology: The systems and equipment utilized in the process of delivering patient care.

More than information technology (IT), medical technology includes the full range of systems and equipment that healthcare staff and clinicians touch and use in the process of delivering patient care. In essence, medical technologies are the "tools of the trade" in healthcare delivery.

Traditionally, medical technology has been treated as three separate disciplines: medical equipment, low-voltage and communication systems, and IT. Today, successful healthcare project delivery requires a unified approach to these three disciplines. The primary reason for treating these as a single discipline during construction is the extensive integration required for these technologies to be installed and function effectively from a clinical and operational standpoint.

Unlike building technologies that create and maintain the environment of care, medical technologies are the backbone of care delivery itself.

To a large extent, healthcare delivery is an information-based exercise. Medical information is the raw material upon which clinicians base their diagnoses and care planning. Unlike the past, when most information was gathered manually through physical observation, touching, timing, and listening, today most medical information is generated or captured by medical technology. It is medical technology, from highly sophisticated imaging systems to laboratory analyzers to smart diagnostic sets, that clinicians use in diagnosing conditions, planning, tracking care, and treatment.

Unlike building systems that primarily rely on networks to communicate with their own components or directly related systems, medical technologies communicate with each other feeding information into, and utilizing information from, patients' medical records. To the casual observer, the variety and complexity of that medical information appears endless, with new technologies generating new data formats almost daily.

Medical information can vary from simple numbers and values captured by a diagnostic set at the bedside or exam table to the hugely complex files generated by physiological monitors, high-definition imaging systems, surgical video systems, etc. All must be captured, frequently converted to a format compatible with the EMR systems, secured to ensure patient privacy, and backed-up in equally secure, preferably off-site, storage locations.

Some medical information, such as patient monitoring and surgical video, must be accessed in real time for immediate use as part of care delivery. Most medical information is captured and stored for use in tracking patients' care progress, billing purposes, and insurance claims, or, in some cases, for protection from litigation. More broadly, information is captured and stored for research purposes directed at improving both the quality and efficiency of healthcare delivery.

Clearly medical technology, and the necessity of integrating all of the technology into a single-functioning care delivery platform, presents a significant challenge for hospital planning, design, and construction.

One study conducted by a medical technology planning firm revealed over 200 technologies routinely found in healthcare facilities today. More than half are commonly referred to as "medical equipment," and another third referred to as "special systems." Taken together, medical technology can account for one-third of an overall project budget.

Virtually all of these technologies generate information pertinent to the diagnostic and treatment process—much of which must be captured in the patient medical record.

Technologies falling into the category of "low voltage" systems, although providing the physical backbone through which other technologies communicate, in reality represent a small fraction of the total medical technology picture. Examples include nurse call systems, intercom, public address, closed circuit video, television distribution, and the structured cabling system that provides the basic infrastructure through which most, but not all, technologies connect.

Given the high degree of technology integration, with multiple systems and equipment requiring interoperability, the facility data network, traditionally regarded as part of the IT discipline, today has become a utility shared by many technologies, the majority of which are medical equipment, not IT.

The challenge for healthcare facility planning and design is understanding all of the technologies required to support patient care delivery, and understanding how these technologies must work together to efficiently support patient care delivery. Equally important, is understanding the impact these technologies have on the planning, design, and construction process.

The challenge for healthcare facility construction is to understand how and when these many technologies must be finally selected, purchased, installed, and integrated to deliver a working care delivery platform—and to manage this process recognizing that the majority of medical technologies are traditionally owner furnished.

What Is Medical Equipment—and How Is It Changing?

Traditionally medical equipment has been planned in three groups

1. Group 1—fixed equipment; equipment physically attached to the building
2. Group 2—movable equipment requiring dedicated building services, such as dedicated power, medical gas, or plumbing connections
3. Group 3—portable equipment that might be used anywhere in the facility

Some of the major equipment types and typical locations are explored in the section that follows.

Imaging, Image-Guided Procedures, and Surgery

Nowhere is the advancement of technology more apparent than in healthcare, and imaging and surgery represent some of the most sophisticated of the changes. Complex imaging technologies like CT (computed tomography) and MRI (magnetic resonance imaging) have become commonplace. Digital radiology has replaced older film-based modalities. Newer technologies like PET (positron emission tomography), PET/CT, SPECT (single photon emission computerized tomography), and PET/MRI are becoming common in hospitals. All of these modalities are digital, generating images stored in PACS (picture archive and communication systems), and connected through IT networks for download of patient information and upload of clinical information to the patient record.

Interventional radiology/special procedure rooms, including angiography and cardiac catheterization modalities, have become commonplace throughout healthcare. Since anesthesia is often utilized during these procedures, spaces are required nearby for patient preparation and recovery. Although invasive, these procedures are not considered surgery, therefore, environment and utility requirements are less stringent than for surgical suites. Video streams generated during these procedures are captured in PACS. Similar to other imaging modalities, systems are connected through IT networks.

Typical imaging modalities are identified below along with their common use and construction considerations:

Radiographic/tomographic systems (Rad/Tomo). Rad/Tomo is the backbone of imaging, the basic imaging modality, seen in every hospital, commonly used for capturing images of skeletal structure, as well as abdominal radiography. The system requires a dedicated room that must have lead-lined walls and doors to prevent radiation exposure, as well as a protective view window, for staff to safely monitor the patient during the radiographic procedure. A substantial rail and structural support system must be provided above the ceiling to carry the equipment, and to allow for movement of the overhead radiographic tube. These systems require significant electrical power and, frequently, a dedicated transformer to ensure power purity and stability. A typical Radiographic suite is shown in Fig 11.1.

Radiographic/fluoroscopic (Rad/Flouro). Rad/Flouro is commonly used for visualization of hollow organs (i.e., esophagus, stomach, small and large intestines, and bladder),

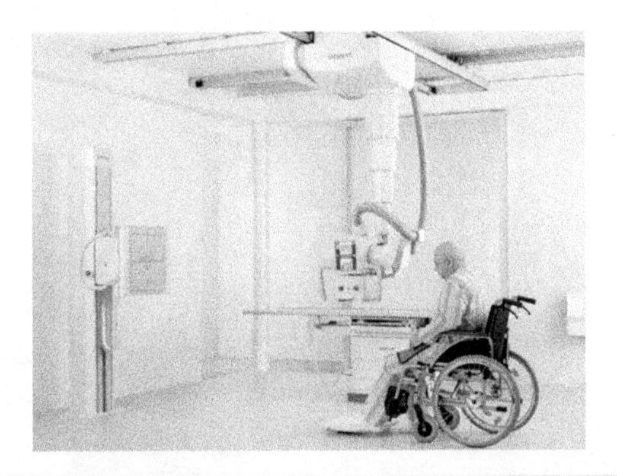

Figure 11.1 Radiographic suite—Siemens Yios. (*Courtesy of Siemens Healthcare.*)

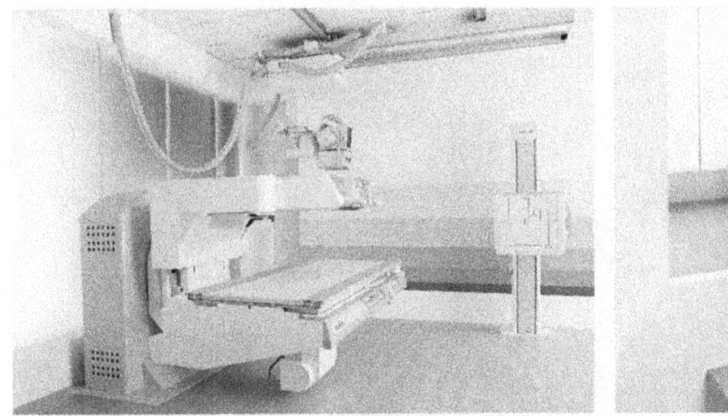

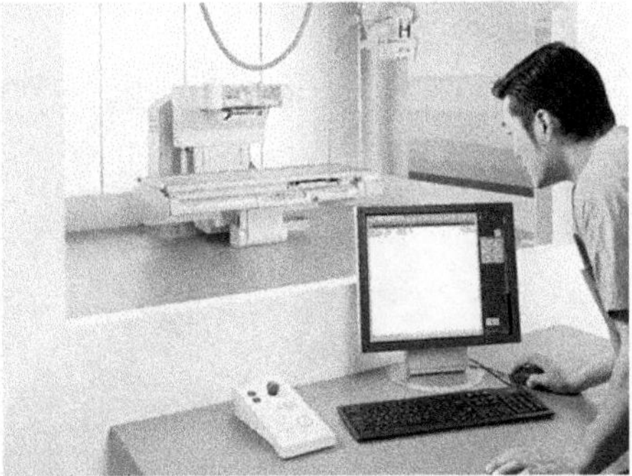

Figure 11.2 R & F modality—Siemens Luminos. (*Courtesy of Siemens Healthcare.*)

other soft tissues. The system is capable of capturing both still and moving images, allowing medical professionals to see movement of organs and fluids, to show internal flows for digestive and other functions. Similar to the Rad/Tomo room, lead-lined walls, a viewing window, overhead structural support, and a dedicated clean electrical power source is required. Visualization of the patient from the control room through a viewing window is required at all times during the exam, especially when the table is angulated to the vertical (standing) position. Figure 11.2 shows a R & F modality and control room.

Mobile x-ray. Two types of mobile x-ray systems are commonly found in hospitals. mobile C-arms and mobile radiographic systems.

Mobile C-arms can easily be moved around the facility, as needed, to capture either still images or moving fluoroscopic studies. These are frequently used in surgery, the emergency department, and intensive care units. A 110-volt power outlet is required at the point of use, and space is required for storage when not in use. Images are typically captured in the unit, then uploaded to PACS through a network connection at the point of use or the storage location.

Mobile radiographic systems provide radiographic visualization of the lungs and skeletal structure. Unlike mobile C-arms, where use is typically confined to a specific

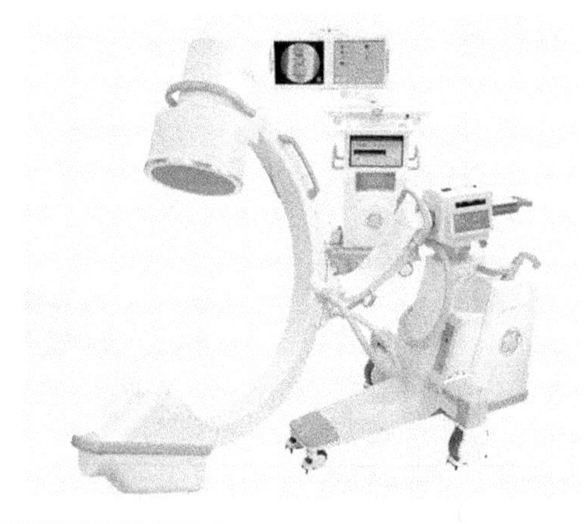

Figure 11.3 Mobile C-arm—GE Healthcare. *(Courtesy of GE Healthcare.)*

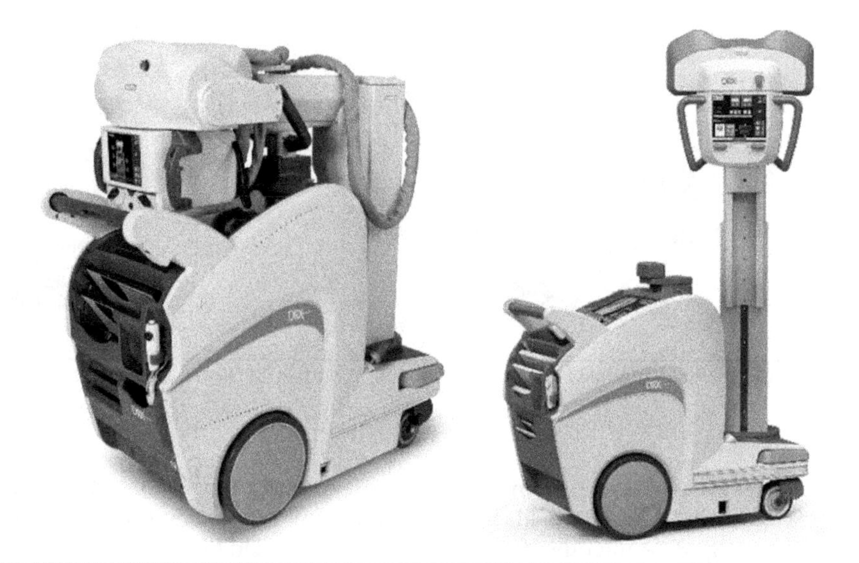

Figure 11.4 Mobile radiographic system—Carestream DRX. *(Courtesy of Carestream Health.)*

department, such as surgery, a mobile radiographic systems may be used anywhere in the hospital (i.e., ICU, NICU, CCU, patient rooms, emergency room, etc.), hence they typically are motorized allowing them to be "driven" to the point of use. System elements usually collapse or fold down for ease of movement through doorways and other tight spaces, and to provide improved visibility for the driver. Most systems in use today are digital. Images are wirelessly transmitted through the hospital network into the PACS system for radiologist interpretation. Common mobile systems are shown in Figs. 11.3 and 11.4.

Computed tomography—"CT scanner"—also called computed axial tomography or "CAT scanner." This modality uses a spiral approach, with a movable table, to capture radiographic images or "slices" of the body, as the patient is moved through the gantry. "Tomos" in Greek means slice. Once multiple slices are captured, the CT scanner uses digital geometry

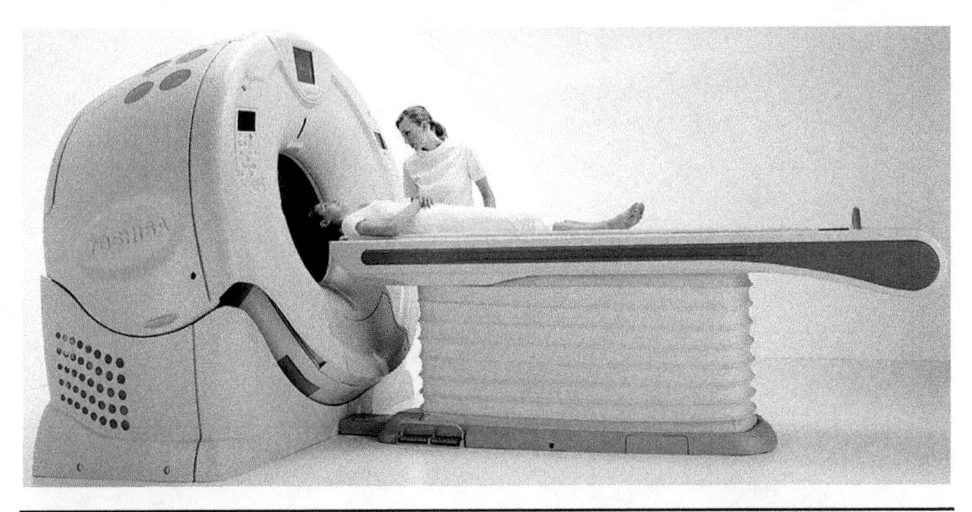

FIGURE 11.5 CT scanning system—Toshiba Aquilion ONE. (*Courtesy of Toshiba America Medical Systems.*)

processing to generate a three-dimensional image of the body. The three-dimensional image is assembled from multiple two-dimensional x-ray images or slices taken around a single axis of rotation.

CT is often used for organs such as the liver and brain, and for the cardiovascular system. Separate control and computer rooms are required, with dedicated power and cooling for the computer. Because x-radiation is used to create the images, the procedure room must be shielded. Similar to most imaging modalities, an observation window is required for staff to view patient from the control room. Closed-circuit video cameras are frequently placed on the opposite side of the gantry, with a monitor in the control room, to provide staff continuous viewing as the patient passes through the gantry. Figure 11.5 shows a typical CT scanner with the movable table.

Nuclear medicine—"Nuc Med"—positron emission tomography—"PET." Nuc Med and PET refer to imaging modalities that employ injection, inhalation, or ingestion of radioisotopes. Unlike other imaging modalities, where radiation is produced outside and then passed through the body to create an image on a receptor placed on the opposite side, Nuc Med and PET utilize radiation sources introduced to the body as described above. Different radioisotopes are used, depending on the targeted organ or system of interest in the body. For example, the thyroid gland needs iodine for production of hormones. During a thyroid study, the patient is given a radioactive iodine compound, which then accumulates in the thyroid. Different isotopes can be utilized to provide visualization of specific body areas, such as the gallbladder, liver, thyroid, lungs, heart, or brain. Physiological functions can be viewed in addition to anatomical detail.

Although PET scans can be used to diagnose a number of health conditions, such as heart disease and Alzheimer's, most are currently used for oncology diagnosis. Nuc Med and PET rooms may require lead shielding, and both will require a dedicated high voltage power circuit. One type of nuclear medicine gamma camera is shown in Fig. 11.6.

Positron emission tomography/CT scanner—"PET/CT." Coupling the capability of PET to visualize physiological functional with the ability of CT to identify precise anatomical structure provides a powerful diagnostic tool. Similar to CT, images are captured in

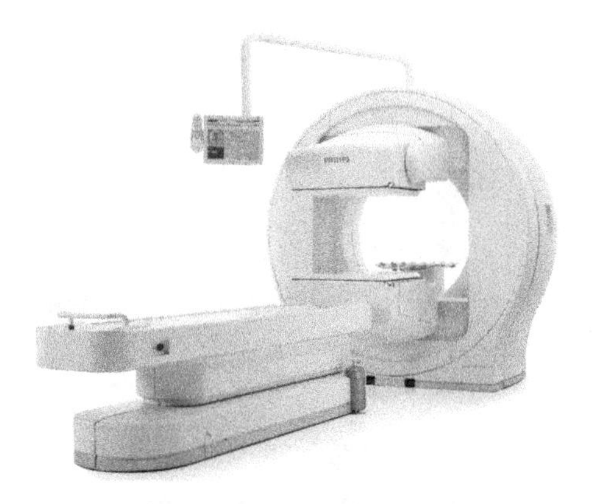

Figure 11.6 Gamma Camera—Philips Brightview. (*Courtesy of Philips Healthcare.*)

Figure 11.7 PET/CT scanner—Philips Brightview XCT. (*Courtesy of Philips Healthcare.*)

slices, then reconstructed using computer analysis to provide three-dimensional images for diagnostic purposes. Figure 11.7 shows PET/CT scanner equipment.

Magnetic resonance imaging (MRI) uses a powerful magnet producing low-intensity radio frequency pulses, and computer technology, to create detailed images of the soft tissues, muscles, nerves, and bones in the body. Nuclei of the body's hydrogen atoms are momentarily drawn into alignment using a powerful magnet. When the magnetic field is removed, radio waves are produced and emitted as the atoms realign to their natural positions. The scanner gathers data produced by the atoms, movements, and a computer is used to create images from the data. MRI scanners can create pictures of nearly any part of the body. Even parts of the body that are surrounded by bone may be seen with an MRI scan, so it is useful for imaging the spinal cord and brain. Because of MRI's inherent flexibility, systems are frequently placed near the emergency department, in order to easily access them. MRI is also used for cardiovascular studies.

MRI systems are large and heavy—hence they are frequently placed on the ground floor of the building, near an outside wall. When placed on an upper floor or below grade, care should be taken during design and construction to ensure adequate size and structure are provided along the path of travel. Often it is easier to remove a large section in the wall, called a *knock-out panel,* to move the machine as required for installation or replacement. We have also seen systems inserted through an opening in the roof with the use of a crane.

MRI requires ducting to the exterior, for exhaust of gases from the cryogenic liquid used to cool the large super conducting electromagnet. If the cooling system fails, the liquid helium used to cool the magnet turns to gas, in a near explosive fashion—therefore, it must be quickly exhausted to the exterior. This causes an unplanned shut down of the magnet, called a *quench.* There is also a danger that helium gas can be released into the room, where it can displace oxygen, causing unsafe conditions for patients and staff. For this reason, oxygen monitors and exhaust fans are also required.

Another danger with the MRI is the strength of the magnet; it can literally pull metal objects through an open door, slamming them into the bore of the magnet. Injuries can easily result to patients, staff members, or even construction workers—as well as damage to the MRI itself. In one documented case, a child was killed by a ferrous oxygen gas cylinder being pulled into the machine's core. In another case, a floor cleaning machine was slammed into the magnet damaging the machine, and injuring the worker. Care must be taken in design to avoid inadvertent entry into the MRI procedure room. Metal detectors should be placed along the path of travel, providing an alarm if metallic objects approach the room entrance. Door frames are now available with built-in metal detection. Security card readers can further limit access to authorized personnel only. Routing access to the MRI procedure room through the control room provides an extra level of safety for both patients and staff.

MRI shielding typically employs copper mesh to create a "clean" RF environment. Mechanical, electrical, and data/communication connections entering the room must pass through waveguides, required to maintain the integrity of radio frequency (RF) shielding. Occasionally, additional magnetic shielding is required to avoid interference from nearby moving ferrous structures (e.g., elevators, dumbwaiters, etc.).

Similar safety and shielding issues occur with movable intraoperative MRI systems, where the magnet is temporarily moved into the surgical field for the capture of images, then moved back to an adjacent storage "garage." In cases where an intraoperative MRI is shared between two or more surgery suites, structural design must create an identical ferrous signature for each room served. Figure 11.8 shows a 3.0 Tesla MRI.

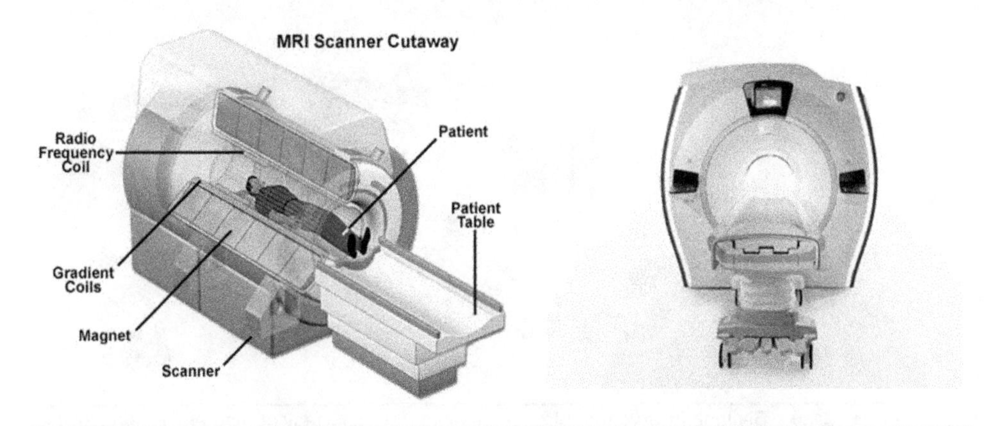

FIGURE 11.8 MRI—GE Discovery 3.0 T. (*Courtesy of GE Healthcare.*)

Catheterization labs/interventional cardiology and angiography. Catheterization labs are typically used for cardiac and other diagnostic and treatment procedures. Interventional radiology is minimally invasive, supporting targeted treatments that are performed using imaging guidance. They are often used rather than open heart surgery, for example, as they are less risky, the recovery times are much shorter, and they are effective for many conditions. Some interventional procedures, such as angiograms, are for diagnostic purposes, while others, such as angioplasty, are therapeutic. Some rooms are set up as electrophysiology (or EP) labs, which study the electrical functions related to the heart and other physiological processes.

These rooms have similar features to imaging rooms and operating rooms. There must be radiation protection and overhead rail systems, as in other imaging rooms, but there also must be seamless flooring and features to support a clean environment, as these are minimally invasive procedures. Typically, a separate space is needed, with power and air-conditioning for the computer hardware needed to support these modalities. Figure 11.9 shows Bi-plane interventional suites with monitors on booms and on rails.

Other diagnostic imaging systems. Smaller diagnostic systems include mammography, ultrasound, and bone densitometry. These may be mobile or fixed, but typically do not require the specialized rooms and support spaces of the more sophisticated imaging systems discussed earlier. Ultrasound does not require shielding, nor does bone densitometry. Mammography, on the other hand, does. Dedicated power and data connectivity are required for each modality. Figure 11.10 shows typical mammography unit installations.

Cancer treatment. In some cases, a hospital may have specialized systems and equipment dedicated to cancer treatment, which require significant space and planning. One of the most common is the linear accelerator (or "LINAC"), which uses radiation beams to target and destroy the cancer. These are very large systems, requiring specialized rooms with thick concrete walls and steel doors, to provide protection from the radiation. Because staff cannot enter the room during a treatment, video cameras are used for visualizing the patient on displays placed in the control room.

Another cancer treatment system finding increasing acceptance is proton therapy. These are unique, complex, and expensive facilities utilizing a cyclotron to accelerate nuclear particles targeted at the cancer. Currently, these are less than 50 such systems in use around the world, yet interest continues to grow. Structure and video needs are similar to a LINAC.

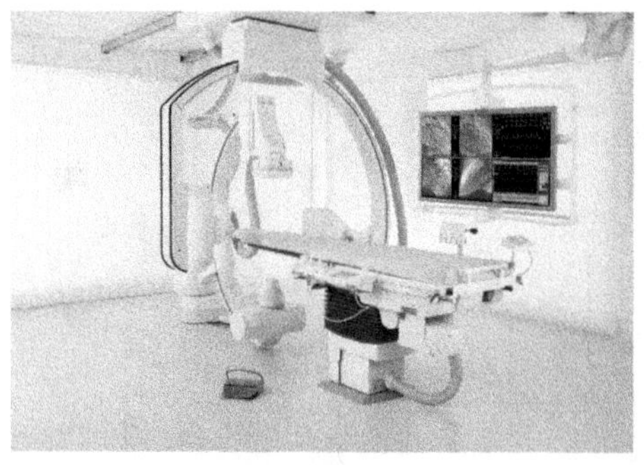

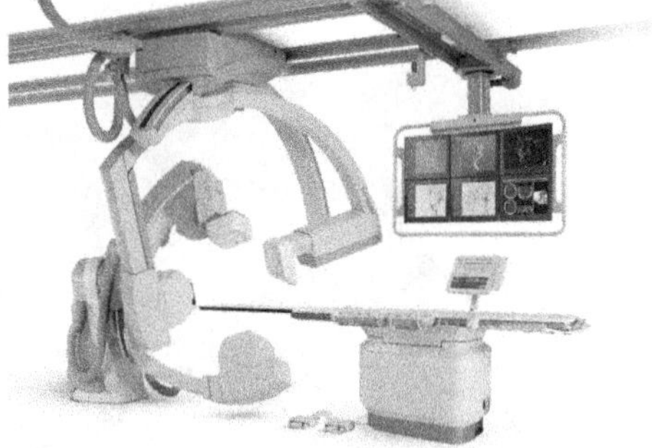

FIGURE 11.9 Bi-plane interventional suites—Artis Zee (*courtesy of Siemens Healthcare*) and Allura XPer (*courtesy of Philips Healthcare*).

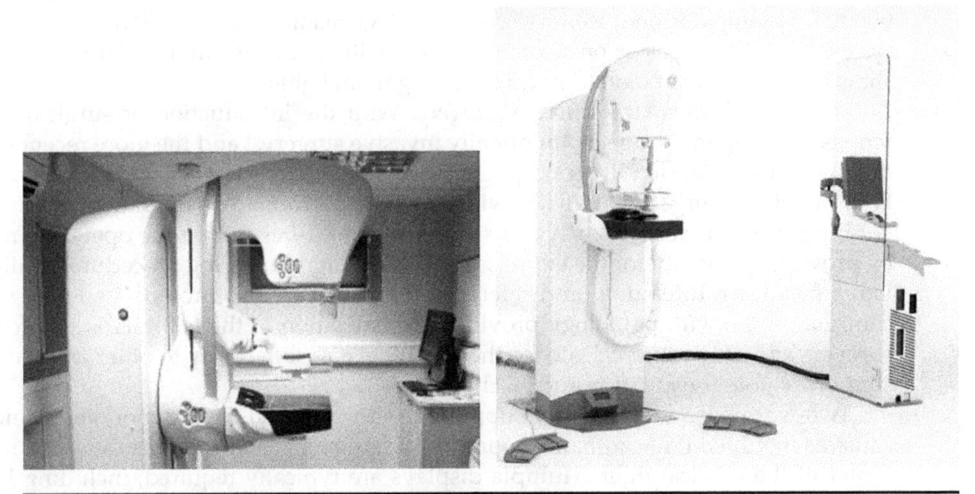

FIGURE 11.10 Mammography—GE Senographe. (*Courtesy of GE Healthcare.*)

Another system commonly used for cancer therapy is the Gamma Knife. These are similar to traditional x-ray systems, but are only used for therapy and not for diagnostic purposes. Radiation protection in the walls and doors is required, but not the extensive rail system.

With all imaging and treatment modalities, careful coordination is required with the manufacturers, to provide the appropriate mechanical and electrical services to the rooms. Typically, these services, such as power and air-conditioning, must be operational prior to the vendor installing the systems, so careful scheduling is required.

In some cases, the component weights are so great that additional structural support is required. Consideration should be given as to how to physically move the imaging or treatment equipment to the room, as it is often large, may not fit through doors, or may be too heavy to move through corridors.

Requirements for lead shielding vary depending on department layout and the modalities deployed. Design teams should employ the services of a radiation physicist to determine the need for, and amount of, shielding required.

Typically, imaging department rooms must be finished with floors, walls, and ceilings complete before equipment installation can begin. Similarly, rooms must be secured; with equipment values well into six and seven figures, inadvertent damage resulting from inappropriate entry can prove costly.

Surgical booms and lights. Surgical lights, anesthesia columns, and equipment booms are required, in varying quantities and configurations, depending on the types of procedures to be performed, and the equipment to be supported and connected. Perfusion booms may be required for such procedures as open-heart surgery.

Despite the industry-wide trend to move equipment off the floor and onto booms, equipment is still frequently mounted on movable carts, moved into the room when needed, and positioned to support the specific procedure to be performed. In addition to providing support for a variety of equipment, booms also provide connection points for movable equipment, including network connections, oxygen, medical air, suction, and nitrogen outlets, as well as multiple electrical outlets with dedicated circuits. Care should be taken to check for possible interference between technologies colocated on a common boom. For example, in one case, an electrophysiology manufacturer would not support their system's function if placed on a common boom with an electro-surgical unit. Figure 11.11 shows boom mounted monitors and operating room lights.

Surgical video/minimally invasive surgery. With the introduction of surgical video scopes ushering in the age of "minimally invasive surgery," and the more recent introduction of the "Hybrid OR" concept, surgery today is experiencing significant change. Four out of five surgeries, industry wide, are now performed on outpatient basis.

Integrated surgical video systems support procedures within the operating room, by providing visibility for the entire surgical team, and capturing procedures in digital video files for future study and reference. Systems can also provide two-way video communication with pathology providing visualization of the lab microscope for the surgeon, and of the surgical site for the pathologist. Connectivity to other facilities can support remote consultation and teaching.

Two-way video is frequently supported with wall and ceiling mounted cameras, cameras mounted on head-bands worn by the surgeons, and cameras mounted in the center of the surgical light. Multiple displays are typically required, including large, wall-mounted screens, and smaller movable displays suspended on flexible arms extending from the ceiling-mounted boom and light hubs. As part of procedure setup, displays are moved and positioned, as necessary, to provide the entire surgical team with easy visibility.

A computer-based video control station is typically placed on a documentation desk or workstation, from which surgical circulating nurses control the video displays, while documenting the procedure through a computer connected to the EMR system. A third PACS workstation may be required to access images needed to support the procedure. Direct line-of-sight visibility is required between the surgeon and the

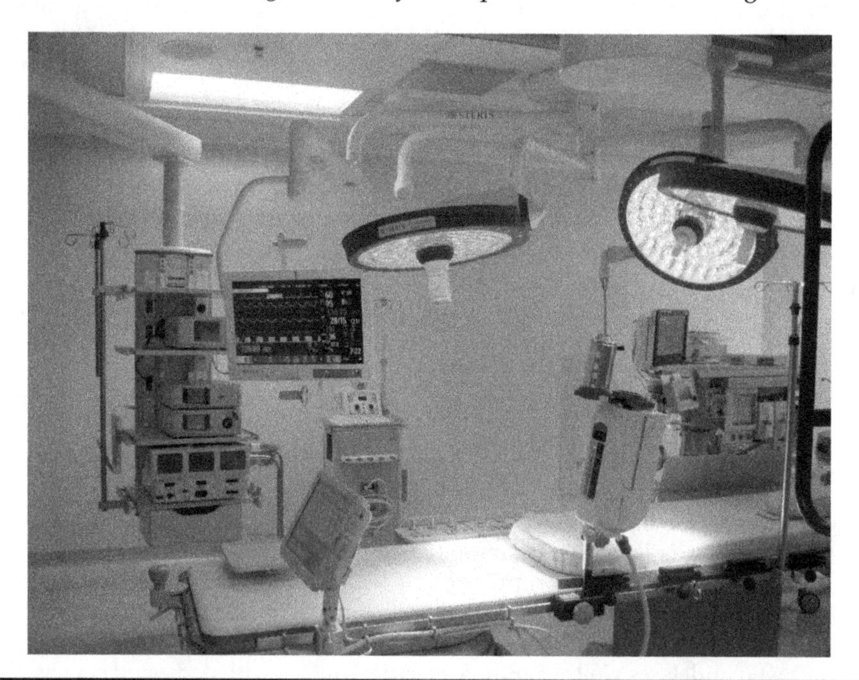

FIGURE 11.11 Surgical lights, equipment and anesthesia booms, and monitor arms. (*Courtesy of Steris.*)

documentation station, to ensure that continuous eye-to-eye contact is maintained with the circulating nurse.

The hybrid OR. Although the utilities, structural needs, and preparation/recovery space requirements of surgery and image-guided procedure rooms have always been similar, these services have traditionally been placed in different areas and departments within facilities, largely because of the different clinical disciplines utilizing the spaces. More recently, image-guided procedure rooms are being built fully equipped and conditioned to support "open" surgical procedures. The term adopted in the industry for these combined spaces is the "hybrid operating room" (or hybrid OR).

Surgical procedures frequently have been supported by imaging studies, previously captured, and viewed during the procedure through PACS workstations. More recently, images have been captured during the procedure using portable "C-arm" x-ray units or intraoperative MRI systems moved into the surgical field for image capture, providing "near real time" images.

The hybrid OR concept takes imaging support to the next level, by providing true real time imaging for guidance throughout the surgical procedure. Today, hybrid operating rooms represent the highest density of technology (and technology cost) found in hospitals. Although a relatively new concept (the first was built at Vanderbilt University Medical Center in 2005), most medical center construction projects today include one or more hybrid rooms. Many facilities are now renovating existing rooms to include hybrid capabilities. Figure 11.12 shows a "hybrid OR" with the imaging equipment in place over the surgical table.

Considerable space is required for a hybrid OR—typically 1200–1500 square feet—including equipment and control rooms. The procedure room itself must frequently house teams of 20 or more staff and clinicians. Similar to traditional surgery suites, hybrid

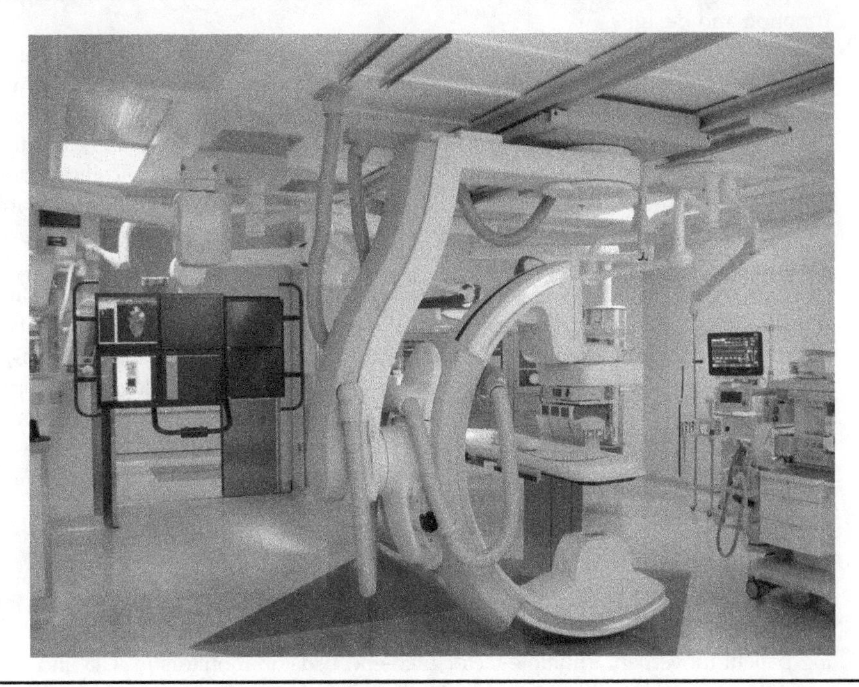

Figure 11.12 Hybrid OR. (*Courtesy of Steris.*)

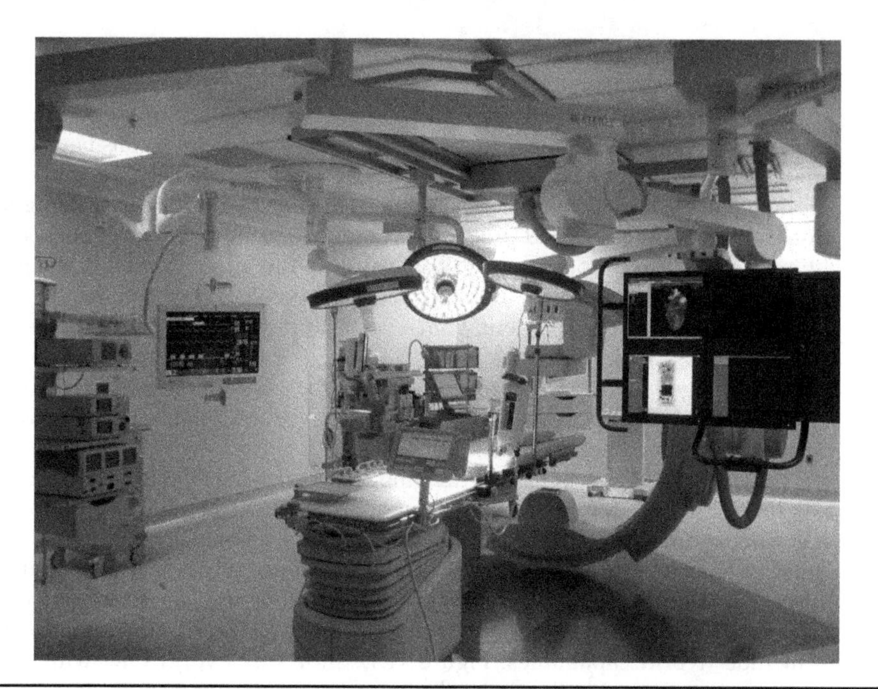

Figure 11.13 Complex ceiling coordination in hybrid OR. (*Courtesy of Steris.*)

rooms are highly customized to support the specific types of procedures to be performed. Frequently, two or more rooms planned within a single facility might vary significantly in function and design.

Imaging, image-guided procedures, surgery, and hybrid surgery rooms all present a considerable challenge to the facility designer, including significant structural, HVAC, electrical, and space requirements, as well as special efforts to coordinate above-ceiling structure and utilities. Figure 11.13 shows the large amount of ceiling mounted equipment required in a "hybrid OR." Increasingly, pre-packaged above-ceiling structural support systems are finding application in these areas, to preserve design flexibility while equipment selections are finalized. Figure 11.14 shows one type of ceiling support system used in operating rooms.

Due to the high level of technology and technological integration required in these rooms, efforts to finalize early selection of system vendors for major elements, including the imaging modality, booms, and lights, and integrated surgical video systems, will prove valuable during design. Although clinicians typically prefer to delay technology decisions as long as possible, to ensure selection of the most current products available, a well-planned and managed procurement process can accelerate the selection process, while ensuring implementation of the latest state of the art products on the first day of clinical use.

Patient Monitoring/Telemetry

Patient monitoring has become a staple of care throughout hospitals. Modalities include relatively simple vital signs monitors found in observation rooms, transport monitors used during patient movement, multiple-factor high-end bedside monitors in ICU and CCU, mobile telemetry monitoring utilized in step-down units, seizure monitors in neurological ICU, fetal

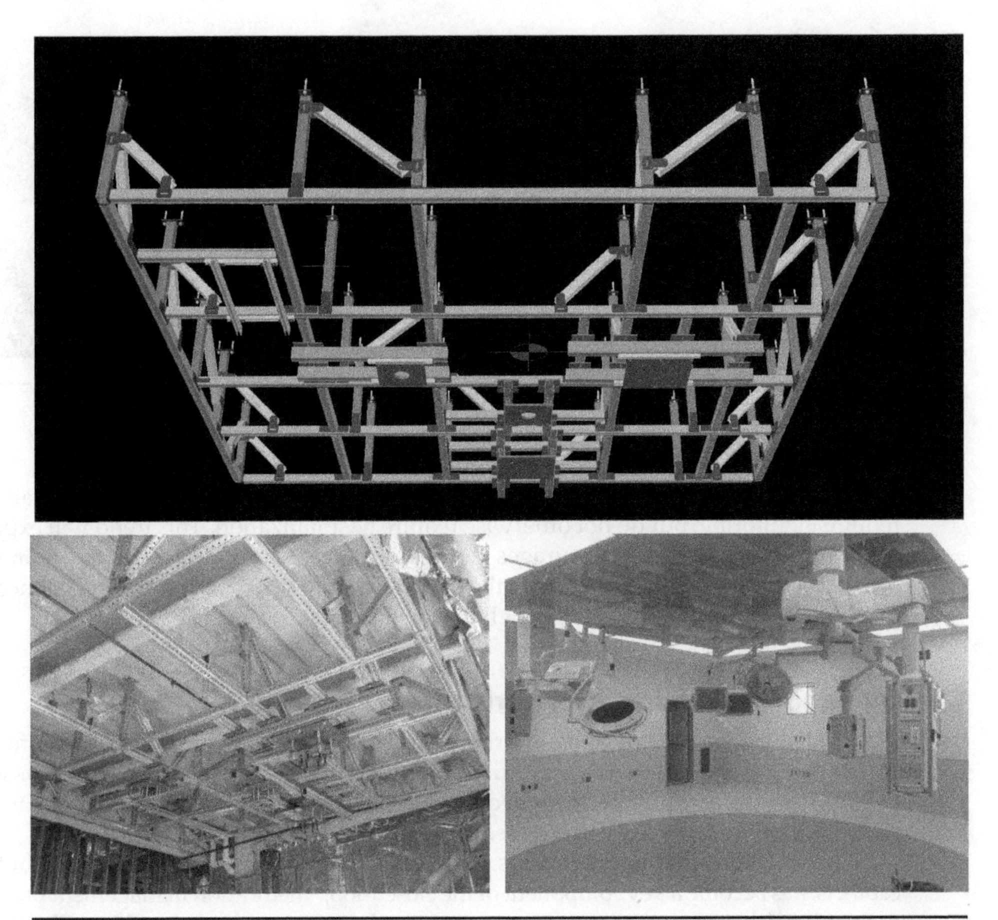

FIGURE 11.14 Above-ceiling structural support—Hilti MI/MQ Strut System. (*Courtesy of Hilti.*)

monitors in ante partum and labor/delivery, and the many specialty monitors found in sleep labs and other specialty treatment areas. Monitoring is utilized and is essential in virtually every patient care area of the hospital. Monitoring systems today are digital, communicating through the facility network to central monitors located at nurse stations, or dedicated monitoring rooms, and to file servers located in the data center. In some cases, central monitoring for an entire hospital is located in a single room where the most highly trained monitor technicians can be located. This is commonly known as the "war- room" concept (see Fig. 11.15).

One industry trend we should expect to see more of is the "eICU" concept, where monitoring for multiple remote facilities is centralized miles away in a single monitoring center. A high level of network integrity is required between the facility and the central monitoring site, with redundant pathways for both monitoring signals and instant communication between central monitoring technicians and facility staff. In this way, smaller facilities, lacking the specially trained monitor technicians and volumes necessary to justify a full time central monitoring workstation, can still provide quality intensive care patient monitoring.

FIGURE 11-15 Central Monitor/"War Room." (*Courtesy of Philips Healthcare.*)

Another trend involves deploying remote monitors in the home when patients require monitoring, but do not otherwise require hospitalization. This trend will expand as the Affordable Care Act encourages more cost-effective care delivery. Mobile monitors are also becoming available, worn by the patient wherever they might go, and connected to a central monitoring station through their cell phones.

Clinical Support—Laboratory, Pharmacy

Automated tracks, controlled by information systems, have transformed the laboratory by connecting multiple analyzers with computerized physician order entry systems (CPOE), to automatically perform necessary tests, while sending results directly to the patient's electronic medical record (see Fig. 11.16).

Similar automation in pharmacies, coupled with automated medication distribution cabinets, has become a key component of the closed-loop medication management process, required to demonstrate "meaningful use" of the electronic medical record (see Fig. 11.17).

A variety of hoods are required in both lab and pharmacy to address specific needs such as bio-safety, preparation of chemotherapy drugs, etc. Recently, facility adoption of the latest USP 797 regulations for hospital pharmacy design has created the need for more dedicated spaces and specialized equipment.

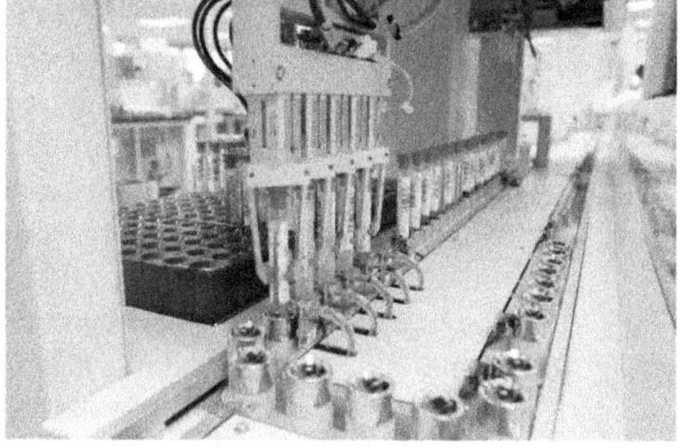

FIGURE 11.16 Laboratory track system. (*Courtesy of Beckman Coulter, Inc.*)

Figure 11.17 Medication Distribution System. (*Courtesy of Omnicell.*)

Central Sterile Supplies

Although CS equipment is primarily supported by the mechanical, electrical, and plumbing systems, even here, technology is gaining utility. Cart washers (see Fig. 11.18) and sterilizers are digitally controlled, network connected to monitor activity and function, and increasingly are automated, ensuring proper operation, while reducing department labor requirements. Surgical instruments with embedded micro chips are becoming available, providing automatic tracking and inventory control through wireless networks. Surgery scheduling systems are becoming a basic tool for managing activities-related equipment and materials control.

Similar to larger imaging modalities, site preparation and planning the path of travel for cart washers and other large sterilization equipment should be considered early in construction. Equipment delivered early must be secured and protected.

Integration with Electronic Medical Records

Driven by the government mandate for hospital compliance before January 1, 2015, the roll out of electronic medical record systems has become the number one target of capital spending in healthcare. While the information gathered in medical records has always been the backbone of medicine, the movement to electronic record-keeping promises to provide vastly improved access to information for speedier, more accurate diagnoses on the individual patient level, and improved access for research and analysis at the community and national level.

Recognizing that the majority of information stored in patient records is captured and generated by medical technology, such as medical equipment, devices, and systems,

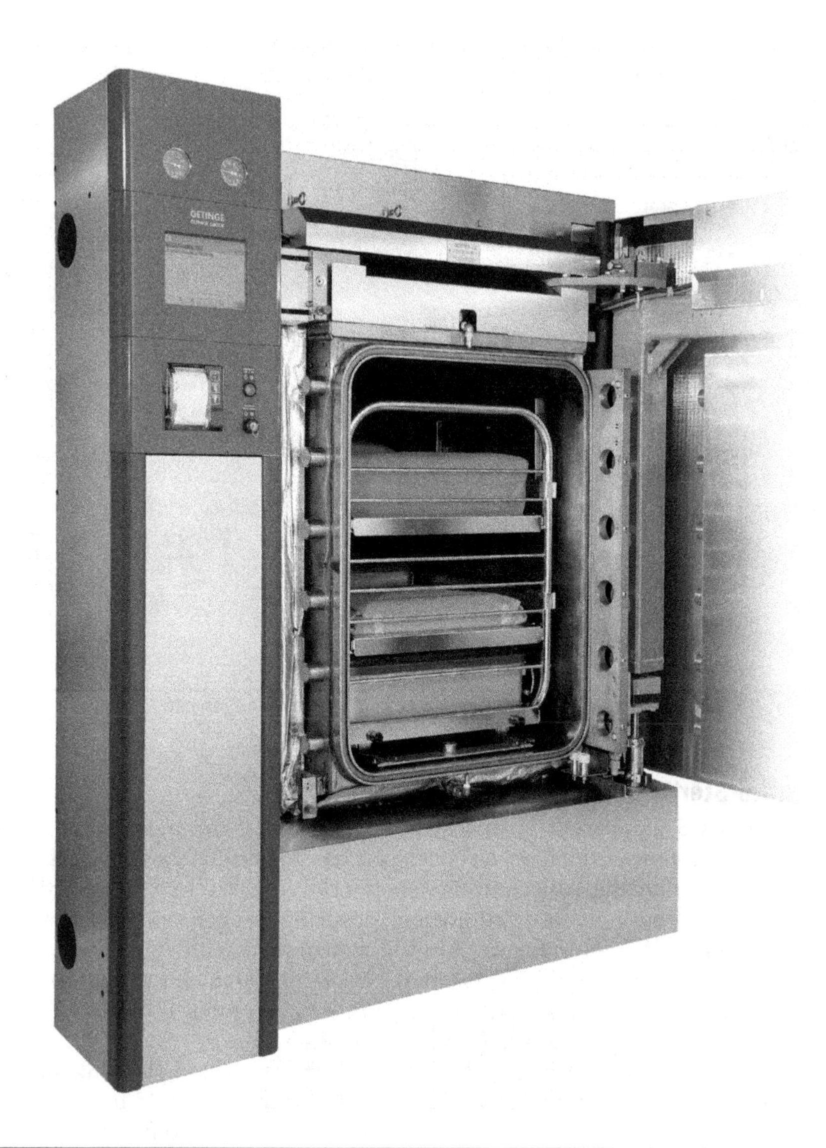

Figure 11.18 Cart washer. (*Courtesy of Getinge.*)

the integration of these technologies to efficiently and accurately communicate information on the correct patient has become a critical element of any successful EMR roll out.

Low-Voltage Systems

What are low-voltage systems, how do they support other medical technologies, and what is their role in supporting care delivery? Low-voltage systems fall into four categories: infrastructure, data network, communication systems, and security systems, as follows

Low-voltage infrastructure—the "structured cabling system": Structured cabling systems (SCSs) consist of passive infrastructure, primarily intended to support the hospital's information, telephone, time and attendance, dictation, security, nurse call, and other communication systems. Additionally, the structured cabling system supports patient monitoring, telemetry, imaging systems, PACS, pharmaceutical and materials distribution systems, laboratory instruments, surgical and pathology scopes, and other medical equipment that communicate using standard data cabling and network protocols.

Where it is appropriate, building systems (including energy management, fire alarm, synchronized clock, pneumatic tube control, medical gas, and other alarms), that communicate using standard cabling and protocols, may be connected through the SCS infrastructure. Although most technology systems today can utilize the structured cabling systems for connectivity, there are still a number of systems requiring proprietary or dedicated cabling systems. It is worth noting that, in many jurisdictions, the fire alarm system must be supported through a separate, dedicated cabling system.

The SCS consists of station cable, outlets, patch panels, termination blocks, copper and fiber riser cable, and associated hardware.

SCS outlet locations typically include all staff workstations, offices, workrooms, conference rooms, medical equipment locations, each bedside and family area in patient rooms, waiting and lounge areas, and charting workstations in and near patient rooms.

For planning purposes, most facilities establish standard SCS outlet configurations for use in various applications, such as workstations, equipment locations, patient monitors, and other key spaces. Similar to power outlets, the exact complement of equipment and devices that might be connected at any one location will vary in daily use, therefore, standard outlet configurations should have multiple connections that can be flexibly assigned for any device. Devices that might be connected at any one outlet might include telephones, fax machines, PCs, printers, scanners, and/or medical instruments.

The variety and mix of devices that might be connected at any one outlet may vary between locations, but will rarely exceed a total of four. Where more than four devices are located at one position, the counter or desk space required for that number of devices is large enough that a second voice/data outlet should be provided. Given the variation in connectivity requirements, the following outlet configurations should be considered:

- Workstation outlets, including all offices, unit secretary, lab and pharmacy work positions, nurse charting positions, dictation cubicles, and other work places should be provided with multi-position single gang outlet plates configured with the quantity of voice/data jacks (see Fig. 11.19) called for by the facility IT standard. Unused jack openings should be blanked-off, hence available for future.
- Bedside and patient room workstation outlets will be similar to general workstation outlets described above.
- Similar outlets with two voice/data jacks should be placed adjacent to copiers and medical equipment requiring data connectivity—to provide for primary network connectivity and secondary connection to a telephone line.
- Patient monitor outlets should also be equipped with two voice/data jacks—providing primary connectivity and a secondary connection for backup.
- Wall telephone outlets should be provided with a single data jack and integral wall telephone hangers.
- Outlets should be placed above ceiling at regular intervals to support wireless network access points. Density of wireless access points has increased over time as more devices have come to rely on wireless connectivity. Several telemetry

Figure 11.19 Multi-position voice/data outlet.

> manufacturers recommend placement of wireless access points on 30-foot centers. Outlets should be equipped with a minimum of two jacks, to allow separation of clinical systems, such as telemetry, from other wireless systems. Outlets should be installed with a minimum 20-foot service loop to permit fine-tuning of access point locations to attain adequate coverage.

Data cable speed and capacity has continued to rise over time, in response to higher network speeds and increased information volume. The current industry standard is Category 6A, which supports gigabit ethernet connectivity to the desktop. The industry standard data jack configuration is the RJ 45 eight connector modular jack.

Data station cables should run directly from each jack to the nearest local technology equipment room (see Fig. 11.20) (also commonly known as a *communications room* or *comm closet*, *data closet*, *Tel/Data*, or *IDF room*), and terminated on 48-port patch panels. Patch panels typically are mounted in 77-inch-high, 19-inch EIA standard data racks, provided with cable management above and between racks, to allow neat and orderly placement of the multiple patch cords required for connectivity.

For ease of identification and safety, cables from patient monitor outlets should be of a unique color, and terminated on dedicated patch panels, mounted in a separate data rack dedicated for monitoring and related medical equipment. Most facilities have established their own color coding standards for cabling, typically blue for data cable, yellow for patient monitoring, or something similar.

In general, fiber-optic station cables and outlets are not required; however, they should be provided where specific equipment or system requirements call for fiber connectivity. The best approach is to have them terminated on dedicated rack-mounted fiber patch panels in the nearest technology equipment room.

Placement and sizing of technology equipment rooms is critical to technology infrastructure design. Data cable lengths are strictly limited to 90 meters (approximately 295 feet). However, that total cable length must include run distance up and down walls,

Figure 11.20 Technology equipment room.

over and around columns, beams, and other structural elements, and run neatly along the corridor walls. Given these factors, a good "rule of thumb" is to plan no cable run to any location served by a technology equipment room longer that 80 meters (approximately 262 feet), when measured from the outlet to farthest point in the technology equipment room. A simpler guide is to place Tel-Data rooms so that no space served is greater than a 150-foot radius from a technology equipment room.

Technology equipment room placement should also take system configurations into account. For example, all patient monitors on a unit should be served from a single technology equipment room. Similarly, many integrated surgical video systems require all components to be served from a single technology equipment room.

Wherever possible, technology equipment rooms should be stacked for ease of cable installation during both initial construction and future upgrades. If practical, at least two technology equipment rooms should be placed on each floor in order to provide a redundant riser cable pathway.

Technology equipment rooms should be sized to house all medical technology control equipment required for the area served by the room, plus space for growth. Equipment placed in technology equipment rooms will include rack-mounted copper and fiber patch panels, network switches, and control equipment for medical equipment and low-voltage systems. Some medical equipment and low-voltage systems may also employ wall-mounted power supplies and control equipment.

Copper riser cable is typically terminated on wall-mounted termination blocks, and then cross-connected to patch panels located in equipment racks. This arrangement allows both for direct connections to dedicated telephone lines when needed, and cross connection via patch cord to station cables serving work spaces.

Where allowed by the IT department's protocols, building management and fire alarm systems panels should also be placed in technology equipment rooms, for ease of connection to the network and telephone lines; however, it should be noted that most facilities frown on this, as the staff and system vendors supporting building systems frequently are unaware of the protocols required to safely work around medical technology.

A study should be performed, prior to designing the technology equipment rooms, to determine the sizes required for station and riser cable terminations, network equipment, medical equipment, and low-voltage systems, and should allow for future growth and build-out of any shelled spaces. Room layout should consider ease of installation where multiple technologies must cross-connect, and ease of access for maintenance as well as for future upgrades/replacements.

Technology equipment rooms serving strictly nonclinical areas, such as administration and business offices, can be sized slightly smaller, but care should be taken to allow for future addition of clinical services. The following provides typical lists of systems and equipment that must be supported, in both clinical and nonclinical area technology equipment rooms.

Vertical copper and fiber riser cabling from each technology equipment room should be terminated in a main technology equipment room, which can also serve as the facility data center (or file server room), and minimum point of entry (MPOE) for cables coming from the Telco ("Telephone Company"), cable TV service provider, and/or other network service providers. Riser cabling should include both multipair copper and multistrand fiber-optic cable.

Although most telephone service today is digital VoIP (Voice over IP—or Internet protocol, the language of networks) served through the network, some amount of copper riser cable (typically 25 to 50 pairs) should be installed to each technology equipment room, to support systems and equipment that require direct connection to Telco lines. Typically, these would include telephone lines for the fire alarm systems, elevator telephones, and maintenance lines for other systems and equipment.

Typically fiber-optic riser cabling should include both *multimode* and *single-mode* fiber cabling to serve the variety of technologies that must be supported. Strand counts should support both initial build-out and considerable future growth. A study should be performed prior to specifying fiber-optic cable strand counts and types, including both the network requirements and the dedicated fiber requirements of medical equipment and systems. It is not safe to assume that all communication will pass through the network. It should be noted that some medical equipment is FDA approved with only specific cable types. Where two or more technology equipment rooms are placed on a floor, a similar strand-count of fiber-optic cables should be run between the rooms to serve as a redundant pathway in the event primary connection to the main technology equipment room is interrupted.

Main technology equipment room sizes vary widely, depending on the facility, the complement and size of systems and equipment to be housed, whether the space serves as a primary or secondary data center, future growth and upgrades, and the potential for future build-out of shell space. An itemization of typical technology equipment room supported systems is shown in Fig. 11.21. Figure 11.22 shows a technology equipment room serving approximately 15,000 square feet. A study should be performed, prior to commencing with design, to determine the configuration of anticipated information and

Systems/Equipment	Areas Served	
	Clinical	Non-Clinical
Structured Cabling System Terminations	Yes	Yes
Data Network Equipment and Terminations	Yes	Yes
Wireless Network (Wifi) Equipment and Terminations	Yes	Yes
RTLS Equipment and Terminations	Yes	Yes
Telephone Equipment and Terminations	Yes	Yes
Nurse Call Equipment, Power Supplies, and Terminations	Yes	
Intercom Equipment and Terminations	Yes	Yes
Public Address Equipment and Terminations	Yes	Yes
Dictation/Transcription Equipment	Yes	Yes
Radio/Wireless/DAS Equipment and Terminations	Yes	Yes
Television Equipment and Terminations	Yes	Yes
Security Equipment, Power Supplies, and Terminations	Yes	Yes
CCTV Video Equipment, Power Supplies, and Terminations	Yes	Yes
A/V Equipment, Power Supplies, and Terminations	Yes	Yes
Infant Security Equipment and Terminations	Yes	
Patient Monitoring Equipment and Terminations	Yes	
Telemetry Equipment and Terminations	Yes	
Imaging Modality Equipment and Terminations	Yes	
PACS & Tele-Radiology Equipment and Terminations	Yes	?
Surgery/Pathology Video Equipment and Terminations	Yes	?
Intergated Surgical Video Equipment and Terminations	Yes	?
Teleconferencing/Tele-Medicine Equipment and Terminations	Yes	?
Pharmacy/Supplies Distribution Terminations	Yes	?
Laboratory Analyzer Equipment and Terminations	Yes	?
Uninteruptible Power Supplies for Above	Yes	Yes

Figure 11.21 Technology equipment room—systems typically supported.

other systems required. A typical list of systems and equipment to be housed in the main technology equipment room might include

- File servers and host computers
- WAN equipment for data connectivity to the outside world
- Internal data network equipment (core network switches)
- Telephone system equipment
- Dictation system recorder
- Cable television head-end equipment
- Interactive television systems equipment
- Nurse call control equipment
- Intercom exchange

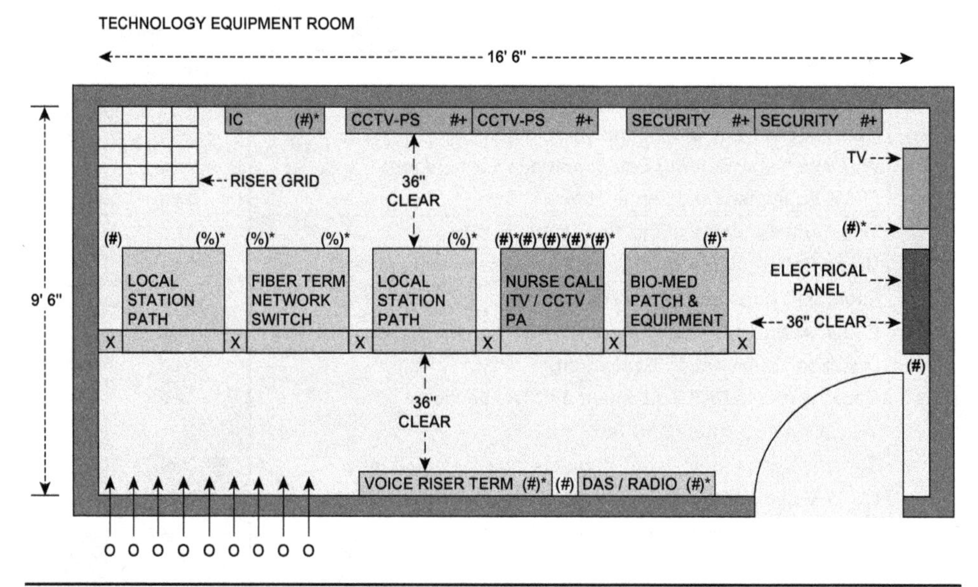

FIGURE 11.22 Typical technology equipment room, serving approximately 15,000 square feet.

- Public address amplifiers and control equipment
- Patient monitoring database file server
- Telemetry system control equipment and file server
- Cardiology information systems files servers
- PACS file servers and deep archive storage
- Integration engines and/or middleware servers
- Radio/DAS (distributed antenna system) systems equipment
- System administration terminals for all of the above
- UPS (uninterruptible power supply) supporting all of the above
- Computer room air-conditioning (CRAC) equipment serving all of the above

All power fed to the main and on-floor technology equipment rooms should be connected to emergency generators. Power requirements should be determined as part of the predesign study, and sized to support future growth, upgrades, and shell-space build-out. HVAC should be sized based on the assumption that all power entering these rooms will be converted to heat that must be exhausted from the rooms.

Cable trays, or a similar hook system, should be sized to support cable required by all technologies, including space for future growth, upgrades, and shell-space build-out. Cable support systems should extend from each technology equipment room along corridors leading to the areas served.

Care should be taken to avoid interference with electrical, HVAC, and other systems, and to allow adequate space for cable support and protection. Use of building information modeling (BIM) is recommended for coordination purposes, and to avoid conflict (i.e., "collision") with other systems and structures. Minimum separation should be maintained between technology cabling and devices and structures that potentially produce electromagnetic interference (EMI)—recommended separations would include

- Minimum distance from 120-volt ac electrical power 8 inches
- Minimum distance from 208-volt ac electrical power 12 inches
- Minimum distance from 480-volt ac electrical power 24 inches
- Minimum distance from fluorescent light ballast 24 inches
- Minimum distance from motors and transformers 48 inches
- Minimum distance from relays and motor controllers 24 inches
- Minimum distance from MRI space 96 inches
- Minimum distance from transformer vault 96 inches

Riser cables should be run in conduit for protection of the cable and for ease of installation. Sleeves between floors are acceptable where closets are stacked. Spare riser conduit/sleeves should be installed for future expansion and technology upgrades. Fiber-optic cable should be run in inter-duct, for ease of installation and protection. Planning for spare riser conduit and sleeves is recommended, allowing for future growth and upgrades. Sleeves through rated walls and floors should be properly fire sealed. Use of reenterable sleeves is recommended, wherever possible.

Cable should be run in conduit where required by code or the authority having jurisdiction (AHJ). Cable should be plenum-rated when run open above ceilings. Riser cable should be riser rated, unless run in conduit. Cable run underground should be moisture resistant—and copper cables protected at both ends with lightning protection. Where conduit is required, the following limitations should be adhered to:

- No more than 180 degrees of bend between pull boxes, manholes, hand-holes
- No single pull longer than 250 feet
- No conduit more than 40% full

Design should include redundant telephone service entrance conduits, based on the following requirements:

- Two sets of entrance conduits from the MPOE to two separate telephone company manholes at two separate locations on the property line.
- The two sets of entrance conduits should be separated by a minimum of 25 feet at all points.
- Typically, telephone companies recommend a minimum of two 4-inch diameter conduits at each entrance point.
- Duplicate redundant entrance conduits for secondary network service provider.
- Two additional 4-inch conduits should be run parallel to one of the telephone company entrance conduits for cable TV.
- All underground conduits should be Schedule 40, installed with no low spots where water can accumulate, and should slope away from the building at an even rate of 0.125 inch (⅛ inch) per foot.

- Where possible, two separate MPOEs should be established, with redundant connectivity to the main technology equipment room, and technology equipment room risers.

All technology equipment rooms should be lined with painted ¾-inch fire retardant A-B plywood (installed "A" side out) from the floor to 96-inch AFF. Power conduits in technology equipment rooms should be concealed within the walls, except where power is piped directly into equipment cabinets. Accessible raised flooring is recommended for main technology equipment rooms. Structured cabling numbering schemes and labeling should comply with Building Industry Consulting Service International (BICSI) standards.

Data Network

Data networks primarily consist of core switches located in the main technology equipment room and edge switches located in each technology equipment room (see Fig. 11.23). Networks are typically connected in a redundant arrangement with redundant core switches—and with half of the edge switches in each technology equipment room connected to each core. Additional redundancy is achieved by connecting half of the edge switches in each technology equipment room through the primary direct riser, and the other half through an adjacent technology equipment room on the same floor.

Network equipment selection and configuration should be based on facility IT standards, with port counts calculated to satisfy requirements of all PCs, telephones, printers, scanners, fax machines, wireless access points, medical equipment, and other technology systems and equipment that require connectivity.

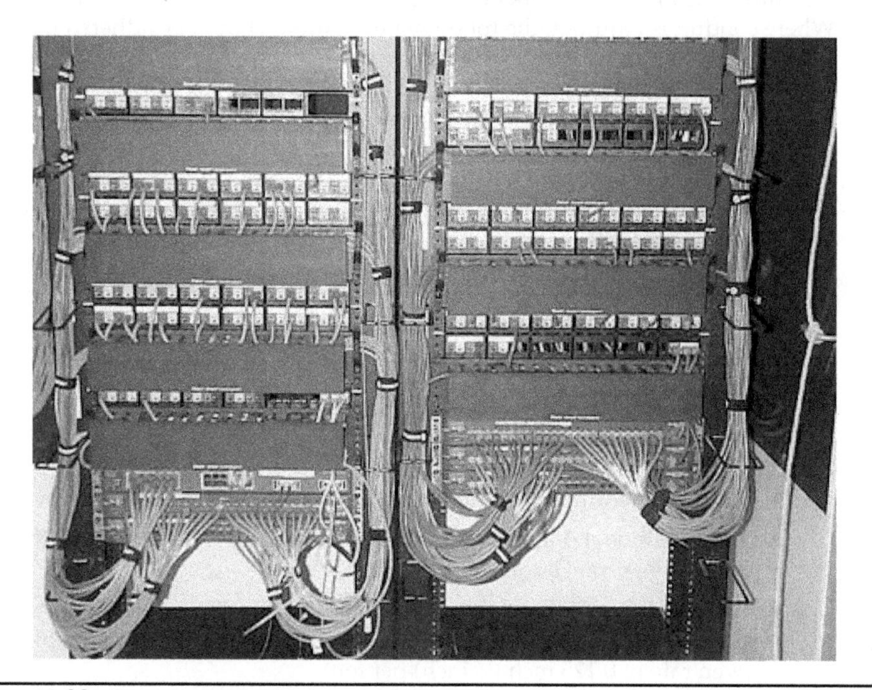

FIGURE 11.23 Data center distribution panel.

An IP addressing scheme should be developed allowing for static addressing and/or address range limitations that might be required for any systems or equipment—and coordinated with any existing networks' addressing schemes.

Planning the network should take into consideration any systems or equipment that might require middleware for communication of data to other technologies.

Today, WiFi provides wireless network connectivity in every hospital network design. As the number of wireless technologies, and the resulting network traffic, has continued to increase over the years, the density of WAPs (i.e., wireless access points) has increased. Additionally, FDA-governed technologies, like telemetry patient monitoring, have strictly defined WAP densities and typically require their own dedicated WAPs. Prior to commencing with design of the network and supporting structured cabling system, a study should be performed to determine technologies that may be connected through wireless, and determine the anticipated traffic, WAP density, and requirements for dedicated WAPs.

Scheduling of network installation should be coordinated with installation, testing, and certification/commissioning of systems and equipment connected to the network. Frequently the network must "go live" 4 to 6 months prior to receiving a temporary certificate of occupancy, in order to support installation and testing of the many connected technologies. Early completion of the technology equipment spaces required to support technology installation, integration, configuration, and testing, requires similar early completion of electrical and HVAC systems, to provide the permanent power and cooling required for technology equipment. Similarly, early security of these spaces is required, to protect expensive technology.

Defining quality of service (QoS) is an important factor in network configuration, to ensure near real time delivery of voice and live video—and to accommodate heavy data users, such as imaging, surgery, patient monitoring, and others. Network configuration should also consider requirements for a mirrored off-site data center, or for cloud storage, which is ideally connected through redundant MPOEs. Connectivity to WAN (wide area network) service providers, such as the Telco, should include redundant connectivity along diverse pathways, preferably connecting to two different central offices, for added redundancy.

Strategies for establishing redundant WAN connectivity for multiple sites might include such structures as a synchronized optical network (SONET) arranged in a self-healing ring configuration, providing dual parallel connectivity. This can be arranged in a ring, passing through every service point, and configured to fold back on itself in the event of a cable cut or equipment failure.

Communication Systems

Telephone system. Current technology telephone systems use a VoIP network, which transmits sound through the network in digital data packets sent. This arrangement allows a telephone instrument to be connected at any location served by the network. This is especially valuable when remote facilities served by the network must be connected to the telephone system.

VoIP systems consist of file servers and software connected to the network and Telco service provider. Network edge switches required for VoIP must be PoE-type (Power over Ethernet) to power telephone instruments through the network cabling.

When VoIP architecture is used, providing a second set of dedicated redundant core network switches for telephony is recommend for additional redundancy, and to more effectively separate real time voice and high-volume data connectivity needs.

Private wireless telephones are widely used in hospitals, to provide easy communications with highly mobile staff. In many cases, private wireless telephones interface with nurse call systems, for signaling and answering of patient calls. Most private wireless telephone systems use VoIP technology connected through the wireless network.

Nurse call/code blue. Nurse call systems (see Fig. 11.24) provide a means for patients, and others, to signal for help and communicate with care-delivery staff. These systems fall into two categories: (1) tone/visual systems that signal the need for help through audible alarm tones and signal lights, strategically located outside doorways and along pathways through which responding staff will travel, and (2) audio/visual systems that signal calls, similar to tone/visual systems, but that also provide for two-way voice communication between caregivers and patients.

A typical complement of the system devices would include

- Bedside patient stations with voice capability located in patient rooms equipped with pillow-speakers (see Fig. 11.25) and/or connection points for beds equipped with call and communication capabilities. These allow patients to signal and communicate the need for help.
- Pull-cord call stations located adjacent to patient toilets, showers, and bathtubs, and in patient dressing rooms in areas, such as imaging, that allow patients to signal the need for help. These stations typically provide for tone/visual signaling only. Increasingly these stations are being provided with voice capability.
- Code blue and staff emergency call stations located at the bedside and other patient care/treatment locations. These stations typically provide for tone/visual signaling only. Increasingly these stations are provided with voice capability.
- Staff call stations equipped with voice capability for staff communication in treatment bays, exam rooms, and other care delivery locations.
- Staff/duty stations located in clean and soiled utility rooms, medication and nourishment rooms, equipment storage rooms, and other locations where staff will be working, to signal calls and the type of calls placed on the system.
- Audio/visual master stations and tone/visual call annunciation panels, placed at locations from which call response will be dispatched.
- Staff terminals located near patient doors (see Fig. 11.26), which display room status, medication and treatment reminders, and other information helpful to care delivery.

Systems are typically configured on a unit by unit basis, with master stations and/or call annunciation panels at the local nurse station or team workroom. Alternatively, some facilities have successfully deployed systems where all patient calls, building wide, are answered at one central location—the term "centralized nurse call" is used to describe this concept.

Four methods are commonly used for dispatching patient calls to nursing personnel, as follows

- *Audio paging.* If the proper person is not located near the central desk, the person is paged throughout the unit or floor, and then must call the desk to learn the

Figure 11.24 Nurse call master station. (*Courtesy of Westcall.*)

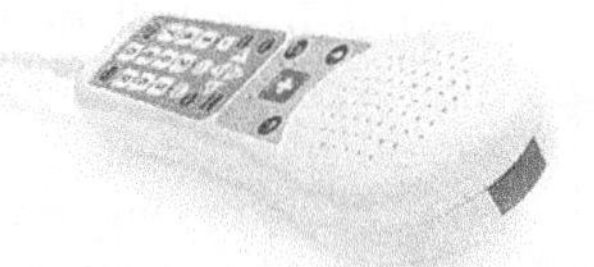

Figure 11.25 Bedside Control Device—Pillow Speaker. (*Courtesy of Crest Electronics.*)

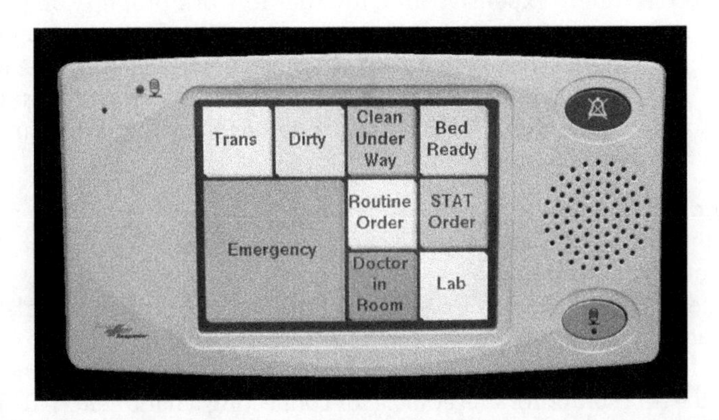

Figure 11.26 Room Terminal. (*Courtesy of Rauland-Borg.*)

location and nature of the call. Audio paging is not generally acceptable in most hospitals because it is noisy and disruptive.

- *Radio paging*. The location and nature of the call can be transmitted directly to the proper caregiver through the use of alphanumeric pagers. Radio paging is silent, efficient, and inexpensive. The primary drawback is that radio paging is one way. The caregiver may still need to call the central desk to confirm reception of the call, or to request additional help or clarification. Radio paging is slightly more expensive than audio paging, but it is silent, and most nurse call systems with radio paging interfaces can deliver calls to caregivers automatically, when no communication clerk is on duty.
- *Wireless telephone*. Similar to radio paging, the location and nature of calls can be delivered through alphanumeric displays on wireless telephones carried by the caregivers, caregivers can be called directly by the communications clerk, or caregivers can speak directly with patients. Wireless telephones also improve communication, because physicians can call caregivers directly without waiting on "hold" while the caregiver is located. Wireless telephones are slightly more expensive than radio paging, but they provide for immediate confirmation of calls, and streamline communications on the unit. As with radio paging, calls can be delivered automatically when no communication clerk is on duty.
- *RTLS/staff locator*. Real time locating systems (RTLS) require all caregivers to wear locator "badges"—small electronic devices that continuously transmit signals to receivers, or to WAPs located in ceilings throughout the hospital. Hence, when a patient call is answered, the proper caregiver can be easily "located," or found through a computer screen. Being able to quickly find staff certainly provides some benefit, but for dispatching patient calls some additional technology (i.e., radio paging or wireless telephone) must be used to contact the person and deliver the message. With some systems, patient calls are automatically diverted directly to the room where the caregiver is located; however, communication is provided directly through the nurse call loudspeaker, which may be disruptive to others in the room where the nurse is located. This approach also raises issues concerning patient privacy, which should always be considered with communication systems. RTLS systems are exceedingly expensive and, in order to work properly, require a unit secretary or communications clerk to be on duty 24 hours a day, to locate staff and dispatch calls. RTLS systems do offer the benefit of being able to track equipment locations. Many studies have shown that staff can spend an inordinate amount of time searching for equipment, rather than spending time with patients, so there is value to this capability.

Prior to commencing with systems design, a study should be performed to determine the anticipated staffing and workflow, and how that might change at different hours of the day and night.

Audio/visual nurse call systems should include patient room stations, located on the headwall beside each bed, code blue stations easily reached from either side of the bed (in most room configurations, placing two code blue stations, one on each side of the bed is the best solution), a receptacle for connecting a handheld patient control unit with call-placement button and loudspeaker, a receptacle for connecting speakers and call buttons located in the bed rail, pull-cord stations located where easily reached from patient toilets, showers, and bathtubs, staff/duty stations in work areas, corridor signal lights, and master consoles located at the unit desks. Monitoring jacks should be located at each bedside station for connection of equipment alarms. Handheld patient control devices, and controls built into bedside rails (see Fig. 11.25), typically include television and light controls. Light controls should be low-voltage only.

Pillow speakers should be provided at each bed location, for signaling patient's need for help, and for controlling the television, room lights, and other devices that might connected. Television controls should include a 10-button keypad for direct selection of television channels, a volume control on-off switch, a selector for alternative audio program (i.e., language selector), and an internal loudspeaker. Broadcast of television audio is typically limited to the handheld pillow speaker, to reduce overall noise level in the room and nearby spaces.

Some similar controls may be included in the bedside rail, however, space is typically not available for the full complement of controls.

Tone/visual nurse call systems should include pull-cord stations, located where they can be easily reached from patient showers and bathtubs, in patient dressing rooms, at staff/duty stations in work areas, corridor signal lights, and call annunciation panels located at the team workstations.

Code blue (i.e., emergency call) capability should be part of the nurse call system, with code calls annunciated both locally and at a central dispatch point, such as the hospital's PBX operator location.

Call-placement pull cords should be trimmed to within 2 inches of the floor, thereby ensuring they can be easily reached, regardless of the patient's position, but avoid becoming soiled.

Corridor signal lights should be color-coded and provided with flash-rates, as necessary, to visually indicate call types. Corridor signal lights should also be placed directly outside the room from which the call is initiated, and duplicated at the head of each corridor along the pathway of staff response.

Depending on the staffing and workflow plan, it may be required for calls from one unit to be annunciated and/or answered from another unit. For this reason, it may be necessary to provide parallel master stations, or annunciation panels, and/or transfer switching capability. It is recommended that all nurse call systems in the facility be networked, to easily support call transfer, and call signaling and answering between units, as census and staffing vary.

All floors and systems should be networked to provide common annunciation of code blue calls at the PBX operator's position, common system administration, and gathering of call activity data in a common database. Terminal equipment and power supplies should be installed in telecommunication equipment rooms. Systems should be provided with a UPS and connected to emergency power.

A quantity of special call devices (i.e., geriatric pads, "breath-call" devices, etc.) should be provided with the systems (see Fig. 11.27).

The nurse call system should be provided with a file server and software for collection of activity data, and for providing management reports. The system should be network connected, and provided with adequate software licenses to permit access to management reports, where required for administration.

Nurse call systems should be integrated with IT systems for download of HL7 patient information and upload of system activity.

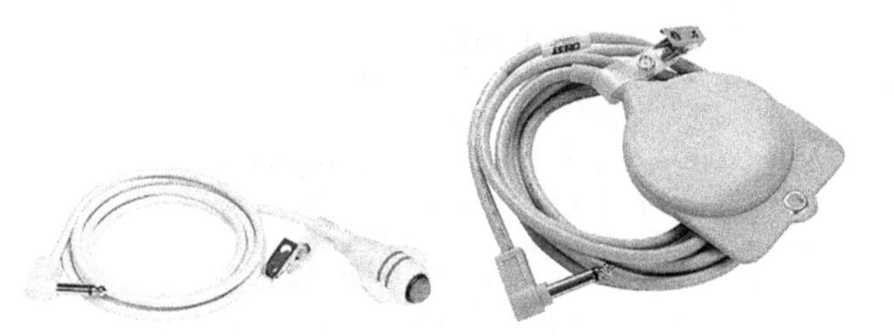

FIGURE 11.27 Specialty, call devices. (*Courtesy of Crest Electronics.*)

Some nurse call systems are available with data ports at the bedside through which various medical devices can be connected, for communication of clinical information to the EMR system.

Intercom. Most internal intercommunication is provided through telephones. However, several departments may require the additional benefits of a dedicated intercom system. Features provided by an intercom system include hands-free loudspeaker communication, flexible departmental group paging, and "meet-me" paging. Telephone systems generally do not perform these functions well.

Departments typically requiring intercom include surgery and imaging. Both have highly mobile staff, who must communicate quickly and efficiently. Hands-free communication is especially useful in surgery rooms, where the ability to communicate without touching the device is important. Surgery intercom typically extends through pre-op holding, PACU, anesthesia office and workroom, Stage 2 recovery, surgeons' and staff lounges, and central sterile supply.

Hands-free intercom (see Fig. 11.28) allows staff to answer calls instantly by simply speaking. Once a call is placed, even the calling party can move away from the station and still communicate hands-free. Meet-me paging allows staff to answer pages by merely touching one button on the nearest intercom station, whereupon two-way hands-free communication, between the paging and responding parties, is established immediately.

Many hospitals install intercoms between the emergency department and imaging, blood bank, surgery, and intensive care, to provide instant communication in emergency situations. In some jurisdictions, the AHJ may require this capability.

Once an intercom system is in place, it can provide internal communication in the event of a telephone system failure. Additional stations placed at unit desks, in the pharmacy, administration, and other key locations can provide internal communication even when the telephone system isn't working.

The intercom exchange should be located in the main technology equipment room, for ease of interface with other systems, and to provide the proper environment. Stations can be connected through the structured cabling system. This system should also be provided with a UPS and connected to emergency power.

Public address and music systems. Building-wide public address paging should be provided for emergency announcements.

Most hospitals find paging disruptive. Therefore, efforts should be made to minimize the use of paging by shifting communication to wireless devices.

Public address systems consist of loudspeakers, local volume controls, amplifiers, wiring, and termination hardware. Public address systems should be interfaced with

FIGURE 11.28 Stentofon "hands-free" intercom master stations. (*Courtesy of Zenitel.*)

the telephone system, to permit initiation of pages from operators' consoles. A microphone should also be provided for operators to use in the event of a telephone system failure.

Public address amplifiers and control equipment should be placed in the main technology equipment room, for easy interface with other systems and to provide the proper environment.

Loudspeakers should be provided with individual volume controls or multiple-tap matching transformers, for setting individual speaker volume levels. The system should be balanced at the time of installation to provide adequate volume evenly throughout all departments.

Once a paging system is in place, background music can be easily broadcast over the system. Speaker cabling should be zoned with public areas, where background music might be desirable, zoned separately from clinical areas, where music could be disruptive or interrupt patient rest.

Local volume controls should be provided in conference rooms, classrooms, selected office areas, and other areas where it might be desirable to reduce or turn off paging volume level. Local volume controls should be equipped with emergency page override.

Automated volume monitoring and control equipment might be required in areas where ambient noise levels vary widely at different times of day. Examples would include kitchens and serving areas, dining areas, lobbies, and other large public gathering places.

Public address systems should be zoned according to department and unit, as well as being connected to local intercom and nurse call systems for initiation of local paging. Zones where local paging is required, and the locations from which pages can be initiated, should be determined during the planning process. Systems should be provided with UPS, and connected to emergency power to ensure broadcast of public announcements in emergency situations.

Television. Television system requirements and functions vary widely between facilities, depending on desired entertainment service levels and educational support needs. Typically, televisions and cable TV outlets should be provided for patient rooms, waiting areas, conference and classrooms, and staff and physician lounges.

Combination TV-DVD units are still recommended for locations where educational programming will be viewed. However, educational programming is increasingly distributed through the facility network, and smart televisions are available for direct network connection. Most televisions have inputs for connection of external signal sources, such as a laptop computer.

Television outlets should be configured with both the traditional coaxial cable TV connector and a data outlet to provide future flexibility.

Signal sources for entertainment programming typically include antennas for local broadcast channels, cable TV service, satellite services, internal video playback systems, or some combination of all four. Control and selection of entertainment programming is typically provided through handheld patient controls or bed-mounted controls connected through the nurse call system.

Signal sources for enhanced entertainment, like movies or premium program channels, typically include internal video playback systems and/or satellite services. Control and selection of educational programming is accessed through telephone interface, handheld remote control, or bedside keyboard, depending on the vendor and system selected. Educational programming may be broadcast throughout the facility on a pre-scheduled basis, ordered on demand, or delivered to a specific location at a pre-scheduled time as part of the patient treatment plan. Systems may include interactive elements to confirm completion of viewing, capturing response to quizzes and surveys in the patient and staff records, and providing access to the Internet.

Educational programming may be provided for patients, visitors, physicians, or other staff. Systems should be configured to prevent access to physician and staff training by patients and visitors.

Planning the complement of required televisions services, for both entertainment and education, should be studied and determined prior to start of design. Patient and visitor privacy and noise levels in public spaces should be considered in the design.

Security systems. Security systems in hospitals typically include an access control/distress alarm, video surveillance, and infant security.

Access control systems. Although frequently referred to as a building system, access control is more correctly considered part of medical technology, as it directly supports the delivery of patient care. Controlling access to certain areas, and the movement of patients, staff, visitors, materials, supplies, medications, and equipment, are key elements of patient care workflow.

Card systems are recommended to control and record access to the building from outside, and to selected departments and spaces within the facility, where sensitive or valuable materials or information are stored. These areas include the pharmacy, medication rooms, materials supply and storage, medical records, technology equipment rooms, cashier and business offices, and others. Card access is also utilized to control movement of staff and visitors within the facility into sensitive areas, such as imaging, surgery, ICU, PICU, NICU, pediatrics, lab, and other areas.

Card access may also be a component of parking control. Most systems include distress alarm buttons that can be discreetly placed under counter for activation by staff in an emergency, or when security should be alerted of a situation. It is recommended to place distress alarm activation controls at all locations where staff interface with the public, including reception desks, cashier locations, and nurse stations. The emergency room is an important area to consider, as there have been numerous cases of domestic and community violence in this area. Human resources and sensitive administrative offices, where staff or visitors may become uncooperative or combative, should also be evaluated.

All activities of the access control system are recorded and displayed at a central workstation, typically located in the facility security command post. System activities and distress alarms can be displayed immediately, thereby allowing security staff to properly dispatch and document response.

Access control systems are highly integrated with electronic locking devices and automatic door operators, which are typically planned and designed as part of the building hardware package. Additionally, systems typically include door position switches—to monitor doors that are open, closed, or ajar—and motion detectors that sense movement, unlock doors as people approach from the secured side, and relock the door when reclosed.

Typically, electronic locking devices placed in the path of emergency egress must be provided with time-lapse release; thereby providing access for only a limited period of time (typically 15 seconds in most jurisdictions). Locking and egress plans should comply with the means of egress provisions of local building code, and be cleared by the local authority having jurisdiction before commencing with design.

Access control must be integrated with infant security systems to properly allow or prevent infant movement, depending on security levels assigned to specific staff functions, and stored in staff members' security cards.

Access control can also be integrated with video surveillance to automatically direct moveable cameras to view and capture movement and events.

Current technology access control systems use proximity cards embedded with unique identifiers, which can be assigned differently to each user, depending on the areas they are allowed to access.

A necessary system component is a badge production workstation, where badges are produced and programmed with the clearances allowed for the individual badgeholder. Most badge production workstations are equipped with a camera, to support production of picture ID badges. Badge production workstations are typically placed at the facility human resources department, and may be integrated with the human resources information systems, to automatically capture staff information for inclusion in the access control system database, and to immediately remove access rights in the event of an employee's termination. Badge design is also frequently coordinated with time and attendance systems, to permit use of a common card for both functions.

System control equipment and workstations typically connect and communicate through the data network, thereby allowing central monitoring and control of access activities in multiple remote facilities.

In locations where control of visitor entry is required, such as ICU, NICU, PICU, etc., door intercoms can be utilized. Systems are typically equipped with two-way audio and a video camera at the visitor call-in station; thereby allowing staff to observe and communicate with visitors, before allowing entry through activation of a door-release control at the master station.

Access control system terminal equipment should be placed in technology equipment rooms, provided with UPS or battery backup, and connected to emergency power. Terminal units typically communicate with the main system control, located at the security command post, through the facility network.

Closed circuit video surveillance. Video surveillance is typically provided with cameras located to observe sensitive areas—both inside and outside—for monitoring and recording at the security command post. Current technologies typically utilize PC-based digital video recorders for capture and playback of video images. Systems typically employ time-lapse recording techniques, with the capability to speed up or slow down playback, for later review and observation of events.

Cameras may be "fixed" to observe a specific location—or equipped with pan/tilt/zoom capability remotely controlled from the security command post, thereby allowing security staff to select and observe multiple locations.

Video surveillance may be required for nonsecurity-related purposes, such as for patient observation in ICU, seizure monitoring, or sleep lab. Video surveillance is also sometimes used as an efficiency tool in surgery and procedure rooms, allowing control staff to track the procedure progress, allowing early start of preparations for the next case, hence, expediting room turnover.

Most cameras today are rated for low-light level usage, and are equipped with automatic iris control, thereby supporting observation in both broad daylight and lower night-time light levels. Cameras utilized in special applications, such as seizure monitoring and sleep lab, are typically infrared sensitive and equipped with infrared light sources, thereby allowing patient observation in darkened rooms

Most cameras today are IP-based, communicating video in digital format through the facility network. IP-based cameras are typically powered through use of POE (Power over Ethernet) network edge switches, similar to those used to power VoIP telephones. In some cases, cameras installed on the building exterior or parking lot locations may be too far from the POE switch, thereby requiring a local power outlet and low-voltage power supply.

Emergency power should be provided for system components. Central equipment at the security command post should be provided with a UPS.

Infant security. Infant security systems have come into common use in recent years. Systems typically consist of ankle bands or umbilical tags placed on the child, which activate and alarm as the child is taken near an exit from the nursery or postpartum care unit. Additionally, these systems may track patient movement within the unit, and can sense and alarm removal of the band or tag.

System enrollment and control workstations are typically placed at nurses stations in the units where used, and interfaced with the access control/distress alarm systems to display and record alarms at the security command post. System control workstations are also used for allowing temporary removal from the unit for clinical and other authorized purposes.

Temporary movement of a child through an alarmed exit can also be controlled, through entry of a valid code on a keypad located near the exit, or through presentation of an authorized card to a card reader located near the exit.

Extensive integration of infant security with access control and building locking systems is required to ensure proper function. Infant security is also typically integrated with elevator controls to stall an elevator with door open on the protected floor during an alarm.

Infant security systems are typically standalone hard-wired systems; however, they are typically integrated with access control through network connections.

What are "special systems" and how are they changing. Systems, such as ambulance and public safety radio, audio-visual systems, interactive and educational television systems, wireless networks, real time locating systems, distributed antenna systems, and others, represent the second largest group of technologies commonly found in healthcare facilities. These can significantly impact design and construction of a hospital, so they need to be considered early in the process. Too frequently, these systems are not included in early planning, or not even identified as needed until later in construction. To avoid changes to construction, a study should be performed prior to commencing design to identify the broad range of special systems required to support facility operations. The following represent just a few of the systems most commonly found in hospitals today:

Radio systems. Radio systems typically include a variety of one-way paging systems and two-way radios, used by a variety of departments and services.

Radio paging is typically handled through outside paging services, but may also operate utilizing a local transmitter, which is typically located in a rooftop technology equipment room. Radio paging may be interfaced with the telephone system or facility network for manual initiation of pages, or with nurse call systems for automatic transmission of patient calls and alarms. Pagers typically provide an audible tone or vibration to announce message arrival, and incorporate a digital or LCD display to deliver written messages.

Localized "coaster" type paging systems, similar to those used in restaurants, are frequently used in family waiting areas and cafeterias, where visitors may be waiting for a family member in surgery or undergoing a procedure.

Two-way radios typically are traditional "walkie-talkie" units, allowing mobile staff to communicate directly with each other, both within the facility and over larger distances, through rooftop repeater transmitters, which automatically receive and resend radio signals. Increasingly, facilities are using private wireless or cellular telephones in place of two-way "walkie-talkie" systems. Where this is planned, care should be taken to ensure coverage is provided throughout the facility, including basements, mechanical rooms, central utility plants, and tunnels where other staff might not travel.

Where required to penetrate a large building or extensive underground spaces, remote antennas and/or transmitter/repeaters to extend coverage of one-way radio paging, two-way radio, private wireless telephone, and/or cellular telephone signals, may be needed.

Two-way radios may also be required for direct communication between EMT staff in ambulances and emergency room medical staff. Systems typically consist of a transmitter with an antenna, located in a rooftop technology equipment room, and a remote control console located at the emergency department. Remote control consoles are typically connected with transmitters through the structured cabling system. In some cases, ambulance radio remote control consoles are equipped with telemetry monitoring consoles, capable of receiving and displaying waveforms sent through the radio system from monitors in the ambulances.

Depending on the requirements of the local AHJ, radio systems may be required for two-way "civil defense" communication with other facilities or for communication to local public safety (police and fire) services. In some locations, "ham" radios are installed, providing an additional layer of emergency communication.

As buildings and building materials have become more impervious to radio signals in recent years, the need for improved distribution of radio signals within facilities has become apparent. Distributed antennas systems (DAS) are now available consisting of rooftop antennas, two-way amplifiers, and antennas distributed throughout the facility. DAS systems are primarily beneficial for supporting the signals of public safety radio services (fire and police) and cellular telephone service providers. In some cases, DAS systems can improve distribution of the two-way radio and one-way paging signal. At least one patient telemetry manufacturer utilizes DAS to support coverage. DAS is typically not recommended for support of WiFi wireless networks, because of limitations on user density inherent with some DAS technologies, and the loss of the ability to triangulate equipment and staff locations.

Dictation/transcription. Dictation systems today consist of file servers and VoIP instruments connected through the network, and accessed through telephones for recording and playback.

Electronic Medical Records (EMR)

Implementation of EMR is a primary focus and, in many cases, the largest single target for capital spending in healthcare today. In 2006, the federal government passed a law mandating adoption of electronic medical record-keeping practices by all U.S. healthcare providers. Limited funding for early compliance was provided in the American Recovery and Reinvestment Act (ARRA), passed in 2009. Compliance is required by U.S. hospitals no later than January 1, 2015, and by all other healthcare providers by January 1, 2016. Failure to comply may result in reduced reimbursement payments, fines, or both.

Working with the Centers for Medicare and Medicaid Services (CMS), the Healthcare Information and Management Systems Society (HIMSS) established a seven step process by which healthcare providers can demonstrate "meaningful use" of EMR (see Fig. 11.29).

As a result of the EMR mandate, most hospitals are in the process of updating their overall IT systems, installing new software with the supporting file servers, adding fixed and mobile workstations, and upgrading networks to support the additional information flow.

In addition to the necessary software and infrastructure upgrades, two key elements of EMR significantly affect facility planning, design, and construction.

Workflow. EMR has a significant effect on day-to-day operations and, therefore, the facility floor plan. Effective planning for EMR requires understanding of the types of information captured in the patient record, how the information is most efficiently and accurately captured and recorded, and the locations where each type of information can be most efficiently and privately observed and utilized by each member of the care team.

Integration. A second, and perhaps more significant element in successful EMR roll out, is technology integration. It is important to note that most information captured in a patient's medical record is detected or generated by medical equipment and medical devices. Today, understanding medical information and its sources, and integrating technology to automatically, efficiently, and accurately capture information in the patient's record, is an absolute necessity for effective hospital planning, design, and construction.

There are numerous examples of the need for integrated systems including "smart" diagnostic sets on the exam room wall, "smart" beds that capture patient weight, and the position of safety elements, such as side rails and brakes, patient monitoring and telemetry, imaging modalities and image-guided procedure systems, anesthesia machines, integrated surgical video systems, laboratory analyzers, automated pharmacy systems, and many more. These clinical and business support systems must be integrated, to efficiently and effectively deliver quality healthcare, and all must be thoroughly addressed and managed through the planning, design, and construction process, to deliver an effective care delivery platform.

US EMR Adoption Model℠			
Stage	**Cumulative Capabilities**	**2013 Q1**	**2013 Q2**
Stage 7	Complete EMR; CCD transactions to share data; Data warehousing; Data continuity with ED, ambulatory, OP	1.9%	2.1%
Stage 6	Physician documentation (structured templates), full CDSS (variance & compliance), full R-PACS	9.1%	10.0%
Stage 5	Closed loop medication administration	16.3%	18.7%
Stage 4	CPOE, Clinical Decision Support (clinical protocols)	14.4%	14.6%
Stage 3	Nursing/clinical documentation (flow sheets), CDSS (error checking), PACS available outside Radiology	36.3%	34.5%
Stage 2	CDR, Controlled Medical Vocabulary, CDS, may have Document Imaging; HIE capable	10.1%	9.0%
Stage 1	Ancillaries - Lab, Rad, Pharmacy - All Installed	4.2%	3.8%
Stage 0	All Three Ancillaries Not Installed	7.8%	7.2%
Data from HIMSS Analytics™ Database © 2013		N = 5,441	N = 5,439

FIGURE 11.29 Seven step EMR adoption model showing 2013 compliance levels.

The advent of the accountable care organization (ACO), and bundled payment agreements designed to incentivize healthcare providers to keep their clients healthy, are changing the landscape of hospital design. The bundled payment concept is intended to reward ACOs for minimizing the quantity and complexity of corrective treatment required, while employing the most effective treatment when corrective care is required. These goals must all be supported by electronic health records that are accessible anywhere, from the hospital patient floor to the doctor's office.

At the same time, new technologies and methods are being developed to speed treatment and recovery. Stem-cell and gene therapy technologies promise the ability to produce exact replacements for vital organs and other tissues. New lab techniques provide individualized medications, formulated for optimal DNA-specific interaction. New techniques for identifying genetic material provide for development of flu vaccines in 60 days from initial identification of the specific strain, far faster than the standard 12 month turn around accepted as the industry standard just a few years ago.

Consideration of medical technology, more importantly, consideration of how medical technology is changing care delivery and the facilities supporting care delivery, is, today, a key element to successful project delivery.

Planning, Design, and Project Management for Medical Technology—How Is This Changing?

Traditionally, technology planning in healthcare has been treated as a series of diverse disciplines with each discipline planning and designing independently, while coordination primarily focused on building systems and utilities. Each discipline was managed independently during the construction process.

In the past, medical equipment planning consisted of a producing a list of equipment desired by facility staff, obtaining vendor literature (cut sheets) around which other disciplines (mechanical, electrical, and plumbing) based their designs. Once construction started, coordinating final selection and delivery of medical equipment was left to the owner and, because of rapid technology change, too frequently became a source of frustrating and expensive changes, complicating and delaying the construction process.

Low-voltage design consisted primarily of internal communication systems, each functioning independently of the others. Later, data cabling was placed on the electrical drawings, and delivered as part of the electrical contractor's scope of work.

IT and special systems were left to the owner to plan, design, procure, coordinate installation, and integrate, where necessary. Like medical equipment, needs for special systems were too frequently identified late in construction, contributing changes and delays.

Integration of the many technologies was typically left to the owner's IT, facilities, and bio-medical engineering staff to coordinate and perform after occupancy. Too often, many of the integrations that would benefit clinical staff simply were never completed, resulting in time consuming, error-prone, manual information capture that could have been automated through technology integration.

Today, medical technology is far more integrated than in the past, and a far more fundamental component of the care delivery process. Medical technology planning should start as part of the earliest functional programming, and, certainly, before space planning. Unlike other disciplines where planning and design can be completed, and documents turned over to a contractor for construction, medical technology management is an ongoing process that must continue through construction and occupancy.

The healthcare world of today is truly different from the world just last year, and very different from the world just a few years ago. Yet, as planners and builders, we're faced with delivering finished facilities that our clients will use for 30–40–50 years in the future. How must our planning, design, and construction processes advance to accommodate this flood of technology-driven change?

Just as this explosion of technology is rendering old healthcare delivery processes ineffective, technology advancement is rendering the traditional healthcare project delivery model obsolete. As technology change has continued, we have seen similar advancement in many elements of construction: new structural systems, new skin systems, new roofing systems, new flooring and wall coverings—the list goes on. However, if the cost for the new building material cannot be justified, the original product may still be used and serve the owner for years in the future.

The difference with technology is that when a chosen vendor, a catheterization lab maker for example, approaches the owner with a new product, that means the old product around which the design team prepared construction documents is no longer available. Hence, in hospital construction, technology changes cannot be avoided, and the corresponding impact on the building and related systems must be managed and addressed.

This difference explains why a significant percentage of change orders that occur during hospital construction are the result of technology changes. Managing medical technology from earliest project inception through occupancy is an absolute necessity for successful project delivery. The process involves six steps

1. *Planning*—operational analysis and needs assessment
2. *Design*—creating documents to support construction and up-fitting
3. *Technology project management*—coordinating final technology selection, delivery, and installation with construction
4. *Procurement*—finalizing technology procurement configurations and specifications, obtaining vendor proposals, vendor selection, and processing purchasing documents
5. *Technology integration*—identifying and coordinating the many connections and interfaces between technologies to achieve interoperability
6. *Commissioning*–testing, certifying, and managing final configuration of technology

Step 1: Planning. The healthcare technology planning process begins with operational analysis, and moves very quickly through needs assessment—identifying the "tools," the systems and equipment required to optimally support workflow. Medical technology is second only to mechanical systems in space impact—and first in technology dense areas, such as imaging and surgery. And medical technology has the greatest impact on operations and workflow, thus having the greatest impact on adjacencies and floor plan layout. Medical technology planning and management process is illustrated in Fig. 11.30.

The key to successful medical technology planning is understanding clinical care processes, knowing the available systems and equipment, and understanding how they can work together to streamline workflow. Operational analysis is the key. Determining

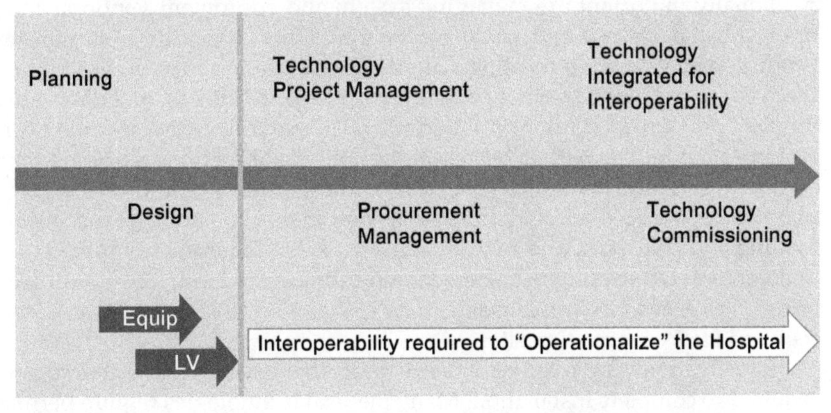

Figure 11.30 Medical technology planning and management process.

the tests, procedures, and treatments to be performed in the facility, and determining the devices, technologies, and systems required to support those activities—i.e., the medical technology.

Step 2: Design. Only after planning, and, more specifically, medical technology planning, is complete can facility design commence. Technology space and utility requirements are primary concerns of both architectural design and mechanical, electrical, and plumbing (MEP) engineering design. Imaging systems and hybrid operating rooms present added challenges for structural design, in addition to special shielding requirements, both traditional lead shielding for systems utilizing x-rays and copper-screen RF shielding required for MRI. Establishing medical technology support requirements, both initially and with an eye toward the future, early in design expedites the process, in addition to ensuring that the design will remain relevant through construction and after. With the use of hybrid operating suites increasing, planning "soft" space near traditional surgery rooms will allow for future addition of control and equipment rooms. Inclusion of planning specialists from all four medical technologies—medical equipment, low-voltage systems, IT, and special systems—ideally as a single discipline, initially in user meetings, and continuing throughout the design process, will ensure good design and minimize future changes.

Step 3: Medical technology project management. For medical technology, completing design is only the start. The rapid evolution of technology means that much of the medical technology will change before a hospital construction project is complete. New products replacing those included in the original planning and design, significant manufacturer upgrades, and totally new products required to support new services added by the owner will constantly appear through the construction time period. Keeping up with product and service changes, establishing realistic on-site delivery dates, scheduling and coordinating procurement, monitoring site preparation to ensure readiness for delivery and installation, all represent ongoing challenges throughout construction. Due to the significant support structures and utilities, and the high degree of integration, required by medical technologies, many fall on the critical path of facility construction.

Critical path systems and equipment represent one of the greatest challenges for construction, and are often the primary cause for delays and change orders in healthcare construction. Ensuring that a realistic procurement schedule has been established, and is being followed, is necessary for successful on-time project completion.

Equally important are managing system and equipment vendors, and their interface with the general and trade contractors. Once a vendor is chosen for a specific product, a coordination meeting with the appropriate trades should be held to ensure there is a clear understanding of rough-in and site readiness, and dates are established for early delivery of vendor provided rough-in materials, as well as the overall delivery and installation. The path of travel through the building for large items, environmental considerations for items delivered early, and security should all be considered and planned. Loading dock and transport equipment availability, responsibility for offloading and transport, and requirements for safety equipment on the part of delivery and vendor staff should be discussed and fully understood. It is not unusual for deliveries to be turned away because trucking company staff lack the basic safety equipment required for a construction site.

Equally important is the consideration of other systems and technologies that will be required to complete installation. Many medical technologies require permanent power early in installation, and some require a live data network, which also requires permanent power, as well as network security. Frequently, this requires early completion of technology equipment spaces, early ordering and installation of network equipment, and early assembly of the IP addressing scheme, sometimes as much as 4 to 6 months before project completion. Installation of network equipment in a technology equipment rooms requires work in that room to be substantially complete, with permanent power and HVAC, and the rooms must be secure. An equal level of completion is required for the main technology equipment room and the MPOE, as well as having outside connectivity established. Clearly, the connected nature of technology has a snowball effect on construction readiness. Early involvement of, and commitment by, facility procurement, bio-medical, and IT staff is required.

Ongoing monitoring of site readiness, and ongoing fine-tuning of vendor delivery and installation dates for every medical technology, is necessary for successful project completion.

Step 4: Medical technology procurement management. Most medical technology, both systems and equipment, is still owner furnished. Yet much of the medical technology falls on the critical path of construction. Many products must be selected, plans finalized, purchase orders placed, and equipment delivered and installed on time, in order to keep construction on schedule. And many technologies must be selected early, site-specific vendor drawings produced and provided to the contractor, so that construction can commence.

Many owner furnished systems and equipment items fall on the critical path of construction. More than simply establishing on-site dates, a comprehensive procurement schedule must be developed addressing every step of the procurement process. LEAN procurement management starts with the commissioning start date, and backs out times for vendor testing and FDA certification, integration and testing, delivery and installation, final negotiation and purchase order processing, selection of options and software packages, proposal review and vendor selection, RFP/RFQ response time, and final configuration and specification development for issuance with RFP/RFQs. For many complex technologies, the procurement process can easily require months, or even years, from start to finish.

Dates of industry trade shows must be factored into the schedule. For example, requiring final selection of an imaging system in the weeks or months prior to the annual Radiological Society of North America (RSNA) meeting held in Chicago at the end of November each year, will probably not be acceptable to the users. Decisions on imaging technology will be delayed until after they have attended the RSNA meeting, seen the latest developments, and determined with administration if funds are available for purchase. Consequently, requiring a final decision in January or February is much more realistic.

A legitimate concern aired by many clinicians is that medical technology is advancing so rapidly that purchasing decisions must be delayed until the last possible moment to ensure that only the very latest technologies are obtained. For critical path systems and equipment that requires significant space and utilities, decisions can be delayed only so long before they start to impact the overall project schedule. Procurement strategies that can address this concern include requiring vendor commitment to provide their most current product on the "first day of clinical use" as part of the RFP/RFQ response. Vendors know what their product development plans are, and they have a good idea which products will sunset during the life of the project. In reality, for many modalities, if the latest platform is proposed, most future upgrades and product enhancements will be software driven.

Performing medical technology procurement for a hospital construction project requires considerable commitment of staff time. Owner's procurement groups are not always adequately staffed and equipped. Therefore, it is important to start procurement planning early—to inform the group of the schedule and volume of work required—and to determine whether they can handle this significantly increased workload, or if outside help will be needed. Levels of outside assistance could include

1. Help with establishing and managing the procurement schedule
2. Direct assistance with procurement of critical path group 1 equipment
3. Direct assistance with other noncritical path group 1 and group 2 equipment
4. Direct assistance with all groups, 1, 2, and 3, equipment

Another significant factor to consider in assembling a procurement schedule is the owner's procurement policies. Many organizations have requirements for bidding. Procurement through a bidding process requires considerably more time than simply identifying a product and issuing a purchase order. Part of the procurement scheduling process is to determine the various ways systems and equipment must be purchased—methods typically include

- Competitive bidding
- Procurement through GPO contract
- Compliance with existing standards
- Sole source availability of specific products

Competitive bidding requires far greater time and effort than other methods. It's important to determine the facility's procurement policies, and to factor the time impact of each into the procurement schedule.

A common misperception is that membership in a group purchasing organization (GPO) will address all purchasing needs. First, it's important to understand that most GPO pricing is typically based on single-unit sales—not the volumes required for a construction project. Hence, volumes that typically occur in construction projects qualify for greater discounts—which, if missed, can significantly impact the overall project budget. Pricing obtained through competitive RFP/RFQ-based competition typically provides enhanced results, and stronger commitment for future support, both during and after construction.

Another common misperception is that the GPO's catalog manufacturers and products can meet the clinical preferences and needs of all practitioners, while, in fact, GPOs typically limit the number of approved vendors. This increases the odds that some preferred vendor's products will not be available through the GPO.

In other cases, established facility standards must be met. Glove box holders must match those currently in use, sharps and medical waste containers must comply with requirements set by the service provider handling those materials, surgical video systems must be compatible with existing systems, many technologies must be compatible with the facility's EMR system. While excellent pricing may be available for a specific product, quality ongoing service and field support may not be available at a reasonable price and on a timely basis, thereby rendering overall cost of ownership for the product higher over time.

It is clear that, closely coordinating the owner's procurement process and schedule with construction is necessary for successful project completion.

Step 5: Medical technology integration. Medical technology forms the backbone of patient care delivery, and medical information forms the backbone of the diagnostic and treatment process. The diagnostic process relies on information from the patient's record. Development and monitoring of the patient's care delivery plan is equally reliant on information. Most of this information is detected or generated by medical technology.

We see this in our everyday lives when we visit the family doctor. The physician turns first to our medical record.

With the 2006 government mandate for facilities to utilize electronic medical record keeping, medical informatics has become the focus of capital planning and spending for facilities across the country. Facility's EMR roll out plans can have significant impact on successful construction project delivery.

Any time the information gathering process can be automated through direct connection of medical devices to the EMR, efficiency and accuracy are improved, allowing care givers to focus on the patient care process, rather than on data collection and entry. Healthcare quality, and, of perhaps equal importance, patients, perceptions of healthcare quality, improves.

Connecting medical technology to the patient record, and managing integration of the many medical technologies routinely used to support healthcare delivery today, requires the following steps.

Before the integration process can begin, it's necessary to determine which technologies can and should be integrated. As mentioned earlier, a study, conducted by one well-known medical technology planning firm, defined over 200 technologies routinely used in hospitals that might be integrated. Perhaps more significantly for construction management, that same study identified over 150 integrations that might be required between those technologies. Integrations range from routine connection of imaging modalities for passage of standard DICOM compliant images to PACS, to complex custom integrations, requiring middleware for translation of unique information formats to EMR acceptable file structures.

Although the bulk of integration is between owner furnished technologies, increasingly contractor furnished systems and equipment are also integrated. Nurse call systems communicate through the data network, downloading HL7 patient information, uploading information captured in the patient room, displaying care plan reminders for staff entering the patient room, and documenting many patient care activities. Access control systems communicate through the data network, interface with human resources systems to allow or disallow staff access to different rooms at different times, document entry into sensitive areas, such as pharmacy and medication rooms, all while documenting activities for future review in the event of a security breach.

Today, there is literally no clear line that can be drawn between owner furnished and contractor furnished systems and equipment, as they are all too highly integrated. Successful project completion is impossible without a comprehensive integration plan, including both owner and contractor furnished medical technologies.

Once a complete list of systems and equipment has been assembled, integration implementation must consider the following five factors:

1. *Physical connectivity.* More than simply providing data outlets for computers, printers, and telephones, the design of the data physical plant for a hospital requires understanding clinical workflow, the multiple equipment and devices that may be employed in the healthcare delivery process, and locations where those technologies, both fixed and mobile, might be connected.

Unlike provision of a power outlet, where any 110-volt device can be connected to any 110-volt outlet, each data outlet must be configured to accept a specific data type and format, generated by the specific medical technology to be connected. Therefore, a significant first step requirement of technology integration is to understand the types and quantities of devices and where they will be used, in order to design the structured cabling systems with adequate connectivity where required.

Given that a variety of users will connect through the structured cabling system, such as contractors, vendors, owner's staff, a single database of outlet assignments must be developed and maintained through construction and activation.

2. *Logical connectivity.* Most technologies are designed to communicate directly through the data network, while some others require interfaces and/or middleware to convert data to network acceptable format. Still others merely utilize the physical cabling for connectivity between components, without communicating through the network. Identifying data formats produced by each technology, which systems and equipment must share information with other technologies, whether data sharing is one-way or two-way, whether data produced by one technology must be shared with other systems, where interfaces or middleware will be required, and the nature of that middleware are all requirements of a successful integration effort.

   It's also important to understand that some technologies may communicate in multiple ways between different components, within the system and with other technologies. For example, most patient monitoring systems utilize proprietary data protocols between bedside monitors and control equipment on the care unit; but utilize standard IP network connectivity between units and central data base file servers, and utilize standard IP connectivity for communication with EMR, for download of HL7 patient information or upload of clinical data to the record.

   Identification, understanding, and planning for each connectivity format is a necessary part of successful integration management.

3. *Security review.* Network security has always been an important focus for healthcare IT. The latest HIPAA regulations, released early in 2013, raise the bar even higher. Most facilities already have security review processes in place, but those who don't will undoubtedly be introducing their requirements.

   Each system connected to a health facility network should go through a security review process to determine if the equipment meets facility standards, and that proposed applications properly track and log user access to data elements. It must also be determined if remote access is required and whether remote access will penetrate the network firewall, thereby opening the system to outside intruders. Every system and each item should be examined and approved before connection to the network is allowed.

4. *Data format.* Once network access is approved, the real work begins on data integration. Each data type produced by each connected product, the data format, cell sizes, and targets must be considered and configured to ensure smooth information flow. Are products compatible with the facility EMR and patient information systems, what existing interfaces can be used, where must new interfaces be written, which interfaces require middleware, and who is

providing each interface or middleware—all of these issues must be determined before integration can be completed.

5. *Operational configuration.* Even after all of the systems and equipment are connected and successfully passing data, operational configuration is still required to achieve interoperability supporting care delivery workflow. What information is needed, by whom, where are they located, what tools will be utilized to access and update information, and how will this change at different times—all are factors that should be considered during final systems configuration. The earliest planning process starts with establishing workflow, but, with the total project time frames often spanning years, the likelihood that workflow will change is very high. Revisiting the facility's workflow and fine-tuning technology configurations are necessary to ensure proper support of care delivery.

Step 6: Medical technology commissioning. Medical technology commissioning primarily includes performing, managing, and documenting final technology testing and certification for each medical technology. Commissioning also includes documenting delivery of O&M manuals and warranty information, scheduling training coordinated with the overall facility activation training schedule, and securing on-site support during start-up.

Most clinical technologies are tested by the installing vendor as part of their FDA certification process. Others require hands-on field testing. Structured cabling system installers typically provide extensive test reports, documenting the performance of each cable and connection. Nurse call systems are typically tested by the owner or the owner's agent, traveling from room to room, literally pressing every button, trying every call type from every location, observing every signal light, conversing through every communication link, and documenting the proper function of each element. Signal strength readings can be taken on television systems, but ultimately testing is not complete until clear pictures are observed on every channel, viewed on each television.

Failure mode testing is recommended for the interaction between access control, infant security, elevator controls, and fire alarm. Similar testing is recommended between such systems and equipment as beds and nurse call systems, nurse call systems and television systems, etc.

Ideally, O&M manuals should be collected and catalogued, along with as-built documents and information related to warranty access, terms, and conditions. These should all be delivered to the owner in a single package, including both electronic and hard copy formats, as appropriate. This process can be further complicated on additions and renovations, with phased occupancies, where parts of the systems must be utilized before the entire system is turned over.

Training for medical technology systems and equipment including, where appropriate, both on-site and off-site technical training, regular user and key user training should be coordinated with the overall facility activation training.

Invariably, questions and issues will arise on the day of occupancy. Vendors of high-use systems like EMR, telephone, nurse call, etc., should be present during initial occupancy, and available to provide additional ongoing training and support, as needed.

Conclusion

Technology is advancing at a rapid pace throughout the world, and will continue at an accelerating pace in the future. It has been said that hospitals represent the most complex structures of mankind, and medical technology certainly represents the greatest collection of unique and complex systems and equipment found in hospitals today. Medical technology represents the primary tools used in healthcare delivery, and must be a primary consideration at every step of the planning, design, and construction process.

CHAPTER 12

Safety and Infection Control

Hospital construction projects vary widely, so the steps to manage safety and infection control are equally broad. The project may be an addition to a critical access hospital with 20 beds in a rural area or a complicated expansion and renovation of an academic medical center in the heart of a major city. Assessment of risk to patients, visitors, the staff and the construction workers begins in the planning stages of the project, and continues throughout the project. A process called preconstruction risk assessment (PCRA) is often used as a tool for assessing risk. Projects vary greatly in their scope and the potential risks that are created. Due to these variations, this chapter addresses only general topics related to safety, the PCRA, the infection prevention processes to help mitigate these risks, and the steps necessary in the event of an accident.

One of the first questions to address is who or what is at risk on a hospital project? There are a number of different categories of people, as well as property and equipment, that can be exposed. As with any construction job, the workers need protection due to the dangerous nature of the work with historically high incidents of accidents when compared to manufacturing or other industries. According to the Bureau of Labor Statistics 2011 Census of Fatal Occupational Injuries, 17.5 percent, or 721 of the 4114 worker fatalities in 2011, were in construction. This is the second highest of all occupations, listed behind transportation/material moving. Fifty-seven percent of the deaths in construction were attributed to the so called "fatal four," which are falls (35 percent), electrocutions (9 percent), struck by an object (10 percent), and caught-in/between (3 percent). These types of dangers would be similar on a hospital project or any other type of construction project. Similar approaches to reduce and prevent accidents should be used in all construction work, such as training, task-specific work plans, full-time safety inspectors, enforcement of all protection measures, lock-out procedures for electrical work, appropriate signage, and the proper use of safety equipment.

A laborer demolishing existing walls without dust partitions, an electrician who lifts an old ceiling panel with no protection, or a subcontractor who allows paint fumes to enter an intensive care unit are all potentially jeopardizing patient and staff safety. Release or transmission of infectious organisms, noise, vibration, and debris can impact a patient's safety and wellbeing. As such, hospitals face critical issues whenever contractors and vendors work within their facilities. Certainly, the stakes are high. Of the estimated 90,000 deaths a year from hospital-acquired infections in the U.S. healthcare system, some may well be linked to construction and facility maintenance issues.

This chapter was written by Tom Gormley.

Hospitals must systematically address risks to patient and staff safety from projects as large as a major facility expansion or as small as a single telecom worker called in to replace a cable or switch. Hospitals have done an excellent job in adopting and monitoring fire prevention policies, as shown by the numbers—less than one death per year over the last several years in hospitals and inpatient hospice facilities. However, patients and healthcare staff can be exposed to many other significant risks during construction or renovation if appropriate mitigation measures are not implemented and strictly followed.

The groups of people that need to be considered during hospital construction include

- Construction workers
- Patients
- Physicians/staff
- Visitors/family
- Vendors

Construction Workers

There are some areas where a hospital project may have additional or different types of risks for workers than other types of construction. Some examples of these are explored below.

Biohazards

In a hospital setting, some patients have infections and contagious diseases that could lead to exposure of the workers. This might occur when workers are close to isolation rooms, where these patients are typically housed. The air from these rooms is exhausted directly to the exterior, so another area of potential exposure could be near the exhaust fans serving these rooms. There is also the potential for infectious exposure to workers if, for some reason, they accidentally open or mishandle biohazard trash bags, also known as "red bags." These contain fluids, needles, soiled bandages, and other hazardous material that are disposed off either using methods that reduce the risk or through disposal in designated medical waste landfills. Construction worker should never handle "red bag" trash. Many health facilities have needle boxes, plastic containers used to dispose of needles, mounted in patient care areas. Construction workers should not handle or remove needle boxes that have been used.

Radioactivity

In the imaging department, there are several uses for radioactive material. Radionuclides are combined with pharmaceutical compounds to form radiopharmaceuticals. These are used for diagnostic purposes. Through ingestion or injection, the physiological functions of systems can be viewed as the compounds move through the body. They can also be used for therapeutic purposes. Nuclear medicine is typically used for scanning or treatments of lungs, gastrointestinal problems, hearts, and the brain, where the radioactivity is internal and imaged as processed. Traditional x-rays are used for anatomical imaging, such as chest, abdomen, head, and bones. In this case, the radioactivity is external, and passed through the body to form an image.

Since the nuclear medicine material is stored in the hospital, a worker could be exposed if they were to mistakenly access or handle it. Construction personnel could be exposed to x-rays if they were to accidentally walk into a room when the x-ray is activated, or if the lead shielding was not installed properly and they were working in an adjacent space. In the past, there have been several cases where old x-ray machines were improperly recycled, resulting in injury and death to people exposed to the radioactive material.

Magnetic Resonance Imaging

This technology uses a strong magnet to align the nuclei of the body's hydrogen atoms, and a resonance frequency to create computer-enhanced three-dimensional images of systems and parts of the body. This provides a good contrast for soft tissues, and is typically used for diagnostic purposes for the cardiovascular system, the brain, muscles, and cancers. The magnet is cooled using cryogenic liquids, typically helium, which poses some risk. If it is accidentally released into the scanner room, it may displace the oxygen, causing asphyxiation. It also undergoes near explosive expansion when changing from a liquid to gaseous state. In the event of an accidental shut down of the electro magnet, called a *quench*, the helium can boil rapidly and must be discharged through a specially designed cryogen vent to prevent an explosion. The other risk posed with an MRI is with the strong magnetic field produced in the scanner room. This field can cause pacemakers to malfunction. It can also pull metallic objects violently into the core of the magnet. There is one case where a metal oxygen cylinder was mistakenly brought into the scan room, and it was "sucked" into the core, killing the patient. Construction workers need to be trained to avoid bringing metallic objects, such as tools, into the scanner room. They also need to be trained on the risks of the cryogens to be prepared to evacuate in the event of a quench.

Patients

Typically, patients will only be present during additions and renovations, but there are projects where a new building may be built on an existing campus, resulting in patients and staff being near construction activities. As discussed in earlier chapters, some patients need special protection due to their compromised position from a health and self-preservation standpoint. Many may have suppressed immune systems due to medications or illness, and are particularly susceptible to dust and other contaminants. Also, fire/life safety issues are important, because many cannot flee the building as a result of medical conditions, age (both very old and very young), surgery, or other issues. Other, more general, construction safety issues are also involved with hospitals, such as crane accidents, water intrusion, overhead structural work, electrocution, scaffolding failures, and more.

Physicians/Staff

As with patients, physicians and staff will typically only be involved in additions/renovations. The staff is generally understanding of the construction disruption and is often excited about the new space they will soon be using. Good communication with this group is very important for the construction manager. Reviewing the infection control risk assessment (ICRA) and the interim life safety measures (ILSM) with the staff helps them understand and support the plan to manage the project. They also need to know their key support spaces will be maintained, such as locker/dressing space, parking, and food service. Safe access to these areas should be verified on a regular basis.

Visitors/Family

Given the purpose of a hospital, it is understandable there are numerous people coming to see the patients or bringing patients for outpatient services. Typically, visitors are ambulatory and capable of evacuation in the event of an emergency. A good life safety plan, with signage for exits, should provide for emergency response. Key issues with this group are clear directional signage, parking, availability of restrooms, and food service. Maintaining clear access on the exterior of the facility during additions is important. There have been lawsuits caused by visitors tripping over materials or hoses and other accidents related to the construction.

Vendors

With the intense activities at a hospital, there is a constant need to resupply, maintain, and support various functions. This ranges from food to the kitchen/cafeteria, to technicians for the imaging equipment, and a steady stream of medical supplies. Access to the loading dock is a key element to be considered in any expansion project, and adequate signage is a must. Many facilities have bulk oxygen systems external to the building. Access is required for filling these systems. Hospitals that use a bulk oxygen system will have an emergency oxygen connection on the outside of the building. Access to this connection must also be maintained.

Preconstruction Risk Assessment

Using a formal process for evaluating the risks on a project is a recommended approach. Avoiding the potential pitfalls can prevent injuries and claims, save costs and time, and preclude other issues, such as disgruntled customers. It is always better to avoid than react to a problem. There are several different approaches used by companies in the industry.

By using a diversified team to identify possible risk issues, mitigation measures can be implemented, and in some cases contingency budgets assigned to each item, to better forecast the overall financial picture. Risk tools that assign frequency and severity can be used to give more weight to contingencies assigned. Monte Carlo simulations can be done to develop contingency budgets needed for the specific project risks. It is wise to track the actual contingency expenditures related to the projections for the risk model to provide indicators on the overall budget.

Infection Control Risk Assessment

A Hungarian obstetrician, Ignaz Semmelweis, is widely credited with being the first to discover, in 1847, that health care providers could transmit diseases to patients. His introduction of a chlorinated lime for hand washing by staff drastically reduced mortality rates and was one of the first well-documented cases of infection control. Florence Nightingale continued to improve hospital conditions in the 1850s with her focus on good hygiene and proper ventilation. These were some of the earliest responses to what are now commonly known as HAIs, or hospital acquired infections. The U.S. Center for Disease Control defines "healthcare associated infections as infections acquired while in the healthcare setting with the lack of evidence that the infection was present or incubating at the time of entry to the healthcare setting." In Nightingale's time, more soldiers died from typhoid, typhus, and cholera than from battle wounds. The modern HAIs often include hepatitis, HIV (human immunodeficiency virus), and *Staphylococcus aureus*. Some pathogens have developed resistance to antibiotics. The *Oxford Journal* (volume 5, issue 1) attributes methicillin resistance to *S. aureus*, also known as MRSA, with causing up to

60 percent of infections in ICUs. OSHA released the Blood Borne Pathogen Standard in 1991 to help focus attention on protection of healthcare workers. The 1999 Institute of Medicine report "To Err Is Human: Building a Safer Health System" suggested that over 90,000 deaths per year in hospitals are attributable to preventable medical errors. This placed more focus on patient safety and hospital acquired infections. In 2008, the Centers for Medicare and Medicaid Services (CMS) actually stopped reimbursing hospitals for patients readmitted with certain HAIs, adding financial as well as public pressure on reducing preventable infections.

One of the tools used to implement patient safety is an infection control risk assessment or ICRA. These are routinely used by hospitals and surgery centers in a variety of ways, from evaluating the risks associated with new surgical procedures to responding to an outbreak or increase of hospital acquired infections. An ICRA is commonly defined as a multidisciplinary, systematic, and documented process to define the risks to patients, as well as steps to mitigate the exposure. A typical hospital risk assessment would consider the types of care, the services provided, the community demographics, the population served, and the environment of care in the facility. In the 2010 Guidelines for Design and Construction of Healthcare Facilities, the ICRA addresses design elements, construction elements, and compliance elements. The design elements include

- The number, location, and type of airborne infection isolation and protective environment rooms
- The number, location, and type of hand-washing stations
- Special HVAC needs
- Water systems to prevent *Legionella*
- Finishes and surfaces

The construction elements address

- "The impact of disrupting essential services to patients and employees"
- "Specific hazards and protection levels" for areas involved
- Patient locations relative to their susceptibility to infections and "the definition of risks to each"
- "Impact of movement of debris, traffic flow, spill cleanup, and testing and certification of installed systems"
- An evaluation of internal and external construction-related activities
- "Location of known hazards"

The compliance elements require a written plan included throughout construction commissioning. This plan "shall describe the specific methodology by which transmission of air and water having biological contaminants will be avoided during construction." It must have a monitoring plan, and changes must be updated and documented.

The focus in this chapter is on the utilization of the Infection Control Risk Assessment—Matrix of Precautions for Construction and Renovation. (Figure 12.1 – Steps 1 to 3 adopted with permission of St. Luke Episcopal Hospital and Steps 4 to 14 adopted with permission of Fairview University Medical Center.)

Step 1 of the ICRA matrix of precautions for construction and renovations identifies four different types or levels of activities, from "non-invasive" to major demolition and construction, and four different levels of risk in patient groups, from "office areas" to the most sensitive areas for patients, such as surgery, ICU, and burn units. These levels of activities and patient groups are contrasted or charted in the matrix to show the risk profile and define the class of precautions required during and after construction. There are four classes of precautions ranging from minimizing dust and cleanup afterward to elaborate measures including isolating the HVAC system, maintaining negative pressure with high-efficiency particulate air (HEPA) filters, and clean up with disinfectants.

The different levels of construction activity are Types A, B, C, and D. Type A would apply to visual inspections above the ceiling and minor maintenance work that does not involve creating dust. Type B would also typically be maintenance work, very minor cosmetic renovations, and installing phone/computer cabling above the ceiling, creating minimal dust which is easily contained. Type C would be construction activities beyond

Infection Control Risk Assessment
Matrix of Precautions for Construction & Renovation

Step One:
Using the following table, *identify* the Type of Construction Project Activity (Type A-D)

TYPE A	**Inspection and Non-Invasive Activities.** Includes, but is not limited to: ▪ removal of ceiling tiles for visual inspection only, e.g., limited to 1 tile per 50 square feet ▪ painting (but not sanding) ▪ wallcovering, electrical trim work, minor plumbing, and activities which do not generate dust or require cutting of walls or access to ceilings other than for visual inspection.
TYPE B	**Small scale, short duration activities which create minimal dust** Includes, but is not limited to: ▪ installation of telephone and computer cabling ▪ access to chase spaces ▪ cutting of walls or ceiling where dust migration can be controlled.
TYPE C	**Work that generates a moderate to high level of dust or requires demolition or removal of any fixed building components or assemblies** Includes, but is not limited to: ▪ sanding of walls for painting or wall covering ▪ removal of floorcoverings, ceiling tiles and casework ▪ new wall construction ▪ minor duct work or electrical work above ceilings ▪ major cabling activities ▪ any activity which cannot be completed within a single work shift.
TYPE D	**Major demolition and construction projects** Includes, but is not limited to: ▪ activities which require consecutive work shifts ▪ requires heavy demolition or removal of a complete cabling system ▪ new construction.

Step 1: ______________________________

Steps 1–3 Adapted with permission V Kennedy, B Barnard, St Luke Episcopal Hospital, Houston, TX; C Fine CA
Steps 4–14 Adapted with permission Fairview University Medical Center Minneapolis, MN. Forms Modified/updated; provided courtesy of Judene Bartley, ECSI Inc. Beverly Hills, MI, 2002. jbartley@ameritech.net Updated, 2009.

FIGURE 12.1 Infection control risk assessment (www.ashe.org/advocacy/organizations/CDC/pdfs/assessment_icra.pdf).

Step Two:
Using the following table, ***identify*** **the Patient Risk Groups that will be affected.**
If more than one risk group will be affected, select the higher risk group:

Low Risk	Medium Risk	High Risk	Highest Risk
▪ Office areas	▪ Cardiology ▪ Echocardiography ▪ Endoscopy ▪ Nuclear Medicine ▪ Physical Therapy ▪ Radiology/MRI ▪ Respiratory Therapy	▪ CCU ▪ Emergency Room ▪ Labor & Delivery ▪ Laboratories (specimen) ▪ Medical Units ▪ Newborn Nursery ▪ Outpatient Surgery ▪ Pediatrics ▪ Pharmacy ▪ Post Anesthesia Care Unit ▪ Surgical Units	▪ Any area caring for immunocompromised patients ▪ Burn Unit ▪ Cardiac Cath Lab ▪ Central Sterile Supply ▪ Intensive Care Units ▪ Negative pressure isolation rooms ▪ Oncology ▪ Operating rooms including C-section rooms

Step 2____

Step Three: *Match* the

Patient Risk Group (***Low, Medium, High, Highest***) with the planned ...
Construction Project Type (***A, B, C, D***) on the following matrix, to find the ...
Class of Precautions (***I, II, III or IV***) or level of infection control activities required.
Class I-IV or Color-Coded Precautions **are delineated on the following page.**

IC Matrix - Class of Precautions: Construction Project by Patient Risk

Construction Project Type

Patient Risk Group	TYPE A	TYPE B	TYPE C	TYPE D
LOW **Risk Group**	I	II	II	III/IV
MEDIUM **Risk Group**	I	II	III	IV
HIGH **Risk Group**	I	II	III/IV	IV
HIGHEST **Risk Group**	II	III/IV	III/IV	IV

Note: Infection Control approval will be required when the Construction Activity and Risk Level indicate that Class III or Class IV control procedures are necessary.

Figure 12.1 *(Continued)*

Description of Required Infection Control Precautions by Class

	During Construction Project	Upon Completion of Project
CLASS I	1. Execute work by methods to minimize raising dust from construction operations. 2. Immediately replace a ceiling tile displaced for visual inspection	1. Clean work area upon completion of task.
CLASS II	1. Provide active means to prevent airborne dust from dispersing into atmosphere. 2. Water mist work surfaces to control dust while cutting. 3. Seal unused doors with duct tape. 4. Block off and seal air vents. 5. Place dust mat at entrance and exit of work area 6. Remove or isolate HVAC system in areas where work is being performed.	1. Wipe work surfaces with cleaner/disinfectant. 2. Contain construction waste before transport in tightly covered containers. 3. Wet mop and/or vacuum with HEPA filtered vacuum before leaving work area. 4. Upon completion, restore HVAC system where work was performed.
CLASS III	1. Remove or Isolate HVAC system in area where work is being done to prevent contamination of duct system. 2. Complete all critical barriers i.e. sheetrock, plywood, plastic, to seal area from non work area or implement control cube method (cart with plastic covering and sealed connection to work site with HEPA vacuum for vacuuming prior to exit) before construction begins. 3. Maintain negative air pressure within work site utilizing HEPA equipped air filtration units. 4. Contain construction waste before transport in tightly covered containers. 5. Cover transport receptacles or carts. Tape covering unless solid lid.	1. Do not remove barriers from work area until completed project is inspected by the owner's Safety Department and Infection Prevention & Control Department and thoroughly cleaned by the owner's Environmental Services Department. 2. Remove barrier materials carefully to minimize spreading of dirt and debris associated with construction. 3. Vacuum work area with HEPA filtered vacuums. 4. Wet mop area with cleaner/disinfectant. 5. Upon completion, restore HVAC system where work was performed.
CLASS IV	1. Isolate HVAC system in area where work is being done to prevent contamination of duct system. 2. Complete all critical barriers i.e. sheetrock, plywood, plastic, to seal area from non work area or implement control cube method (cart with plastic covering and sealed connection to work site with HEPA vacuum for vacuuming prior to exit) before construction begins. 3. Maintain negative air pressure within work site utilizing HEPA equipped air filtration units. 4. Seal holes, pipes, conduits, and punctures. 5. Construct anteroom and require all personnel to pass through this room so they can be vacuumed using a HEPA vacuum cleaner before leaving work site or they can wear cloth or paper coveralls that are removed each time they leave work site. 6. All personnel entering work site are required to wear shoe covers. Shoe covers must be changed each time the worker exits the work area.	1. Do not remove barriers from work area until completed project is inspected by the owner's Safety Department and Infection Prevention & Control Department and thoroughly cleaned by the owner's Environmental Services Dept. 2. Remove barrier material carefully to minimize spreading of dirt and debris associated with construction. 3. Contain construction waste before transport in tightly covered containers. 4. Cover transport receptacles or carts. Tape covering unless solid lid. 5. Vacuum work area with HEPA filtered vacuums. 6. Wet mop area with cleaner/disinfectant. 7. Upon completion, restore HVAC system where work was performed.

Step 3 ______________________________

Adapted with permission V Kennedy, B Barnard, St Luke Episcopal Hospital, Houston, TX. Forms modified/updated; provided courtesy of Judene Bartley, ECSI Inc. Beverly Hills, MI, 2002. jbartley@ameritech.net Updated, 2009.

Steps 1–3 Adapted with permission V Kennedy, B Barnard, St Luke Episcopal Hospital, Houston, TX; C Fine CA
Steps 4–14 Adapted with permission Fairview University Medical Center Minneapolis, MN. Forms Modified/updated; provided courtesy of Judene Bartley, ECSI Inc. Beverly Hills, MI, 2002. jbartley@ameritech.net Updated, 2009.

FIGURE 12.1 *(Continued)*

Step 4. Identify the areas surrounding the project area, assessing potential impact

Unit Below	Unit Above	Lateral	Lateral	Behind	Front
Risk Group	Risk Group	Risk Group	Risk Group	Risk Group	Risk Group

Step 5. **Identify specific site of activity e.g., patient rooms, medication room, etc.**

___

Step 6. **Identify issues related to: ventilation, plumbing, electrical in terms of the occurrence of probable outages.**

___

Step 7. **Identify containment measures, using prior assessment. What types of barriers? (E.g., solids wall barriers); Will HEPA filtration be required?**

___

(Note: Renovation/construction area shall be isolated from the occupied areas during construction and shall be negative with respect to surrounding areas)

Step 8. **Consider potential risk of water damage. Is there a risk due to compromising structural integrity? (e.g., wall, ceiling, roof)**

Step 9. **Work hours: Can or will the work be done during non-patient care hours?**

Step 10. **Do plans allow for adequate number of isolation/negative airflow rooms?**

Step 11. **Do the plans allow for the required number & type of handwashing sinks?**

Step 12. **Does the infection prevention & control staff agree with the minimum number of sinks for this project?** (Verify against FGI Design and Construction Guidelines for types and area)

Step 13. **Does the infection prevention & control staff agree with the plans relative to clean and soiled utility rooms?**

Step 14. **Plan to discuss the following containment issues with the project team. E.g., traffic flow, housekeeping, debris removal (how and when),**

___

___

___

***Appendix:** Identify and communicate the responsibility for project monitoring that includes infection prevention & control concerns and risks. The ICRA may be modified throughout the project. Revisions must be communicated to the Project Manager.*

Adapted with permission V Kennedy, B Barnard, St Luke Episcopal Hospital, Houston, TX. Forms modified/updated; provided courtesy of Judene Bartley, ECSI Inc. Beverly Hills, MI, 2002. jbartley@ameritech.net Updated, 2009.

Figure 12.1 *(Continued)*

Infection Control Construction Permit

	Permit No:
Location of Construction:	Project Start Date:
Project Coordinator:	Estimated Duration:
Contractor Performing Work	Permit Expiration Date:
Supervisor:	Telephone:

YES	NO	CONSTRUCTION ACTIVITY	YES	NO	INFECTION CONTROL RISK GROUP
		TYPE A: Inspection, non-invasive activity			GROUP 1: Low Risk
		TYPE B: Small scale, short duration, moderate to high levels			GROUP 2: Medium Risk
		TYPE C: Activity generates moderate to high levels of dust, requires greater 1 work shift for completion			GROUP 3: Medium/High Risk
		TYPE D: Major duration and construction activities Requiring consecutive work shifts			GROUP 4: Highest Risk

CLASS I	1. Execute work by methods to minimize raising dust from construction operations. 2. Immediately replace any ceiling tile displaced for visual inspection.	3. Minor Demolition for Remodeling
CLASS II	1. Provides active means to prevent air-borne dust from dispersing into atmosphere 2. Water mist work surfaces to control dust while cutting. 3. Seal unused doors with duct tape. 4. Block off and seal air vents. 5. Wipe surfaces with cleaner/disinfectant.	6. Contain construction waste before transport in tightly covered containers. 7. Wet mop and/or vacuum with HEPA filtered vacuum before leaving work area. 8. Place dust mat at entrance and exit of work area. 9. Isolate HVAC system in areas where work is being performed; restore when work completed.
CLASS III Date Initial	1. Obtain infection control permit before construction begins. 2. Isolate HVAC system in area where work is being done to prevent contamination of the duct system. 3. Complete all critical barriers or implement control cube method before construction begins. 4. Maintain negative air pressure within work site utilizing HEPA equipped air filtration units. 5. Do not remove barriers from work area until complete project is checked by Infection Prevention & Control and thoroughly cleaned by Environmental Services.	6. Vacuum work with HEPA filtered vacuums. 7. Wet mop with cleaner/disinfectant 8. Remove barrier materials carefully to minimize spreading of dirt and debris associated with construction. 9. Contain construction waste before transport in tightly covered containers. 10. Cover transport receptacles or carts. Tape covering. 11. Upon completion, restore HVAC system where work was performed.
CLASS IV Date Initial	1. Obtain infection control permit before construction begins. 2. Isolate HVAC system in area where work is being done to prevent contamination of duct system. 3. Complete all critical barriers or implement control cube method before construction begins. 4. Maintain negative air pressure within work site utilizing HEPA equipped air filtration units. 5. Seal holes, pipes, conduits, and punctures appropriately. 6. Construct anteroom and require all personnel to pass through this room so they can be vacuumed using a HEPA vacuum cleaner before leaving work site or they can wear cloth or paper coveralls that are removed each time they leave the work site. 7. All personnel entering work site are required to wear shoe covers.	8. Do not remove barriers from work area until completed project is checked by Infection Prevention & Control and thoroughly cleaned by Environmental. Services. 9. Vacuum work area with HEPA filtered vacuums. 10. Wet mop with disinfectant. 11. Remove barrier materials carefully to minimize spreading of dirt and debris associated with construction. 12. Contain construction waste before transport in tightly covered containers. 13. Cover transport receptacles or carts. Tape covering. 14. Upon completion, restore HVAC system where work was performed.

Additional Requirements:

Date Initials	Date Initials — Exceptions/Additions to this permit are noted by attached memoranda
Permit Request By:	Permit Authorized By:
Date:	Date:

Adapted with permission V Kennedy, B Barnard, St Luke Episcopal Hospital, Houston, TX. Form modified/updated; provided courtesy of Judene Bartley, ECSI Inc Beverly Hills, MI, 2002. jbartley@ameritech.net Updated, 2009.

Figure 12.1 *(Continued)*

routine maintenance that involve some demolition, new walls, minor mechanical, plumbing, and electrical, with the distinction that the work cannot be completed in a single work shift. Type D would be major demolition and construction activities typically seen on most hospital addition and renovation projects. A key element on these classifications is the amount of dust generated and the ability to contain it.

Dust can cause respiratory problems for patients, as well as provide a mode of transfer for infections. Inhalation of pathogens that have become affixed to particles of dust can cause viral, bacterial, and fungal infections. A variety of diseases can result from airborne transfer, including meningitis, influenza, tuberculosis, and SARS. Another problem with airborne contaminants, particularly from demolition work, is the release of mold spores. Often mold is found inside walls and above ceilings, where moisture and temperature levels promote growth. This is a significant problem in hospitals that have building leaks and moisture issues, or where the air pressure differential inside the hospital is negative with respect to the outside, causing infiltration of outdoor air, further encouraging the growth of mold. During demolition of walls and ceilings, especially outside walls or areas where moisture is present, dangerous mold spores can be disturbed and spread in the hospital. One of the common forms of mold is *Aspergillus*. Some species of this can cause serious illnesses, and are particularly dangerous for patients with suppressed immune systems, such as those suffering from HIV or leukemia.

Step 2 is defining the patient risk group that will be impacted by or adjacent to the construction activities. The groups are categorized more by location in the hospital or services provided than by medical condition or specific diagnosis, so it is easier for nonclinical individuals to understand. In some cases, there may be the need to clarify specific patient types, medical conditions, or procedures involved that may be impacted by the construction. This is one reason the ICRA team should be multidisciplined, so there is clinical staff to provide more in-depth analysis.

Step 3 requires the team to match the patient risk group with the planned construction activities to identify the class of precaution or infection control steps to be implemented. It is best to be conservative in the projected types of construction, so there is a safety factor in the precaution taken. For example, if the project involves some minor demolition on one end of a patient floor, but heavy demolition on the other end, the safest approach, of assuming Type D construction on the entire floor, may be the wisest choice. This must be balanced against the cost and time to implement the additional precautionary steps.

Two items on the matrix should be noted. First, there are a few sections that give the option of Class 3 or 4 precautions. In these cases, it is the responsibility of the ICRA team to analyze the specific conditions and implement the most appropriate response. Second, there is a note clarifying that, for Class 3 and 4 precautions, the infection control officer or department of the hospital must approve the steps being implemented. Thorough documentation of the analysis used in determining the precautions required, the infection prevention approvals, and any changes to the approach should be maintained for review during Joint Commission surveys, and for legal reasons in the event an issue comes up during or after the construction.

The infection control procedures for each of the four classes are self-explanatory, but somewhat broad, given the many different types of projects and the variations in the layout and design of hospitals. These ongoing and completion requirements must be customized to the specific project, and even to the phases of the project. On a multiphased project, there may be different classes of precautions underway at the same time. For example, the scope of work may involve only finish upgrades in administration offices and require Class II precautions while, at the same time, operating rooms are being gutted and replaced, requiring the highest levels of precautions—Class 4. It is the responsibility of the ICRA team to develop a specific set of procedures for their project, and even phases of their project, based on the applicable precautions. On the Class 3 and 4 precautions, the entire design and construction team may need to be involved to comply with the Guidelines. The mechanical engineer would need to help develop a plan to isolate the HVAC system and create negative air pressure work zones. The architect may need to design the "critical barriers" to isolate the work zones, so they do not impact fire exits. The contractor and subcontractors may need to develop work plans to comply with the requirements, while at the same time allowing completion of the construction in a timely, cost-efficient, and safe manner.

Step 4. This effort is typically very straightforward by looking at areas adjacent to, or "surrounding," the work area. It may involve the departments in the same smoke zone, adjacent smoke zones, and floors above or below the work area. At times, this can be more complicated, and the ICRA team may need to look at areas or departments that are not near the work area. For example, an existing hospital may not have adequate or functional water shut-off to isolate the work area. Old valves often rust and cannot be closed, or valves were not installed as strategically in older facilities. In this case, the water for the entire facility could need to be shut down temporarily for a tie-in, which would result in the loss of hand-washing stations in areas not near the work zone. The ICRA team would need to draw on the facilities management member to develop an approach to minimize the impact in this case. Bottled water, alcohol-based hand sanitizers, and the filling of bathtubs with water to flush commodes could be options. The ICRA team needs to take a broad view, with specific assessments of the unique features of their facility, when evaluating the precautions for areas surrounding the project.

Step 5. Identifying the "specific site of activity" is typically more complicated on additions and renovations, because they generally involve multiple departments. Some projects may be simply remodeling a single patient floor, or even a few rooms that would be easily defined. It is a best practice to review the life safety plans to identify the work areas that can then be "marked up," to clearly communicate the impacted areas of the hospital.

Step 6. The ICRA team will need to evaluate and identify potential outages of mechanical, plumbing, and electrical systems that may impact infection control. One of the common issues with the mechanical systems is the loss of proper air pressure relationships. The requirements are defined in Table 7.1 in Chap. 6, Sec. 7, of the Guidelines, which is the ASHRAE/ASHE STANDARD 170-2008. For example, soiled linen rooms must have negative pressure relative to the adjacent space or corridor, to help prevent airborne contaminants from escaping. If the mechanical system is compromised, the air pressure relationships could be impacted, resulting in the risk of transfer of infectious airborne material. This is just one example of many "outages" that could occur impacting infection control.

Step 7. The ICRA document notes that renovation/construction areas must be isolated from occupied hospital space. The ICRA team must use the assessment matrix, and their experience, to define the containment measures required in a specific case. The containment measures will vary with the duration they will be in place, the size of the openings to protect, and the requirements of the local regulatory agencies. While drywall on metal studs with all joints, openings, and cracks sealed is the best barrier for hospital construction, this may not be practical in all cases. For example, in some renovations in operating rooms the work must be done at night, and then the space is cleaned so it is ready for surgeries the next day. In this case, it may not be possible to build drywall partitions each night, and plastic barriers may be a better option. Figure

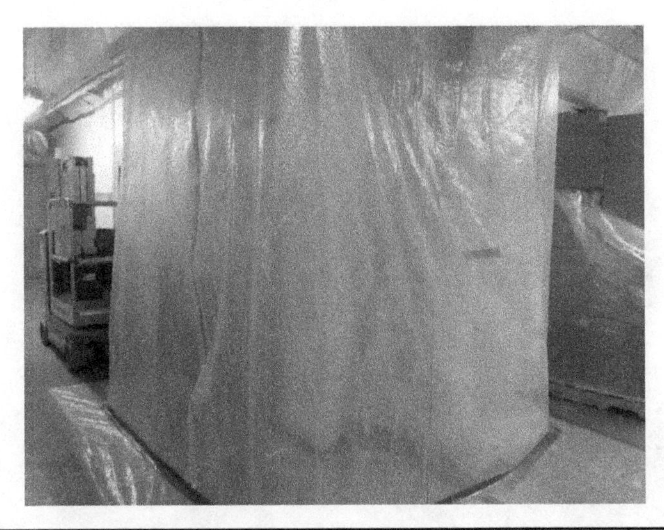

Figure 12.2 Temporary dust enclosure. (*Courtesy of Gobblell Hays Partners, Inc.*)

12.2 shows a plastic dust enclosure. The ICRA form also mentions plywood for barriers in addition to plastic, however, the fire rating and smoke classifications must be considered when using these potentially combustible materials. Fire retardant plastic and plywood may be used for these applications. There are also prefabricated enclosures available commercially.

The other question to be addressed in Step 7 is the need for negative air pressure fans and high-efficiency particulate air (HEPA) filters. The existing HVAC system, in almost all cases, should not be used to create negative pressure in a construction site. The best practice is to isolate the HVAC system and use auxiliary equipment to remove air from the space. If the HVAC system is to be used in any way, such as isolating the supply and return, and allowing an existing exhaust fan to draw air from the work area, a very careful and well-documented evaluation should be performed to identify all negative impacts of using the existing system. A common practice is to use portable fans exhausted directly to the outside to create the negative pressure in the work area relative to the operational hospital space. With the temporary negative pressure system in place, the dust and airborne contaminants are kept away from the occupied areas and patients. Depending on the location of the discharge of air from the construction site, the ICRA team will need to determine if HEPA filters are required to clean the air before it is exhausted. If air is exhausted from the construction site to a location in the interior of the building, HEPA filters must be used to clean the discharge of the temporary fans.

Step 8. This requires the ICRA team to evaluate the potential for water intrusion as a result of additions and renovations. There are waterborne pathogens that could cause infections. In addition, and more commonly, the water trapped inside walls and above ceilings can facilitate the growth of mold. As discussed earlier, several strains of mold can cause infections in patient populations, and are particularly dangerous to those with suppressed immune systems. Drywall is a very good medium to grow mold when water is introduced. One of the most common molds that cause problems in buildings is often called "toxic black mold." The technical name is *Stachybotrys chartarum*. This mold grows

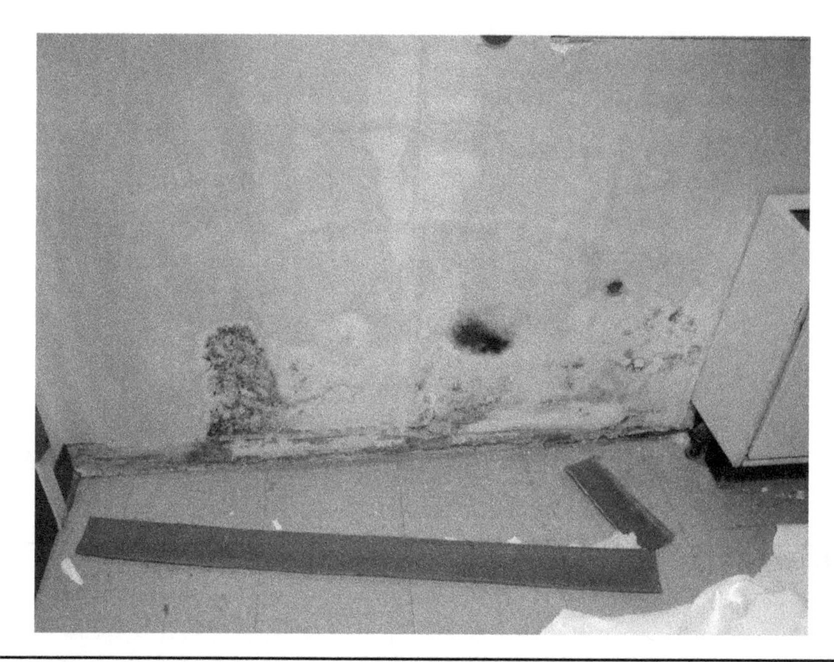

Figure 12.3 Moldy drywall due to moisture. (*Courtesy of Gobbell Hays Partners, Inc.*)

in material with cellulose, such as drywall, ceiling tiles, and wood. Figure 12.3 shows a mold problem at exterior wall due to moisture penetration.

Due to this issue with water and drywall, most exterior walls now are designed with a fiberglass wrapped board rather than paper wrapped to inhibit mold growth. A popular is a product called Dens-Glass. If the ICRA team identifies the potential for water intrusion, then mitigation steps should be implemented. These could include temporary enclosures to protect the building from water, dehumidifiers to reduce the moisture levels, and routine inspections of exterior walls and above-ceiling spaces to identify leaks before significant mold develops.

Step 9. The ICRA team needs to determine the best times for the construction work to be done. Ideally, it could be executed during "non-patient care hours," but this is often not a possibility. On simple projects, there may be the opportunity to adjust construction or patient care times to avoid mixing them in the same area. On complex, multiphased projects, or with areas that cannot be shut down, i.e., emergency department, there may be no option other than working during times patients are receiving care. The ICRA team needs to define the allowable work hours, and document them so they can be effectively communicated to the hospital staff, the contractor, and subcontractors. Ideally, the allowed work hours should be developed prior to subcontractor bidding and included in the contract, as the timing could affect cost. Another important step to include in the planning is a mechanism to stop work if it is impacting patient care, even if it is during the defined allowable work hours. A communication system using radios, phones, or other methods must be established, so the hospital staff can quickly contact the construction manager to have the crews stop work. This system must clearly define the hospital staff member(s) who have the authority to stop work and who they are to contact to avoid confusion. Stopping work can have cost and safety implications. One example where this was important was a project involving driving piles for an operating room expansion. It was agreed that the pile foundation work could be done during the day for safety and schedule reasons; however, the team knew the vibration could impact surgery. A radio system was set up between the OR supervisor and the project superintendent. During a particular eye surgery procedure, the pile driving was vibrating the operating room light causing the surgeon problems during the delicate operation. With the system in place, a quick call stopped the work until the surgery was safely completed. Coordination of work hours, hospital functions, and construction activities is critical to a safe and successful project; however, all members of the construction team must know patient safety is the highest priority.

Step 10. The Guidelines, in Sec. 1.2 to 3.2.1, state that the number, location, and type of airborne infection isolation and protective environment rooms shall be addressed by the ICRA. This is done during the design phase, based on the actual and projected patient demographics, acuity levels, case mix, patient volumes, and types of services offered. During a construction project, the ICRA team must determine if any of the isolation or protected environment rooms will be impacted or put out of service. Based on this, the decision can be made if any additional rooms must be provided, or if the rooms need to be relocated. Due to the cost of building these rooms with the required exhaust system, pressure monitoring, and ante room, it is best to avoid impacting these if possible.

Step 11. The Guidelines, in Secs. 2.1 to 7.2.2.8 and 2.1 to 8.4.3.2, define the requirements for hand-washing stations as far, as the size, anchorage, valves, and other features. The location and number of hand-washing stations is defined in the Sec. 2.2. The specific requirements for each department or functional area may vary. For example, in CCU Section 2.2 to 2.6.2.5 states that "there shall be at least one hand-washing station for every three beds in open plan areas and one in each patient room" and be "located near the entrance." In contrast, in the post-anesthesia care unit (PACU) hand-washing stations "shall be available for every four beds, uniformly distributed to provide equal access from each bed."

During a construction project, the ICRA team must determine the impact on the required number of hand-washing stations for each department and/or area. If the requirements will not be met during the construction, then temporary stations must be built in order to comply. These requirements also come into play when temporary units or spaces are set up. For example, in a surgery renovation project, a temporary PACU was required to be built in some adjacent storage space in the hospital. A hand-washing station was required to be installed for every four beds per the Guidelines.

Step 12. The ICRA team needs to confirm with the hospital infection prevention staff that the requirements are being met per the Guidelines as described in Step 11. There should be a member of the hospital infection prevention department on the construction ICRA team, and that person should coordinate with the hospital department, and provide documentation of their approval.

Step 13. The clean and soiled utility rooms are key elements in a hospital and can have an impact on infection control. The clean workroom is typically used to prepare patient items, and store clean and sterile materials. The Guidelines require a work counter and a hand-washing station if it is used for preparation not just storage. The soiled workroom also requires a work counter and hand-washing station, as well as a flushing rim clinical service station sink with a bedpan washing device, and storage space for covered containers for soiled linens and other waste. These rooms cannot be connected to avoid cross contamination. The soiled utility workroom is an obvious source of contamination, and requires negative air pressure to retain the potentially dangerous air. These rooms are required not only on patient floors, but also in critical care units and surgery suites. The ICRA construction team must evaluate the impact of the work activities on the required clean and soiled workrooms and develop plans to provide access to the existing areas or temporary locations. These plans must be verified and documented as acceptable to the hospital infection prevention staff.

Step 14. The ICRA construction team must develop a plan to address containment of dust, water, or airborne pathogens and other contaminants due to the work. This involves the actual work areas, but also must include other potential factors. These include

- *Traffic flow.* This relates to the movement of the construction workers through hospital occupied spaces to access the work areas. The workers may have dusty and dirty clothes. In Class 4 precautions, an anteroom is required to provide a location for construction personnel to clean up before entering a patient care area. For example, renovation of a surgery room is taking place and there is no option but for construction personnel to pass through the active surgery corridor to get to the renovation site. The ICRA team established a process for workers traveling in and out of the worksite. Workers put on disposable jumpsuits, hair covers, and shoe covers before entering the surgery area. Once they reach the renovation site, they enter the anteroom, remove the cover ups, and enter the worksite. When ready to leave the worksite, the workers vacuum off their clothing using a HEPA filtered vacuum, enter the anteroom, put on clean jumpsuit, hair cover, and shoe cover, and exit back through the active surgery corridor. Traffic flow must also address staff and visitors that may need to safely travel through work areas to insure they do not track debris and spread contaminants.
- *Trash and debris removal.* There must be an effective plan in place to safely remove all unwanted material from the work area. Construction and, particularly, demolition typically create large amounts of trash, excess materials, cardboard boxes, and general debris that must be taken to dumpsters located outside the hospital. Ideally an outdoor trash chute can be installed so the material can be taken directly out without going through the hospital. Figure 12.4 shows a trash chute mounted on the side of a building for debris removal.

Often this is not possible due to the location of the work area or lack of access for the chute. In these cases, the trash and debris must be taken in closed containers through the hospital, often down elevators, to the exterior. It is important that these trash containers be cleaned on the outside, including the wheels, each time before they are moved from the construction site through a clean area. Figure 12.5 shows a rolling trash cart that can be used for construction debris. These also come with covers to contain dust when inside an operational facility.

The role of the ICRA construction team is to develop a plan regarding how and when the material can be transferred safely out of the facility. The location of the dumpsters must also be considered, so contaminants and dust are not pulled back into the facility through outside air intakes.

- *Housekeeping.* The ICRA team must develop a plan to address keeping the hospital clean during construction work. This is often a joint effort involving the construction crews and the hospital housekeeping or environmental services department. The roles and the areas of responsibility for construction and environmental services must be defined so there is a clear understanding as to by whom, when, and how areas will be cleaned. The four classes of precautions

Figure 12.4 Trash chute. (*Courtesy of aerial innovations of TN, Inc. and Turner Construction Company.*)

provide guidance on the cleaning procedures, indicating where disinfectants and HEPA filtered vacuums may be required. Most hospitals also have their own specific requirements for cleaning prior to patient use. The construction team should work with the hospital's environmental services department to comply with any unique requirements, based on the services provided, patient

Figure 12.5 A rolling trash cart. (*Courtesy of Multivista.*)

population, and local conditions. Construction work in the operating rooms also requires special housekeeping considerations. While the contractor or subcontractor does the general cleaning after working in an OR, the hospital typically does what is known as a "terminal" cleaning. This is also sometimes referred to as a "deep cleaning," and commonly involves the following steps:

- Move all removal equipment and furniture to the hallway or opposite side of the room.
- Scrub all kick buckets and racks.
- Vacuum air condition grills.
- Clean all shelves and cabinets.
- Damp wipe with disinfectant solution all overhead lights.
- Wash wall surfaces with disinfectant solution.
- Damp wipe all furniture and equipment, including wheels and casters.
- Empty all soap dispensers, change tubing, disassemble foot pedals, and clean dispensers and foot pedals; refill dispensers and reassemble foot pedals.
- Remove the operating table cushions and removable pads; scrub cushions and pads in sink; thoroughly clean the entire operating table; remake the operating table.
- Scrub all sinks with disinfectant solution.
- Flood floor with disinfectant solution, machine scrub, absorb solution with vacuum.

- Return all furniture and equipment to room when floor has dried.

- *Food in the work areas.* While not mentioned specifically in the ICRA document, the introduction of food, and subsequent scraps and trash, can lead to infestation with bugs and rodents. This can pose infection control issues, as these pests can move into adjacent occupied hospital space, carrying germs and pathogens. It is best not to allow food in the work area during construction in an occupied facility. Separate eating areas can be set up for workers with proper containers for disposal of food scraps and trash. While this is not possible on all projects, the issue of eating and smoking by the construction workers should be addressed.

The ICRA construction team as mentioned before should be a multidisciplinary group that includes qualified individuals from the following groups:

- Hospital infection prevention
- Hospital facilities management department
- Hospital environmental services
- Contractor or construction manager
- Key subcontractors
- Design team—architects and engineers

The team should follow the ICRA 14 step process and thoroughly document their assumptions, the matrix, the steps, and any changes to the plans. They should use the Guidelines, or whatever code is in effect in their state, to define the requirements.

The plan should be communicated in a timely and effective way to the other constituencies in the hospital, such as physicians, nurses, administration, and department supervisors, as well as other members of the design and construction team. Training may be required for new procedures or systems. A monitoring system should be implemented to measure the results of the infection control steps. This may involve air monitoring, daily inspection of barriers, measurement of air pressure relationships, and supervision of workers to verify compliance with procedures.

Some projects have implemented the use of infection control construction permits to help in planning and documenting procedures. This process is a useful step to make sure the risks have been identified, mitigation measures are clearly defined, and the proper authorities have been notified and approval received. While this is practical and positive on some projects, it may not be possible on large addition and renovation projects that may continue for years. These projects may impact multiple areas, involve more than one patient risk group, and require different types of construction activities. On major projects, the permit process may need to be modified to be completed by phase, by department, or some other category. It is the responsibility of the ICRA construction team to develop an approach that is followed by the construction team to avoid significant cost and schedule impacts. This is necessary for the hospital to comply with their specific needs and regulatory agencies, and to protect the patients, staff, and visitors from infections resulting from construction activities. Much of the focus of the discussion of the ICRA process has been centered on renovation projects, where patients may be in the immediate vicinity of construction activities, or where patients are cared for in areas served by HVAC or other systems that are impacted by the renovation. The ICRA process also applies to new construction. Construction of a new patient tower that will eventually tie in to an existing hospital can have a negative impact on patients if precautions are not taken. There have been cases of patients contracting diseases where construction was taking place external to the building in which they were being cared for. Demolition of facilities, and excavation and disruption of utilities, creates a situation in which patients could be negatively impacted. As a new addition comes out of the ground and tie-ins to an existing building take place, the opportunity for the spread of fungal spores and other risks is present. Even construction of a new building that will not connect to an existing health facility can create risks.

Interim Life Safety Measures

An effective interim life safety measure (ILSM) program is critical to protecting the safety of the patients, staff, visitors, and construction workers. In addition, it is required by the Joint Commission, as part of maintaining the hospital's accreditation. The concept behind ILSM is to develop a plan to address deficiencies or noncompliance with the fire/life safety codes during construction or maintenance activities. By definition, this plan is to be implemented for a limited or "interim" period, and is not intended to address long-term or permanent failures to comply with the codes. For example, if the steps on the exterior of a required fire exit in a hospital were to be rendered unusable, the owner should put an ILSM plan in place to reroute the occupants to another compliant exit. This may involve adding signage or other measures to provide safe emergency egress. The steps must be replaced in a reasonable time period, or the hospital is subject to penalties by the local code officials and the Joint Commission.

For deficiencies or code violations that cannot be corrected immediately, a hospital can operate temporarily with one of the options defined in the Life Safety section of the Joint Commission standards. With these, the hospital can evaluate the cost, time, impact, and other factors to determine the best approach for their specific situation. They can remedy the issue within 45 days, repair it over time or, after recognizing that it cannot be resolved without major cost, provide another approach for protection. The options are defined below.

1. A *management process* can be developed to document the problem and corrective actions taken to resolve the issue within 45 days.
2. An *equivalency* can be used if the deficiency cannot "be corrected without major construction. In this case, the hospital must demonstrate "that alternative building features" exist that comply with the intent of the life safety code. This must be approved by the Joint Commission and other applicable regulatory agencies.
3. A *plan for improvement* can be prepared that addresses how the deficiency will be corrected along with a time frame.

No matter which option is used to correct a life safety deficiency, the need for ILSMs must be evaluated. If the evaluation determines that measures are needed, which will usually be the case, the ILSMs must be implemented to provide protection until the plan is fully executed and the deficiency corrected.

The ILSMs generally are approached in three phases. The first phase is planning the measures, the second is implementation and the third is closeout. These are explained further below:

1. Planning: This phase involves reviewing the project scope to assess the areas where the life safety systems or codes may be compromised. A life safety plan provided by the architect is a good tool for the risk assessment phase. The drawing typically shows all fire/smoke compartments, fire exits, exit signage, fire/smoke detectors, fire extinguishers, and fire hose cabinets. By overlaying the construction activities on this plan, it is easy to see if any of these elements of life safety may be compromised. Areas to focus on are fire alarm devices that may need to be shut down or moved, openings in fire/smoke walls that must be made, emergency exit routes that may be blocked or rerouted, and temporary partitions that may be required. Another risk that should be analyzed is from the systems standpoint, such as the fire protection or emergency power. The lack of fire sprinklers poses an obvious threat, and a fire watch may be needed, particularly if there is welding, cutting, or other direct ignition sources. The generator may be an issue, if it is down for tie-ins, because the emergency exit lighting would be down, unless there are battery backups.

   Once the risks are identified, the design and construction team and the hospital staff should develop the ILS measures. These will need to be customized to the specific project, as there are always different situations. The plan will also need to address various phases, as additions and renovations often involve finishing a portion of the work then moving the hospital in, to open up the areas for work. The plan may alternate the areas for work, and must change accordingly. For example, on most emergency department expansions, it is best to build the new area and let the hospital occupy it. There is rarely room in the existing emergency department for remodeling. It is probably crowded and overloaded with patients which is the reasons for the expansion. The ILSM during the construction of the new section would most likely involve rerouting emergency exits to maintain egress around the work, and managing tie-ins to systems. This contrasts with the ILSM after the new section is open and the renovations are underway. At that time, the safety priorities may involve temporary fire walls, to separate the construction areas from the hospital, and a fire watch while the fire sprinkler system is down to relocate the heads. Clear documentation, with life safety plans and a schedule, as well as measurable outcomes for safety, air quality, hospital satisfaction scores, and other project-specific key indicators are imperative. Reviews with the local regulatory agencies are also important during the planning of ILSM.

2. *Implementation*. As the project begins, the hospital and construction teams must implement the defined steps, while maintaining and documenting the results. On almost all additions and renovations, the plans will change based on monitoring feedback, operational needs of the hospital, regulatory agencies, and other factors. A responsible person from the construction manager and the hospital should walk through the project every day to review compliance with the ILSM. Documentation of the results and corrective action on any deficiencies

should be completed every day as well. The daily safety inspection should address key issues such as:

a. Access to fire exits with clear egress to the exterior

b. Fire/smoke doors and smoke compartment walls

c. Access to the building for ambulance and first responders

d. Any "hot" work involving cutting and welding that creates sparks

e. Any training that is needed or scheduled

f. Operation of emergency systems: fire alarm, fire sprinkler, generator, medical gas

g. Any high-risk work activities, such as erecting steel, main utility tie-ins, major equipment hoisting, or others

The plan must be flexible to allow for adjustments and to provide quick response to issues. On multiphase projects, the work will conclude at different times, so it is important to identify the areas, and complete the close-out or closure phase, before the space is occupied by the hospital. The data from the monitoring should be provided to all of the design, construction, and hospital staff as a comprehensive approach considering design cost, schedule, operations, and other key factors.

3. *Closures or close-out.* Prior to occupancy by the hospital, it is typically the responsibility of the design and construction team to verify that the life safety deficiencies caused on an "interim" basis by the construction or maintenance activities have been corrected. Some areas are straightforward to verify, such as the fire exits being clear, exit signs in place, and the space being clean. Verification of the proper air relationships, and the operation of fire alarm, or other systems is best done through a formal commissioning process. This is addressed fully in Chap. 13 of this book.

Most projects should use a checklist to verify that the items on the original plan have been addressed. Depending on the size of the project, this could be completed in 1 day for a quick tie-in to a life safety system, or it could be a complicated process over several weeks for a large addition and renovation. Coordination with the local and state building code officials is important on phased projects, as they often require life safety or other inspections prior to occupancy. There can be a variety of agencies involved, depending on the type of project. The Department of Health is usually involved with food service projects. The elevator inspection is typically performed by an independent state agency. There is often a Fire Marshal involved, in addition to a state hospital regulatory agency, and maybe local building departments. Thorough research on the "authorities having jurisdiction," and the role they will play, can help avoid problems and potential delays.

Hazardous Materials

On additions and renovations, particularly in older facilities, it is not uncommon to encounter some type of hazardous material. There is a wide range of hazardous products that were used in past years for different applications. These include the following:

Asbestos

This may be found in many forms in older buildings. The most expensive and disruptive is sprayed on fireproofing applied to the structural steel. Often the material is "friable," which means that it contains more than 1 percent asbestos and is easily broken or crumbled, by contact or deterioration, releasing asbestos fibers into the air. Breathing asbestos containing material (ACM) can have serious health consequences, such as respiratory issues, lung cancer, and mesothelioma. The tiny asbestos fibers cause scarring in the lungs, potentially leading to inflammation, difficulty breathing, and cancer. Conversely, non-friable ACMs are not easily crumbled by hand or by pressure. These materials might include asphalt roofing, vinyl asbestos tile (VAT) flooring, wallboard, and mastic and insulation on pipes or ducts. Non-friable ACMs can become friable by grinding, cutting, or demolishing the material, causing the release of fibers. The asbestos fibers are so light that they hang in the air for hours after typical dust particles have settled.

Prior to developing estimates or starting work in older facilities, a construction manager should check with the hospital's facility management department to verify if they know of any asbestos containing material in the work areas. If reports are not available, then testing should be done to determine if asbestos is present. All asbestos-related work should be done with guidance and oversight by a certified industrial hygienist (CIH). Only professional firms that are experienced, licensed, and protected by appropriate insurance should be used with ACMs.

There are three primary ways to deal with asbestos containing material:

1. *Removal*. This is the best long-term solution for ACMs. It may be a simple process for vinyl floor tile that can be taken out in whole pieces, avoiding breakage that could release fibers. Removal of insulation or mastic (glue) fibers on pipes can be limited. Sections can be done using what is called the "glove-bag" method which involves wrapping plastic around the pipe enclosing it as it is removed. This method is low cost and causes minimal disruption. See Fig. 12.6 for the "glove-bag" method of abatement. Friable sprayed on fire proofing may require elaborate measures, such as enclosing the entire area with plastic in negative pressure, to contain the material as it is scraped off, after being sprayed to wet the material, to minimize the release of fibers. Workers must wear full respiratory protection and suits, plus have decontamination areas for entering or leaving the work space. This is a very expensive and disruptive process. See Fig. 12.7 for full respiratory protection and protective clothing.
2. *Enclosing*. This involves sealing the ACM in some airtight material or box. Simple examples would be sealing ACM pipe insulation with a plastic wrapping material or laying a new floor over the old ACM floor, so it does not have to be removed. With sprayed on insulation, drywall enclosures can be built around the beams or columns sealing the ACM inside. These enclosures must be caulked to provide an airtight seal, to prevent the dust from escaping into the air stream. On all forms of enclosing the ACMs, bright labels indicating that asbestos is present should be placed on the outside, so maintenance staff or other workers do not accidentally open it and become exposed.

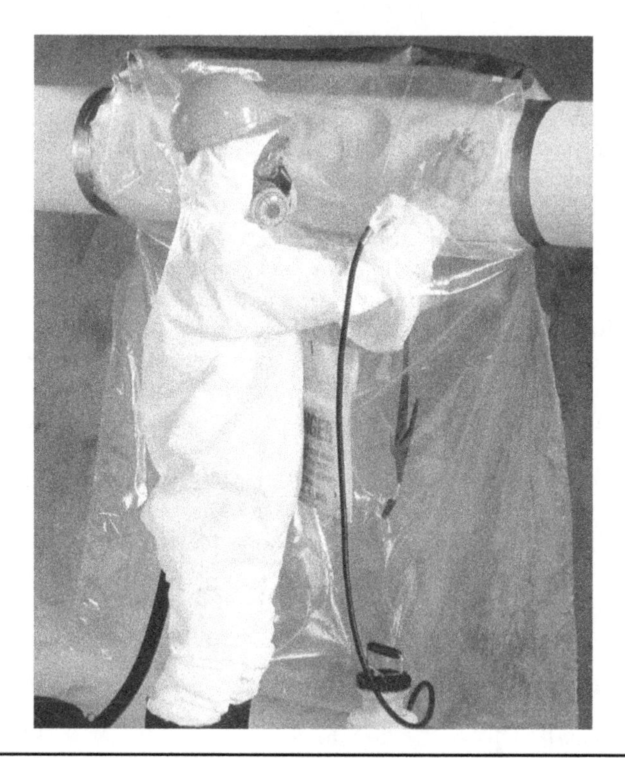

FIGURE 12.6 Glove-bag method. (*Courtesy of Gobblell Hays Partners, Inc.*)

3. *Encapsulation.* This method involves spraying a material (shellac, glue, or a special coating) over the ACMs to hold the material together, so it does not crumble and become friable. A simple explanation is just sealing or painting over the sprayed on fireproofing so it stays in place. It is important to verify that the ACMs are in good repair, and not already falling off, or the encapsulant will not be capable of holding the material in place.

Lead-Based Paints

Lead was added as a pigment to paint for many years. It improved durability, resisted moisture, and accelerated the drying process. This practice was banned in the late 1970s, so it is only found in older buildings. Lead is harmful to adults, affecting "practically all systems within the body" per the Environmental Protection Agency (EPA) website. It is particularly dangerous for children, causing nervous system and kidney damage, delayed development, and other maladies. Dangerous exposure is typically a result of improper removal by "dry scraping, sanding or open flame burning" per the EPA. Lead dust can be created during demolition work in a hospital that can impact workers, patients, and staff. As with asbestos, testing should be done prior to preparing estimates or doing the work.

Removal of lead-based paint should be done a by a professional with appropriate training and safety protection. Scraping and sanding to remove the paint can cause

Figure 12.7 Type C respiratory protection for asbestos abatement of sprayed on fireproofing. (*Courtesy of Gobbell Hays Partners, Inc.*)

dust, so these methods must be used in controlled areas with negative pressure and HEPA filters, to protect the adjacent space, as well as full respiratory protection for the workers. As with asbestos, water or other fluids can be used to minimize dust. Chemical stripping is another choice for lead paint removal, however, it involves the use of chemicals that are also toxic, so professional care is required. Heat guns or blow torches, which are sometimes used for paint removal, are not recommended for lead paint removal, due to the potential for creating lead dust or igniting the material. Painting over lead-based paint is an option, if the material is still bonded well and not flaking.

Polychlorinated Biphenyls

Polychlorinated biphenyls (PCBs) are a group of chemicals manufactured for a variety of uses. They have no smell or taste, and are generally oily and clear to yellowish in color. The chemicals are used due to their resistance to extreme temperatures and pressure. PCBs have been linked to birth defects, as a result of exposure by expectant mothers, liver and reproductive system damage, and skin reactions. The U.S. EPA classifies these chemicals as "probable" human carcinogens. PCBs can be found in caulking, light fixture ballast, lubricants, and other fluids. They were commonly used in electrical transformers and capacitors due to their dielectric and coolant capabilities. Products with PCBs need to be identified during the planning stage of a project prior to preparing estimates or beginning work. Professional environmental engineering firms should be engaged for any remediation efforts involving materials containing PCBs.

Fuel

All medical/surgical hospitals are required to have an emergency generator, which is most commonly powered by a diesel motor. The fuel for these is typically stored in a large underground supply tank, although, in some cases, they are above ground. This fiberglass or, in older facilities, steel tank is generally outside the power house. The fuel is piped to a smaller tank, often called a *day tank*, which is located adjacent to the motor to properly supply the needed positive head pressure. The day tanks will vary in size, depending on the generator as well as the supply tank. Most codes require a medical/surgical hospital to have a fuel supply to operate the facility for 24 hours. There is a system of pipes, pumps, filters, and other devices for this fuel. As with any system, there is the possibility of failures that could result in diesel fuel leaks. This is a potential environmental as well as a fire hazard, plus it can be harmful to construction workers if exposed. This exposure could result from encountering the remnants of old leaks during demolition on an addition or renovation. A spill could also occur on a new project during testing or start-up of the emergency generator and the associated fuel system. The emergency response and cleanup of diesel fuel should only be handled by trained professionals specializing in this field. The scope of this type of leak could vary widely, from a few gallons lost on the powerhouse floor due to a leaky coupling on the pump, to extensive loss of hundreds of gallons, due to a buried pipe leaking underground for many years on an old facility. Fuel that has leaked or been spilled outside can result in environmental damage and contamination of groundwater. The construction manager should make sure proper safety procedures are in place for handling fuel, as well as an emergency response plan in the event of a spill.

Radioactive Material

The imaging department in a hospital often uses radioactive material for some diagnostic testing and therapeutic procedures. While this material cannot be used by terrorists to make nuclear bombs, it could be used with conventional explosives to develop a "dirty bomb" to contaminate populated areas. The materials often used are cesium 137, iodine 131, and cobalt 60. All can be very dangerous. A 1987 accident in Brazil, when a medical device was mistaken for scrap metal, "resulted in four deaths, injuries to many, and widespread contamination of Goiania". From the website article by the International Atomic Energy Agency, http://pub.iaea.org/MTCD/publications/PDF/te_1009_prn.pdf.

During an addition or renovation, construction workers could accidentally be exposed to radioactive materials that are improperly labeled or stored. This material is most commonly found in the nuclear medicine department, but may also be found in chemotherapy and radiation oncology areas.

Prior to construction work in these areas, a risk assessment should be done to identify the potential hazards. The work areas should be searched by trained hospital employees, or third party experts, to verify they are free from radioactive materials. If they are accidentally encountered during construction activities, trained experts in this field should be immediately mobilized to help clean up any workers that have been exposed, and to safely remove the materials. Medical evaluations should be provided for any construction workers exposed to radioactive materials.

Hazardous Construction Materials

In addition to the hazardous materials that may be found in an existing hospital, construction work often involves many different types of materials that will endanger hospital patients, staff, and visitors, as well as workers. These may include paints, sealants, fuel, solvents, adhesives, and other products. Given the variety of materials that may be specified by the designer to be installed in a hospital, or that are used in some way in the construction process, it is impossible to define all of the situations where the materials may be found. The hazards would typically be exposure to breathing the fumes or the risk of flammable materials. Several common examples are provided below

- *Paint.* While most of the paints used today are water based, as opposed to oil based, there are still applications where an oil- or epoxy-based material may be needed. These paints can contain volatile organic compounds (VOCs) that may be harmful to humans. When applying these materials, the work areas must be properly separated from occupied hospital space, and workers must be provided with the appropriate protection. Paint storage can also be an issue as these can be flammable. Large quantities should not be stored in an existing hospital, and material that is combustible should be stored in rooms with rated walls and doors as well as fire sprinklers.
- *Floor covering adhesives.* Many adhesives are now water based; however, they can also have odors or fumes, which can be irritants to patients and staff. Similar precautions should be used as with paint or solvents.
- *Roofing adhesives or mastics.* With the common use of membrane roofs on buildings today, adhesives typically are required to seam the joints in the sheets that are fabricated in 4- to 6-foot-wide rolls. There are several different types of materials used, such as ethylene propylene diene monomer (EPDM), polyvinyl chloride (PVC), and modified bitumen. The adhesives and sealants are also required at the flashing, which is used at intersections to walls, at penetrations, and around roof mounted equipment. In some cases, built-up roofs comprised of liquid asphalt and sheets of bituminous material are used. It is very rare to see coal tar pitch roofs today, but there could be instances where this system is used to tie in to an existing old roof. With both built-up and coal tar, additional safety measures are required, as the process involves heating the asphalt or coal tar to high temperatures so it becomes liquid, and is then applied between and over the roofing sheets, to provide a watertight system. There are odors and gas-fired kettles that add to the risk with these systems. While all roof work is obviously done outside the hospital, there are always fresh air intakes, as well as doors that can allow the odors and chemicals to be introduced into the inside, where they pose risks to patients and staff. The chance of fire must also be considered while installing a roof and when performing "hot" work, such as welding or cutting near a roof.
- *Fuel.* Many construction operations require gas or diesel fuel to power such equipment as welder saws, compactors, and others. As these fuels are highly flammable, they must be stored and used safely, to mitigate the risks to construction workers and hospital occupants.

The Occupational Safety Health Administration (OSHA) and good construction practices require contractors and subcontractors to have material safety data sheets (MSDSs) on all potentially hazardous materials used on a project. These sheets provide information on the material, such as the chemical makeup required, safety measures, health effects, and other information. While OSHA does not have a specific form, the agency does recommend using the 16-section format developed by the American National Standards Institute (ANSI). The OSHA website states that this promotes "consistent preparation of information" that will "improve the effectiveness of the MSDSs by making the information easier for the reader to find …." The 16 sections are

- Identification
- Hazard(s) identification
- Composition/information on ingredients
- First-aid measures
- Fire-fighting measures
- Accidental release measures
- Handling and storage
- Exposure controls/personal protection
- Physical and chemical properties
- Stability and reactivity
- Toxicological information
- Ecological information
- Disposal considerations
- Transport information
- Regulatory information
- Other information

The MSDSs should be obtained from the subcontractors and suppliers by the contractor/construction manager, for all materials to be used on the project, and maintained in a central filing system. There are web-based systems for managing these, as well as a paper copy system. The subcontractors' supervisors and the construction manager should review the plans for all work, to verify that the approach is consistent with the MSDS requirements, and follow up with periodic inspections during the execution, to make sure the project is in compliance with the safety measures. Failure to maintain the MSDS information could result in a violation during an OSHA inspection, which can lead to fines and/or work shutdowns.

The MSDS files should be readily accessible at the worksite, so the appropriate response measures are easily known and taken in the event of an accident. For example, if a worker is accidentally exposed to a hazardous chemical in a paint or solvent, the MSDS should include the first aid measures for the medical responders. Similarly, the MSDS would provide information for cleanup in the event of an accidental spill to minimize environmental damage.

Emergency Response Procedures

In spite of the best efforts to implement risk assessments and use appropriate safety procedures, accidents do occasionally happen. Construction is a risky business with accident rates that exceed other industries. The Bureau of Labor Statistics website shows, in their September 20, 2012 news release, that the construction industry was second in number of fatal work injuries by sector in 2011. This results in the fourth highest fatal injury rate out of 15 industry sectors.

Given the potential for accidents during construction projects, and the increased risk involved with accidents on complex hospital projects, especially additions and renovations, it is important to have a well-defined and comprehensive emergency response plan. One approach to emergency response is to address the incident in two phases, which are often defined as primary and secondary.

Primary. This phase involves saving lives and stopping the immediate threat. For example, in the case of a fire during construction on a hospital renovation, this would include moving patients, staff, visitors, and workers into a different smoke zone or, in extreme cases, evacuating the building, as well as extinguishing the fire.

Secondary. This phase involves stabilizing the situation, providing medical care, assessing the damage, furnishing temporary facilities or housing, and securing the area. In the example above of a fire during a renovation, this would include moving people with injuries to the emergency room, finding temporary rooms for patients that were relocated, providing security, communication with the staff and the press, assessing the damage, and developing plans for interim services, if the space cannot be reoccupied.

Emergency response may be required for numerous reasons, and it may be an issue that is quickly resolved or may involve extended periods of time. The problem could be a construction accident, a fire/life safety problem in the hospital, a natural disaster, such as a hurricane, tornado, or earthquake or a man-made problem, such as a terrorist attack, bomb threat, or a deranged shooter. If one of these occurs during a hospital construction project, the construction manager will need to be prepared to help in the response. It is best to develop an emergency response or emergency management plan for all construction projects. The plan will need to be coordinated with the hospital's current plan on additions and renovations to an existing hospital. On a new hospital, the emergency plan will need to be a stand-alone approach that coordinates with the nearest hospital for emergency services.

An emergency response or management plan needs to be well-documented, with paper copies in the hands of the key individuals for quick response. Electronic copies or records are needed for backup, but there may be times when computers are down due to lack of power, so printed versions will be necessary. Key individuals would include the hospital administration (CEO or COO), director of engineering or facility manager, on-site construction superintendent and/or manager, and possibly others with the hospital, such as the head of security, department heads in the work area, and the emergency department. The emergency plan must also be communicated to the workers through training, signage, drills, and other means.

All project emergency response or management plans will be a little different, given the variation in the scope of work, hospital layouts, community services and capabilities, corporate or hospital management approaches, and other factors. Some of the key elements that should be included are outlined below:

- *Emergency response team and leadership*. The group that will coordinate and direct the response should be clearly identified, along with their roles. The leader, who will be responsible for making timely, critical decisions, should be agreed on and his/her direction should take precedence. It is best to have a multidisciplined team with hospital management, clinical, engineering, and construction personnel. One person should be identified to document the issues that caused the emergency, the decisions made, the timing, and other relevant items to fully record the incident, as these situations can sometimes lead to litigation. It will be much easier to prove the appropriate steps were taken in a timely manner with contemporaneous notes.

- *Command center.* A specific location should be defined where the emergency response team will gather during a situation. The space should be equipped with phones, computers, the emergency plan, flashlights, water, and possibly food. Ideally it would have emergency power, but that may not be practical. There should also be a designated backup location, in case the primary location cannot be accessed or has been damaged by the emergency.
- *Contact list.* The team should develop a contact list, or call sheet, that includes all of the appropriate persons to be notified in the event of an emergency or disaster. This might include the CEO of the hospital, the corporate office, the insurance company, the police or fire department, and others. The contact list should include cell, home, and office phone numbers, emails, or other data that would be helpful. In some cases, if there are numerous people to notify, a "call tree" can be utilized for multiple people to assist in making contact. The communication with the so called "first responders," such as police, fire department, and ambulances, is critical from a time standpoint. This emergency response team should have regular contact with this group and include them in the drills, so they are very familiar with the location, access routes with alternates, evacuation plans, hazardous materials on-site, or other issues that would impact their ability to provide timely and effective assistance.
- *Communication with the press/public.* In order to effectively control the flow of accurate information, one person on the emergency response team should be identified as the point of contact to speak to the press and issue official statements. There have been numerous examples when poor communication and inaccurate information produced disastrous public relations results. One of the worst was the Sago Mine explosion, when the governor of West Virginia, other officials, and the press announced that 16 buried miners were alive, only to later have to confirm all were dead except one. A designated person should develop contact with the local press, so they have a clear system to disseminate information and confirm the accuracy before making public statements. During the training of construction workers and hospital staff, it should be made clear that they are not to speak to the press, to avoid rumors and inaccurate information that could hurt people and damage the reputation of the companies involved and the hospital.
- *Security.* The emergency response team should work with the hospital security personnel on existing facilities, and engage third party security personnel for new hospitals, to develop an approach for controlling access and maintaining order during an event. This may involve a "lock down" of the facility to keep unauthorized people out of the building, controlling "onlookers" so first responders can access the injured, or controlling a disgruntled worker. Proper training, clearly defined rules of engagement and good communication with the emergency response team are essential for positive results.
- *Evacuation points/assembly areas.* As part of the emergency response plan, there must be clearly defined points where the building can be evacuated and assembly areas where the people can be safely gathered and counted. This will help the "first responders" and emergency response teams determine if anyone is missing, and possibly still inside, an unsafe building. The areas need to be a safe distance

from the building and outside of roads that may be used by the emergency personnel. Typically, parking lots or grassy areas outside the emergency egress points, at a stair tower or other fire exits, are used for the assembly areas. In some cases, where people may have been exposed to hazardous chemicals or fuel, a decontamination area with enclosures and showers may need to be provided.

Conclusions

Safety and infection prevention are very important issues when constructing hospitals, especially on additions and renovations where patients, staff, and the public are present. Proper planning and effective preventive measures are essential to avoiding the potential risks that can injure people, disrupt hospital activities, and cause significant financial impact. The construction manager needs to work with the entire design and construction team, as well as the hospital, to safely deliver the project. Good communication, the use of proven tools, and effective leadership skills will be needed to bring together the diverse group that is involved with hospital construction.

Implementing formal construction and maintenance policy and procedures can significantly reduce health and safety risks, minimize disruptions to hospital operations, and ensure compliance with state and federal regulations. The following are best practices to be considered:

Choose the right team. Developing effective policies and procedures is best done by a multidisciplinary hospital team that may include representatives from engineering, security, risk management, nursing, infection prevention, and other departments impacted by the process. A key issue is establishing a construction-specific infection prevention (IP) policy that meets current regulatory and compliance requirements. A contractor experienced in healthcare construction can assist at this stage.

Map risk zones. The team should conduct a hospital-wide infection control risk assessment (ICRA) to map and color code low-, medium-, and high-risk zones. Precautions to be followed during construction or maintenance should be clearly noted for each zone. For example, any construction work in a cancer treatment unit might require such safety measures as isolating the HVAC system, creating negative air pressure, using HEPA filtration, and wearing disposable shoe coverings when entering the work zone. Before a project begins, the risk map alerts contractors and vendors to the level of precautions required.

Provide written policy, procedures. Use a written construction and maintenance policy, drafted by the hospital's team, to clarify the special requirements every contractor must meet when working on hospital property. For small- and large-scale projects, provide contractors and vendors a list of procedures, and task a member of the hospital's facilities team to ensure compliance. The list should begin with the basics, including where to park, any uniform and badge requirements, location of lunch areas and restrooms and which elevators to use. This document would also focus on operational issues, such as timing of the work, noise-reduction measures, and, when applicable, more complex project-specific ICRA requirements.

Train at regular intervals. It is important for the hospital or its general contractor to conduct periodic training sessions for subcontractors, vendors, and appropriate hospital staff on construction-related health and safety issues, including infection prevention procedures to reduce potential airborne and waterborne contaminants. Hospital staff, especially those who oversee contractors, should understand the policies and procedures.

Set up a permit system. To better manage construction work, many hospitals have created a permit system for any cutting, welding, or soldering on a construction project. In this way, facilities management and other relevant departments are informed in advance of any such "hot work" scheduled by the contractor. Interim life safety management (ILSM) procedures should also be set up in advance of the project and communicated to hospital staff. Another safety precaution is setting up a "lock-out, tag-out" control on electrical panels to prevent anyone from turning on a circuit that has been shut down for a project.

Document and monitor. Finally, the hospital should monitor the health and safety steps taken during a construction project, and document findings. For instance, a facilities manager or environmental specialist should monitor air quality at the work site at regular intervals, and compare those results with a baseline reading. Post-construction inspection and air quality testing may also be needed to complete compliance requirements.

Protecting health and safety is extremely important. After all, working in construction involves many risks, such as work involving heights, lifting heavy loads, and working in close proximity with heavy machinery. "Zero accidents" is a term often heard and defines the goal to keep everyone working at or near jobsites 100 percent safe and healthy—including patients, employees, suppliers, and the community. To achieve this, proper planning, training, and enforcement are essential. While the task is difficult, it is not impossible with good team work and support from the leadership in the organizations involved.

CHAPTER 13

Commissioning of Healthcare Projects

What Is Commissioning?

In the context of construction, commissioning is the process of ensuring that building systems are installed and perform in accordance with the design intent, that the design intent is consistent with the owner's project requirements, and that operations and maintenance (O&M) staff is adequately trained to operate and maintain the completed facility.

In the context of healthcare construction, commissioning presents an outstanding way to achieve a facility of which so much is expected. While all owners want buildings that are energy efficient and comfortable to occupants, in a healthcare setting, owners have the added complexities that

- Lives are literally depending on the building's systems
- Building systems are increasingly more complicated, and expected to work interdependently, without fail.

When a failed system can mean the loss of life, the building itself becomes just as accountable as providers under the oath to "do no harm," and, with a population of people whose health is already compromised, health facility commissioning (HFCx) is critical to success.

Construction has traditionally been, and continues to be, primarily a linear process. By contrast, commissioning is circular and iterative. This chapter illustrates commissioning as a process that requires players to revisit plans and designs many times over, working always toward maximizing performance.

Achieving success from the HFCx process requires a clear understanding of individual roles and responsibilities, and early collaboration among a team of people who will be involved in providing patient care, operating the facility, designing the facility, and building it. That team must include representatives from the owner (including the O&M staff), the construction manager and contractors, the design team, and the commissioning authority. To manage the commissioning process, a health facility commissioning authority (HFCxA) will be selected by the owner. The HFCxA will serve as a qualified owner's representative or as an independent third party. Under the leadership

---

This chapter was written by Rusty Ross.

of the HFCxA, the commissioning team should be authorized to work with the project design and construction team to meet the commissioning goals and the owner's project requirements (OPR).

For commissioning to realize its greatest potential, the HFCx team should be in place from the very beginning of a project, and continue through the planning, design, construction, occupancy, operational, and warranty phases. Commissioning is a proactive process, not a "punch list" process. Its focus is to define what is expected (via the OPR), and to advise the team members that what is expected will be measured and documented, whether during the design or construction phase. In the execution of the commissioning process, a list of deficiencies similar to a traditional punch list may be developed, but, more importantly, commissioning will provide documentation that expectations have been achieved. It is critical to the success of the project that this iterative process be applied to the design phase as well as the construction, acceptance, and occupancy phases.

Independent, Third-Party Commissioning Authority

The HFCxA is an advocate for the owner. This person manages the commissioning process and reports directly to the owner. To avoid a conflict of interest, the HFCxA should not be employed in any respect by any member of the design or construction team. (The HFCxA could be a member of the owner's staff, however.) The best practice is for the owner to contract directly with an independent, third-party HFCxA, who will provide the most objective and unbiased services. If this practice is not followed, and commissioning is provided by a design or construction team member, a conflict of interest will inevitably arise.

Further, the primary business of the HFCxA should be providing commissioning services. Members of the commissioning team, such as the contractor or designer, will possess knowledge of the systems involved in commissioning, but their primary focus is their core business—building the facility or designing it. Historically, it has been difficult for design team members to objectively review their own designs, or to critically analyze the effectiveness of specified control sequences. The construction team likewise historically has demonstrated difficulty in objectively assessing the installation of the equipment and systems they have installed and testing the functionality of those systems. As well, the urgency of construction and occupancy schedules often compromise the constructor's objectivity and willingness to engage in detailed testing, or to take the time to correct deficiencies prior to occupancy. Finally, these team members are generally not skilled at developing project-specific documents, such as installation checklists or functional test procedures. They may not be skilled at executing steps on checklists, conducting the test procedures, or troubleshooting or diagnosing integration problems that often arise during functional testing.

Today's healthcare facilities are architecturally challenging, with many complex, integrated, computer-driven systems that control the environment through temperature, humidity, air pressurization, lighting, security, life safety, and other parameters. The criticality of economic realities requires these systems to be as energy efficient as they are life saving. The delivery of the constructed facility in today's market is often the result of multiple subcontractors and vendors delivering their services in "silos." It is difficult to identify who is responsible for the integration of system functionality, integration that is increasingly reliant on computer system integration—systems talking to systems. Achieving this desired integration functionality is a top priority of the commissioning process. It is best achieved by those whose focus is documenting integrated completion, in lieu of completion of individual "silos" of responsibility.

The most urgent focus of many projects, whether in design or construction phase, is meeting the defined schedule for completion. Schedules are important to owners for many obvious reasons; however, nonfunctional facilities are rarely considered acceptable to the owner or its medical staff because the schedule was satisfied. The focus of the commissioning process is to monitor the quality of the design and construction process without respect to the schedule. Commissioning monitors the design process for meeting the owner's objectives, such as maintainability and functionality requirements. Commissioning documents the construction process, ensuring that it meets the requirements for installation, functionality, system integration, and O&M staff preparations, as defined by the design. Should the design or construction team meet the schedule without completely finishing their processes, commissioning will identify many of the shortfalls. This may result in design revisions after the designs are issued for construction, or additional functional or integrated systems testing after occupancy, but, in the long haul, it provides the end user a better product from which healthcare can be delivered to the patient.

A critical aspect of commissioning is that it will be proactive in identifying gaps, gaps that may occur in the design before they become an issue in the design phase, or critical path issues, or budget issues during the construction phase, and gaps in the construction process or owner training process, before they become issues after occupancy.

A third-party HFCxA professional focuses on commissioning, not designing or constructing, and offers objectivity and a wealth of commissioning and team-building experience, required to effectively lead and coordinate the commissioning team. While the commissioning agent will be concerned with meeting the schedule, his tasks will more acutely focus on such items as the design meeting the intent of the OPR, the maintainability of the facility, if the sequences for systems will optimize performance, if the training process desired by the owner is adequately defined, and if the process for the delivery of the O&M documentation and record documents meet owner requirements and are in the format requested. The agent's focus during construction will be similar for system installation, start-up, functional performance, system integration, and the such actual delivery of deliverables as staff training, O&M manuals, and record documents.

Establishing the Commissioning Scope

Establishing the commissioning scope defines the boundaries of the commissioning process. This first step determines the work that will be done (and, by default, the work that will not be done). Defining the scope is the identification of the systems to be commissioned and the commissioning tasks to be performed for each system.

There are tasks to be executed in each phase of a project, tasks that should be defined in the commissioning scope. Each phase of the project will require the HFCxA's involvement, and the extent of that involvement must be defined.

During the design phase, the commissioning process should begin with the development of the OPR. The scope should determine whether the HFCxA facilitates the development of this document, has responsibility for authoring it, or if the HFCxA is simply to review a document prepared by others.

The HFCxA's most effective and proactive service is often the design phase plan reviews. The scope should define the number of reviews and the content of the reviews. Are these reviews to be for commissionability, maintainability, and/or functionality, or should these reviews be somewhat "peer review" focused? The review criteria must be defined. The number of reviews also must be defined, and will be determined by the type of design process used on the project. The traditional process involves packages at each phase: schematic design (SD), design development (DD), and final construction documents (CDs). If a fast-track method of design delivery or a design-build delivery method is utilized, the deliverables by the design team may be different.

The design phase scope should define the number of meetings the HFCxA attends and, specifically, which meetings, usually related to the design reviews required. Projects often require design meetings after the delivery of the design at each major milestone (SDs, DDs, CDs). Some projects will require meetings to ensure the defined sequences of operations have been coordinated. Thus, the scope must define which meetings the owner wants the HFCxA to attend. If the owner desires the HFCxA to attend the pre-proposal meeting, for example, this attendance should be stated.

The commissioning agent should author the commissioning specifications, and coordinate with the design team regarding the impact of commissioning on other sections (identify related sections and additional language needed due to commissioning). The commissioning specifications should contain typical installation checklists and test procedures, in order to establish the rigor of the process of documenting installation and testing. It is critical that the requirements for commissioning be defined for the contractors prior to establishing the construction cost. The commissioning specifications are the vehicle for defining their roles and responsibilities, and because the specifications are a part of the design documents, it is contractually binding to the contractor, subcontractors, and vendors. A design phase commissioning plan is sometimes desired, but generally communicates the same information as the commissioning specifications, and may be considered redundant.

There are numerous activities the HFCxA should perform or be involved in during the construction and acceptance phases, and these activities must also be defined in the scope. The extent of the HFCxA's responsibilities in each of these may include

1. Preparation of the commissioning plan, installation checklists, and test procedures for functional testing and integrated systems testing.
2. Reviews of shop drawings, record drawings, testing records (such as duct pressure testing), O&M manuals, start-up documentation, HVAC test and balance reports, electrical testing reports (such as the emergency generator testing reports), the owner training process, and proposed training programs and supporting literature.
3. Documentation of installation. Should the HFCxA or the installing subcontractors execute the checklists? If the installing subcontractors execute them, should the HFCxA spot check them for accuracy in the field, or simply review to determine if the checklists were completed by the subcontractors?
4. Flushing of hydronic systems. Should the HFCxA attend or witness?
5. Start-up. Should the HFCxA witness the start-up process, review the start-up documentation, or review for adequate maintenance of equipment during the start-up and runtime period, prior to substantial completion?

6. The direction, witnessing, and documenting of all functional and integrated systems testing.
7. Meetings. How many meetings will the HFCxA be required to attend outside of those necessary to plan and execute the commissioning process? Will he or she be required to attend regular owner meetings? (The regularity of the attendance must be defined.)
8. Final report, documenting and summarizing the findings of the commissioning process.

There will still be tasks for the HFCxA after occupancy. These must be defined in the commissioning scope as well. These should/could include peak seasonal testing, involvement in end of warranty meetings, lessons learned meetings, reviews of issues that have surfaced during the warranty period, and other items. One critical service often overlooked is the measurement and verification of actual building/system performance during the warranty phase. It is critical to document the success of commissioning by measuring facility performance during occupancy. This process may be as simple as documenting an Energy Star score, based on energy consumption during the first 12 months of occupancy. Or, it could be a more robust documentation of performance as compared to the energy model developed during design.

A healthcare facility has more than a dozen interactive systems, with 10 times that many subsystems. The main systems include building envelope, life safety, HVAC, controls, plumbing, medical gas, electrical, fire alarm, information technology, fire protection, interior lighting, exterior lighting, refrigeration, vertical transport, and materials/pharmaceutical handling.

Not all commissioning projects will include all systems. A construction project's size, budget, and schedule are likely to shape the scope of the commissioning work. The commissioning on smaller projects may be limited to mechanical systems and critical elements of the electrical distribution system. Or, the effort may be focused on major systems that present the greatest opportunity for operational savings and building performance.

The type of project and the project schedule will impact the scope of the commissioning process. Often renovation projects will have "domino" type phases, which require smaller subsections of scope to be completed, prior to moving on to the subsequent phase. This phasing will impact the scope of commissioning. Projects may require central utility services to be completed prior to the loads they serve. This can require multiple testing of the same systems (once upon substantial completion of the utility, and again after the load is connected to the utility).

When renovation projects utilize existing systems and/or involve additions to existing systems, as they often do, these existing systems should be included in the scope of work. A project that adds or replaces a chiller, cooling tower, or other major elements of a chilled water system will require commissioning of the system as a whole, not just the added piece of equipment. The addition or replacement of automatic transfer switches will impact the essential system, priority loads, load shed schemes, coordination of electrical system protection devices, etc. The commissioning scope should define the extent of testing of these existing systems.

Defining the commissioning scope is not a simple task. It requires clarity and completeness if the owner expects consistent quality proposals by the commissioning agents. Once the commissioning scope has been agreed upon and the team firmly

Figure 13.1 Phases of commissioning.

established and well informed about the commissioning process, the process moves into the design phase (Fig. 13.1).

Commissioning in the Design Phase

The Predesign Conference

The HFCxA organizes and leads a predesign conference prior to the beginning of the design phase. Representatives of the owner, architect, engineer, and contractor (if engaged by this point) will be in attendance. Productivity at this meeting, and all other meetings, is dependent on the authority given to these representatives to make decisions on behalf of their organizations (including the owner).

The predesign conference is a discussion about project goals, scope, team members' roles and responsibilities, and processes. The conference addresses the project schedule and performance expectations.

Successful commissioning will require effective communication, which will typically lead to a number of meetings. Meetings should be used as a tool to facilitate the process and should not be overused. In the early stages of the design phase, the goal of meetings is to firmly establish the OPR and the basis of design (BOD) for systems to be commissioned. These two documents will guide the project through design and construction, into occupancy and beyond.

The Owner's Project Requirements

It is crucial to define and document, early in the design process, the owner's expectations and requirements for the project. The document is referred to as the OPR. The OPR defines the building's functional requirements, and owner expectations for its use and operation, as they relate to the commissioned systems. The HFCxA should work collaboratively with the owner and design team to develop the OPR, soliciting input from key stakeholders on the project team, the owner's medical staff, and O&M staff. This input is key to the quality of the OPR, but, more importantly, it is vital in establishing expectations for the project. Because it defines the owner expectations, it is much more than the design narrative that is traditionally developed by the design team in the pre-design and schematic phases of projects. Often the most effective way to create this document is to conduct an OPR charrette—an organized brainstorming session that gathers input from the key stakeholders on the criteria listed below.

From data gathered during the charrette, an OPR draft is developed for final review and comment by the participants. The *final* OPR is a dynamic and living document, and, as such, will be regularly edited as the project moves through design and construction.

The basic contents of the OPR are as follows

1. *Background.* A narrative description that places the project within a useful context. This description defines the primary purpose of the project and the needs that its programs will meet.
2. *Objectives.* A complete list of measurable objectives, including first cost, change orders, life cycle cost, energy efficiency (which should be consistent with Energy Star rating goal), control of infection rates, patient and visitor satisfaction scores, staffing requirements, maintenance costs, services, floor area, capacity for future growth, and schedule.
3. *Functional program.* Created by the owner and the architect, the criteria contained in the functional program define the specific purpose and use of each space in the building and list any specific design requirements, such as safety, energy consumption, and maintainability.
4. *Life span, cost, and quality.* The expectations of the owner related to the quality of equipment, the expected life span of the equipment, the cost expectations, and the level of system redundancy. Examples include: Should the essential electrical system serve code minimum loads or additional loads? If so, which loads and what are their priorities? Are Uninterruptable Power Systems (UPSs) required to serve some loads? Should transfer switches be closed or open transition? Should stored fuel capacity for generators or boilers exceed code minimum? If so, what capacity is desired? Should HVAC systems have redundant or *backup* equipment (and to what degree of redundancy)? What level of automation and information is expected in the building automation system (BAS)? What level of "smart" systems, such as electrical switchgear components, is desired? What level of integration between the BAS system and equipment control panels (boilers, chillers, etc.) is desired? Which manufacturers are preferred, and are there any that will not be acceptable? What future capacities are expected in systems?
5. *Performance criteria.* The OPR should address minimum performance criteria for systems, such as expected energy efficiencies. The owner may expect an energy model to be utilized, an Energy Star target score to be defined, an Energy Usage Index (EUI) index to be met.
6. *Maintenance requirements.* The OPR should address expectations for system maintenance, the training required for the O&M staff, training archiving requirements, trends that will be required to be set up in the BAS for specific systems, criteria for minimum information related to systems documentation that will be needed for Computerized maintenance management system (CMMS) systems, and O&M documentation format requirements.

The BOD for Systems to Be Commissioned

After the OPR is complete, the next task is to document the methodology to be used to incorporate the OPR into the design of the facility. The design team develops the BOD to define how the design will be developed to meet the requirements defined in the OPR. The BOD should describe the basic design assumptions that would not be described in the design documents. Like the OPR, the BOD is a living document that is regularly updated during the planning, design, and construction process.

The BOD should begin with descriptions of the building (such as floor area and number of floors) and the building envelope (including roof, walls, glazing, vapor barriers, and R values). The BOD should also identify the following:

1. Applicable codes and standards
2. Outdoor design conditions (such as geography and weather)
3. Indoor design conditions (such as temperature, relative humidity, and air changes)
4. Expected building occupancy for all building uses[1]
5. Diversity assumptions
6. Miscellaneous power loads for all building occupancies
7. Illumination levels, controls, and power requirements for all building occupancies
8. Other internal loads
9. Ventilation requirements for all building occupancies
10. Building infiltration and pressure requirements
11. Anticipated maintenance management program
12. Maintenance and service requirements for all commissioned systems
13. Indoor air quality requirements
14. Acoustics criteria for HVAC system
15. Life safety criteria (fire protection, fire alarm, and smoke control)
16. Sustainable design elements
17. Energy conservation measures

Reviewing the OPR and BOD

The HFCxA reviews the OPR and BOD, and verifies that they are comprehensive, specific to the project, and clearly understandable. The HFCxA also verifies that the OPR identifies all appropriate goals and objectives, and that the designs, equipment and systems identified in the BOD are responsive to the goals and objectives established by the OPR. The HFCxA also reviews the OPR and BOD with hospital O&M staff and

---

[1]ASHE, *Health Facility Commissioning Handbook: Optimizing Building System Performance in New and Existing Health Care Facilities* (2012).

documents their comments and concerns. The HFCxA prepares a written list of comments, and forwards the list to the owner and the design team,[2] working in this manner until the OPR and BOD are satisfactory to the owner.

Design Document Review

The HFCxA reviews the design documents produced by the design team for compliance with the OPR and BOD. These reviews should include input from the facility O&M staff. The frequency of reviews should be at the end of each phase of design: schematic design, design development, and construction documents.

The key objectives of all document reviews are the following:

- Verify design document consistency with the OPR and BOD
- Verify design document completion, including all the necessary information for construction, maintenance, and operations
- Verify inclusion of all the features required to execute the health facility commissioning (HFCx) plan
- Suggest alternative designs that will reduce energy consumption, improve maintenance features, and maximize life cycle costs without significant increase to construction budget
- Review all documents with hospital O&M staff and document their comments and concerns

The HFCxA should provide complete written responses after every review, and confirm that HFCxA and owner comments are fully addressed between design stages.

Schematic Design Document Review

During schematic design (SD) document review, the HFCxA should, at a minimum

- Ensure compliance with the OPR and BOD
- Confirm that general concepts comply with "standard" healthcare design practices (see previous chapters)
- Review system capacities against normal rules of thumb (not a detailed load check)
- Confirm that redundancy and future capacity issues have been addressed
- Ensure that adequate space is provided for accessibility, maintenance, and equipment removal
- If the project is leadership in energy and environmental design (LEED), confirm compliance with LEED-related commissioned systems or design criteria, such as non-CFC refrigerants, lighting controls, and daylighting
- If the project was planned to comply with the International Green Construction Code (IgCC), confirm preoccupancy and post-occupancy commissioning details

[2]ASHE, *Health Facility Commissioning Guidelines* (2010).

Design Development Document Review

During design development (DD) document review, the HFCxA should, at a minimum

- Review and comment for space allocation, equipment layouts, and maintainability, and begin reviewing for coordination of systems and general equipment sizing (capacities)
- If the OPR or BOD has been revised since SD, review for compliance with the revisions
- Confirm that the DD concepts match those presented in the SD
- Review of major equipment capacity information provided against experience rules of thumb
- Review for redundancy and future capacity requirements, as outlined in the OPR
- Review the mechanical/electrical room layouts for adequate space for accessibility, maintenance, and equipment removal
- Review architectural sections and reflected ceiling plan information, to spot check above-ceiling clearances to ensure that adequate space is provided for accessibility/maintenance of all above-ceiling MEP equipment (valves, air terminal units, dampers)

Final Construction Document Plan Review

During the review of the final construction documents (CD), the HFCxA should address, at a minimum, operational efficiencies, energy efficiency, maintainability, and commissionability. To accomplish this, in this stage the HFCxA will

- Review that the documents/concepts follow previous submissions (i.e., no major differences from previous submissions)
- If the scope was not adequately defined at the previous design phase, address again for this phase the review elements defined previously for DD
- Provide input on making the systems easier to commission
- Review sequences of operation for clarity and completeness
- Review accessibility to valves, gauges, thermometers, dampers, control components and similar operations, and maintenance equipment
- Review the quantity and location of sensors, flow measuring stations, etc., for maintenance and operations even beyond those required to control the system
- Review for sufficient isolation valves, dampers, interlocks, and piping, so that conditions can be simulated for overrides, failures, etc.
- Coordinate the documents between disciplines for system integration

Commissioning Specifications

Commissioning specifications integrate commissioning into the contract documents for all contractors and vendors. These specifications define the scope of commissioning (systems to be commissioned and the tasks to be performed). The commissioning agent should prepare this specification section, and identify the related sections in the remainder of the contract specifications that will require coordination with the commissioning process. The HFCxA should develop language that will need to be included in these related sections. The commissioning specifications should be in compliance with ASHE Health Facility Commissioning Guidelines or other equivalent guidelines. Another such guideline is ICC G4-2012 "Guideline for Commissioning" published by International Code Council (ICC).

The commissioning specifications should be prepared for inclusion in the SD documents, and updated with each design milestone delivery. The specification format should be coordinated with the design team, to ensure the specifications integrate seamlessly into the construction documents. The specifications also define the roles and responsibilities of all commissioning team members, including, at a minimum, the owner, design team, construction manager/general contractor, such major subcontractors as the mechanical, electrical, HVAC controls (BAS), the HVAC test and balance, and the subcontractors responsible for life safety systems. The specifications should include sample installation checklists, functional performance test procedures, and sample integrated systems tests that define the rigor of the documents that will be prepared and issued as a part of the commissioning plan, after the final design is complete. These checklists are to be defined as examples, and should be clearly noted as such. Project-specific documents will be prepared and issued after the design is finished. As shop drawings are completed, it should be clear that the contractors will be required to follow the project-specific checklists to be issued during the construction phase.

The commissioning specifications will address the following:

- Commissioning team involvement, and roles and responsibilities
- Submittals and submittal review procedures for HFCxA processes/systems
- Operation and maintenance documentation/systems manual requirements
- Project meetings related to commissioning, and the frequency with which commissioning team members are obligated to participate (at a minimum, one at each major milestone: shop drawings, installation, start-up/controls, calibration/test and balance, functional and integrated systems testing, O&M staff training and O&M manual submission, opposed season testing, etc.)
- Construction installation verification procedures
- Start-up plan development and implementation
- The testing plan and ready-to-test procedural requirements
- Functional performance test procedures
- Integrated systems test procedures
- Training requirements, including development of a training plan
- Opposed season testing requirements
- Warranty review and lessons learned site visit/meeting

The Commissioning Plan

The commissioning plan can be developed during the design phase, or it may not be developed until the construction phase. The plan conveys the same information to the project team that the commissioning specifications provide. If developed in the design phase, there will be some project-specific elements of the plan that will remain incomplete until the design is completed, requiring the plan to be updated during the construction phase of the project.

The commissioning plan lists all equipment and systems to be commissioned, includes a narrative defining the process, identifies the key project milestones associated with commissioning, and defines all deliverables required by all members of the commissioning team (Fig. 13.2). It includes project-specific installation checklists, functional test procedures, and integrated test procedures for the equipment and systems to be commissioned. It identifies requirements defined in the contract design documents for tasks the HFCxA will monitor for completion, including O&M staff training, equipment start-up, record documents, O&M manuals, etc.

The basic components of an HFCx plan include

- *Commissioning narrative.* A description of the commissioning process, defining what processes will occur during the design, construction, acceptance, post-occupancy, and warranty phases of commissioning. This narrative will include
  - A description of the roles and responsibilities of each commissioning team member
  - A list of key commissioning milestone activities
  - A list of all work products that will be generated during the commissioning process, and which team member will be responsible for each deliverable
  - A description of the communication protocols that will be utilized

FIGURE 13.2 The commissioning plan describes the commissioning process in *project-specific* terms, as defined by the project design requirements.

  - Matrices defining all specified requirements for start-up, training, O&M manuals, record documents, etc., related to the systems to be commissioned
  - A description of post-occupancy requirements, such as opposed season functional testing, end of warranty meetings/site visits to address warranty issues

- *List of equipment and systems.* A list of all the equipment and systems that will be commissioned, utilizing the equipment descriptions, as defined in the equipment schedule sheets in the contract documents.
- *Installation checklists.* Pre-functional checklists (PFCs) for each piece of equipment and system in the commissioning scope. The PFCs define all elements of installation, related to each piece of equipment, that are required by the contract documents (clearance, mounting height, piping arrangements, control devices, hydronic specialties, insulation, labeling, vibration isolation, etc.).
- *Functional test procedures and integrated systems test procedures.* Functional performance test (FPT) and integrated systems test (IST) procedures for each piece of equipment and system that is to be commissioned. These test procedures will document initial set points, sequences of operation in all modes, all alarm and safeties, start/stop sequences, interface functionality with other systems, annunciation/graphic requirements, etc. These test procedures will document all aspects of the specified performance.

Commissioning in the Construction Phase

Commissioning milestones during the construction phase include

1. Shop drawings
2. Equipment installation
3. Equipment start-up
4. Test and balance
5. O&M manual preparation and O&M staff training
6. Functional testing and integrated systems testing
7. Final commissioning report

These milestone activities chart the process from the design phase through system functionality and the turnover of the constructed facility to the owner. Meetings are utilized to facilitate the process, but are not to be over-used. The commissioning team *kicks off* the process with an initial meeting after the construction phase commissioning plan is completed. This meeting walks the team through the commissioning process that will be utilized, and the commissioning plan serves as a textbook for the process. It defines each step, and includes much of the documentation that will be required. As the construction process approaches each milestone activity, the HFCxA chairs meetings to revisit the commissioning tasks to be executed in the milestone, the communication and cooperation that will be required, and the reporting procedures that will be utilized. The HFCxA utilizes an issues log to document and track items that require action by members of the commissioning team. The issues log tracks all elements of the process that require action from the commissioning team, from shop drawings to integrated systems testing.

Commissioning integrates into the project construction schedule, which will include the milestone activities related to commissioning. The HFCxA works with the contractor in coordinating and scheduling the Cx activities. Shop drawing reviews occur as the design team reviews occur. Depending on the type of Cx review required by the owner, the review occurs either in parallel with the designer's review or after the designer has approved the shop drawings. The HFCxA coordinates his comments with the design team.

The HFCxA reviews installation of systems during regular visits to the site. The goal of the installation reviews is to proactively document the equipment installation and confirm that it conforms to the design requirements (Fig. 13.3). Should the installation deviate from the requirements, such deviations should be noted in the issues log, and resolution should be achieved as soon as possible. The frequency of the visits should be such that the HFCxA can monitor installation of prototypical elements of each piece of equipment as it occurs. All elements of installation should be complete prior to functional testing.

Testing and flushing of systems should be monitored by the HFCxA. As hydronic systems and ductwork systems are pressure tested, the HFCxA should witness a sampling of the testing, and review a sampling of the test reports. The HFCxA should witness a sampling of the flushing process, and review the flushing procedures and documentation. The HFCxA should review similar test procedures for electrical elements to be commissioned.

Figure 13.3 The HFCxA documents system installation, as systems are set in place.

As the project approaches the start-up of equipment, the commissioning team should plan the process. The contractors and vendors should develop a start-up plan, and a plan for maintaining the equipment, prior to turnover to O&M personnel. The HFCxA monitors the execution of the start-up plan, witness start-up of major equipment and systems, and reviews start-up documentation completed during the process by the contractors and vendors.

At this stage of a project, many activities are occurring in parallel. HVAC test and balance (T&B), essential electrical system load bank testing, and preparation of O&M manuals should be completed, O&M personnel training planning should be nearing completion, and preparation and planning for functional and integrated systems testing should be underway. To functionally test an HVAC system that has not been balanced would be an incomplete test. The T&B process establishes the minimum HVAC control set points for devices such as static pressure sensors (airside and hydronic) and air terminal flow stages (maximum, minimum, and interim control points). T&B confirms the accuracy of air flow measuring stations and hydronic flow meters, and other elements of the HVAC control system that are critical to functionality. Consequently, T&B precedes functional testing. The HFCxA should sample the accuracy of the T&B process to confirm the report is accurate.

A testing plan should be developed that defines the sequence for testing systems, based on the completion of construction in the various areas of the facility. For example, if a facility has a patient tower and ancillary areas, the construction would typically progress from one area to the other, that is, from the patient tower to the ancillary areas. The testing plan identifies the systems serving each area of the facility, and sequences the testing of systems to flow with the construction completion in that area. The construction schedule should already indicate that the HVAC T&B process and life safety system testing will follow the construction flow, so too should the functional testing of systems. As individual systems are tested, the testing of integrated systems should follow. Obviously, systems that do not have dependencies (such as the nurse call system and medical gas systems, or domestic hot water systems and sump pumps) can be tested as soon as the construction process is completed on these systems. As the testing process follows the construction completion out of the building, testing of facility-wide systems should be conducted. These would include the testing of systems, such as smoke evacuation systems, "black site" testing of the essential electrical system during which all utility power is disconnected from the facility, smoke evacuation systems, stairwell pressurization systems, and overall building pressurization testing. All testing, except for seasonally dependent testing, should be completed prior to occupancy. Testing of all life safety-related systems should occur prior to demonstration to the local authority having jurisdiction (AHJ), and should be based on the local building code.

At this stage of the project, training of the O&M personnel should be underway, as the staff must be prepared to operate the facility before substantial completion. Consequently, training should be complete before occupancy. The O&M manuals must be completed and available prior to training, as they will be used as reference materials during the training process. The training plan should be developed and executed in coordination with the O&M staff requirements. In addition to the training specified to be delivered to the staff by the contractors and vendors, O&M staff participation in functional and systems testing can be invaluable to their understanding of systems. They should be encouraged to witness the testing as much as possible.

Documentation that will be included in the final commissioning report should be compiled as the commissioning process is executed. This document is a summary of the process, and should include documents that record the commissioning process. The final report should include an executive summary of the process in narrative form, site visit reports, a summary of the issues log and how each item was addressed, all executed installation checklists and test procedures that document the findings of the commissioning process, a summary of the O&M staff training process, plan review and shop drawing review process, and a list of outstanding scope yet to be performed, such as open issue log items, opposed season testing, and other warranty phase activities.

Commissioning in the Post-Occupancy Phase

Commissioning continues after occupancy, with the most common tasks being opposed season testing, an end of warranty review, and documentation of facility performance through measurement and verification. Additionally, the commissioning agent can equip the O&M personnel to operate the facility more effectively by assisting in setting up trends and dashboards in the BAS.

Elements of the HVAC systems cannot be adequately tested unless peak seasonal conditions exist. However, initial occupancy may not coincide with a peak seasonal period, so functional testing of certain aspects of the HVAC system must be deferred until peak seasonal periods during the first year of occupancy. There are two peak seasons: summer and winter. Examples of tests that require peak seasonal testing are full load testing of the capacity of the chilled water system, water side economizers, air side economizers, and humidifiers. The testing must be coordinated with the facility staff, and often must be scheduled for off-peak hours. Participants in the testing should include, at a minimum, the HFCxA, subcontractors responsible for the functionality of the systems (such as the mechanical and controls subcontractors) and the O&M staff. The HFCxA should follow up on any noted deficiencies until resolved. The testing reports should be issued as an addendum to the final commissioning report.

As the project nears the end of the warranty period, the commissioning agent should participate in an end of warranty review meeting, to address any outstanding issues related to the construction process, additional items that surfaced during the warranty period, and lessons learned. Attendees should include all members of the commissioning team. Items that remain outstanding at the end of this meeting should be tracked to resolution. Minutes of this meeting should be issued as an addendum to the final commissioning report.

Measurement and verification of facility performance is critical to documenting that the facility is performing as designed and commissioned after occupancy. It is vital to confirm that the owner continues to receive the ROI from his investment in energy efficient design elements and commissioning. If the design included an energy model, that model will have included an estimate of the monthly energy consumption and an Energy Star target score. The energy model will have assumed certain operating parameters. A straightforward method for documenting performance would be a process that utilizes the EPA's Energy Star program. The HFCxA assists the facility staff in establishing a portfolio manager account on the EPA Energy Star website. Monthly, the HFCxA or the facility staff will enter actual electricity, heating fuel and water consumption, and costs into the portfolio manager account. These figures should be compared each month to the consumption predicted by the energy model. After the first 12 months of data has been recorded and entered into portfolio manager, a baseline Energy Star score will be established, and can be compared against the Energy Star target score identified during the design. If the actual Energy Star rating is less than the Energy Star target score, or the monthly utility bills indicate energy consumption greater than the energy model predicted, the HFCxA should work with the facility manager, design team, and contractor to identify the cause of the disparity and implement corrective action. If an energy model was not created, the Energy Star program can still be used to evaluate energy performance. The HFCxA and the facility staff can still evaluate monthly energy consumption over the first year of operation and maximize energy performance to reduce energy consumption.

Even with a well-trained operations and maintenance staff in place, buildings inevitably experience "performance decay" after occupancy. Performance decay is the natural degradation of building systems that can result in a 30 percent loss of efficiency over time. Performance decay is caused by such factors as sensors gradually falling out of calibration, dirty filters and coils, and "over ridden" set points, or sequences of operation, which were manually changed temporarily, and not returned to automatic mode. Consequently, commissioning should continue throughout the life of the facility, to verify that the building is operating as intended. This work should be primarily accomplished by the facility O&M staff. To assist the staff in monitoring performance, the HFCxA should facilitate the development of dashboards that will readily report critical performance to the staff before an issue arises. These dashboards can be performance-related, such as monitoring operating room temperatures, relative humidity, and pressure, or energy-related, such as monitoring the temperature differential on the chilled water system. Additional dashboards can be created to monitor performance of air terminal units (the greatest waste of energy in a hospital is simultaneous heating and cooling), air handling units, the heating hot water system, the chilled water/condenser water system, isolation room pressurization, and many other systems. While the dashboards present an outstanding tool for monitoring building control, they are only as good as the staff using them and, without proper training, much of their benefit will not be realized. Additionally, the HFCxA can assist the staff in establishing trends of critical elements in the BAS system that will indicate performance issues, such as trends of the chilled water control valves at the air handling units, chiller load factors, air terminal box, and air valve and water valve positions.

Commissioning Standards

While there are many commissioning processes and guidelines available to the industry, including standards created by American Society of Heating, Refrigerating, and Air Conditioning Engineers (ASHRAE), Associated Air Balance Council Commissioning Group (ACG), National Environmental Balancing Bureau (NEBB), ICC, and LEED, the only one that specifically addresses the complexities of commissioning healthcare facilities is a set of companion publications issued by American Society for Healthcare Engineering (ASHE), The ASHE *Health Facility Commissioning Guidelines* (2010) and the ASHE *Health Facility Commissioning Handbook* (2012). Most of the processes discussed in this chapter have described the processes as defined by ASHE. LEED for healthcare is a commissioning standard for healthcare facilities, but it does not address many of the systems critical to the performance of healthcare facilities. Further, many of the tasks in the LEED process are optional, such as plan reviews, shop drawing reviews, monitoring and review of O&M staff training, measurement and verification, and the use of independent, third-party commissioning providers.

Case Study: A Hospital in the Midwest

(Use of commissioning in a large renovation and addition healthcare project)

Project Overview

The scope of this project included renovation of an existing hospital, the renovation and expansion of its central energy plant, and the addition of a new 12-story women's and children's hospital to the existing facility. The completed project expanded the total square footage of the hospital to more than a million square feet, and the construction process spanned a 4-year period. The construction in the existing hospital was phased, beginning in 2008 with the addition to and renovation of the existing hospital's surgical department and central sterilization areas. There were six major phases in the renovation of the existing facility (many had subphases), which included radiology, emergency room, endoscopy, labs, and main entrance, in addition to the surgical suite. The construction of the new tower was completed in parallel with the renovation projects. The central plant revisions included replacing and expanding the existing chilled water system, and adding additional steam boiler capacity. The project was completed in June 2012. The commissioning scope included preparation of the OPR, conducting plan reviews, preparation of the commissioning specifications and a commissioning plan that included installation checklists and functional test procedures, shop drawing reviews, documentation of equipment and system installation and functionality, review of owner training, review of O&M documentation, opposed season testing, and conducting end of warranty reviews. The systems that were commissioned included all major elements of the HVAC system, the electrical distribution system (including all elements of the essential system), the life safety systems (fire alarm and fire protection), nurse call system, domestic water system, and medical gas systems. The commissioning team included representatives from the owner, the design team, the construction manager, all major subcontractors, and the HFCxA team. The HFCxA executed all installation checklists, and directed and documented all functional and integrated systems training.

Commissioning Process and Findings

The process reviewed the design documents at SD, DD, and CD stages, and the review comments were vetted with the design team, and incorporated into the design. The HFCxA prepared the commissioning specifications and reviewed shop drawings. During the construction phase, compliance with the design documents was assessed and documented through the execution of the installation checklists and functional and integrated systems testing. Deviations were noted and recorded in the master issues log, and were tracked to resolution. (A total of 994 deviations were recorded.) It is impractical to list all items noted and corrected during this process; thus, a general summary of findings is provided, as follows.

Surgical Suite Renovation

The operating room (OR) air handling units (AHUs) were scheduled to deliver 55°F and 85°F air into the ORs. With the ORs at the normal set point of 68°F, the systems were required to reach these extremes within 30 minutes, but these systems did not achieve the set point changes initially. Numerous AHUs lacked the required tracking of the supply and return fans, thus impacting building pressurization. The ORs were required to maintain 0.1 inch positive pressure and did not do so initially. Numerous required alarms were not programmed into the systems, and sensors were found to be out of calibration. Leaks were found in the humidifier assemblies. Elements of the fire alarm and nurse call system were not functioning according to design, nor were they addressed correctly.

Electrical Distribution and Essential Electrical System

The load tests on generators serving the women's and children's hospitals failed due to overheating. (The commissioning agent discovered the vendor attempting to cover up the failure.) Work in the existing hospital added 10 new transfer switches, but the priority loads and load shed scheme were not identified for the renovated essential system. Meters were indicating incorrect readings. The alternator heaters on the new emergency generators were "running wild." Skid-mounted fuel tanks for the generators were not vented properly. The phasing on the utility service to the women's and children's hospital did not match the phasing on the utility service serving the existing hospital. Some electrical equipment did not have code-required electrical working clearances.

Chilled Water System

The fuses in the main switchgear serving the 5-kV chillers were oversized. The cooling towers were headed together in a common return water system, yet the cooling towers were not designed with an equalizer line. The 200HP condenser water pumps were designed to start across the line, damaging the contacts in the starters within a short period of time, requiring the addition of "soft start" capabilities. The sequence of operation was programmed incorrectly for adding/dropping chillers based on demand. Pump failure alarms for chilled water pumps (CHPs), secondary chilled water pumps (SCHPs), condenser water pumps (CWPs) were inoperable. The refrigerant monitor was found not to be programmed correctly. The pressure regulating valve (PRV) setting on the makeup water line was found to be set too low.

Heating Hot Water System

Several different independent systems were installed. The control loops were not tuned to maintain heating hot water system (HWS) set points and satisfy the differential pressure set points. On some systems, the PRV setting on the makeup water line was found to be set too low, and relief valves at the heat exchangers were not installed.

HVAC Control System, Building Automation System (BAS)

Graphic screens were found to be incomplete and incorrect. Sensors were found to be located such that they were not able to control properly, and sensors were found not to be calibrated properly. Sequences were incorrectly programmed, and some sequences were omitted. Dead-bands were programmed incorrectly, and safety set points were programmed incorrectly and omitted. Status such of equipment was reported incorrectly. Some equipment was found not to be communicating with the BAS system.

Other Items

The pressure relationship between the spaces in the MRI imaging suite could not be attained. Integrated sequences between the fire alarm system (FAS) and the BAS were not programmed correctly. Elevator recall functions due to fire alarm commands were not programmed as specified, and FAS devices were labeled incorrectly. The nurse call system labeling and programming was found to be incorrect, and devices were found to be missing. When sampling the HVAC T&B report for accuracy, some readings were noted to be incorrect. Numerous pieces of equipment were found not to be installed, including strainers, isolation valves, access doors, check valves, pressure relief valves, dirt legs, drain lines, and crankcase filters on emergency generators. Numerous labels for both mechanical and electrical equipment and pipe labeling were found to be incorrect or incomplete, including arc flash labels for electrical distribution equipment. The scale on gauges and thermometers was incorrect for the type of service being monitored.

Lessons Learned

The commissioning process documented compliance for installation and functionality of equipment and systems, and that owner training and O&M manuals were delivered as required by the contract documents, in many cases without deficiencies. The purpose of commissioning is to document completion, and to confirm that the building reflects the owner's requirements and intentions. It is not a punch list process. However, the commissioning process will also proactively identify noncompliance. For this project, the deficiencies noted were critical to providing the required healthcare environment. They included items related to life safety, healthcare environmental standards, patient comfort, energy consumption, and maintainability. Commissioning saved the owner costs and operational efficiencies, and potentially saved lives with the life safety-related issues. Commissioning also eliminated contractor callbacks, by proactively addressing issues before occupancy, thereby providing savings to the contractor as well.

CHAPTER 14

Occupying the Project

While all phases of a project are important to the overall success, the final completion and occupancy require front-end planning and coordination. It is often an intensely busy time during the transition from construction to use as a healthcare facility. The process of preparing for the owner to take responsibility for the operation of the building is often referred to as "transition planning." With many projects, particularly large healthcare facilities, the owner may hire an independent firm that specializes in planning and managing this complex process. These individuals or firms generally advise the owner to start the transition planning early in the construction phase to facilitate a smooth and timely move into the new facility or addition/renovation. The approach usually involves developing interdisciplinary teams to coordinate the people, systems, equipment, and processes that may be involved.

The scope of transition planning can vary widely due to the vast array of types of healthcare projects. On the small end of the spectrum, it may involve opening a few new rooms related to a newly added imaging modality. On the opposite end, it may involve relocating an entire hospital into a newly constructed facility. The costs and benefits of an external transition planner should be evaluated on each project. On a small project or a phase of a renovation, the coordination could be done by the hospital facility management people and the contractor. Large or multiphased projects would most likely benefit from having a third party planner, so they can focus on this transition, as the hospital staff typically has its own routine duties running the facility, and the contractor's priorities should be safely completing the job on schedule, in budget, and with good quality.

One of the authors of this book was involved with a new hospital built adjacent to the existing facility, and the owner used the move to practice a mock facility evacuation. This provided the opportunity to see in "real time" the challenges and timing involved in moving all the patients and staff out of the building, which would have to be done in the event of a catastrophic emergency. The hospital administration's "lessons learned" included that the amount of time it took was longer than expected, that the staff wanted to bring various types of equipment with them, which further complicated the mass exit, and that it was important to make sure the information technology systems at the new facility were operational.

The transition planning team typically consists of members from several different hospital departments, as well as the contractor, and often subcontractors, equipment vendors, and consultants. The hospital members would typically include infection control, clinical department managers, facilities management staff, administration, and others directly involved with the project. The team would work to develop a detailed plan and schedule for the smooth completion of construction, and the transition to a safe and operational healthcare facility. One approach is to develop a "last 60 or 90 days

This chapter was written by Tom Gormley.

schedule" to graphically illustrate the activities (Fig. 14.1), timing, and relationships. Some of the key activities include

- Completion of construction/punch lists/contractual requirements
- In-service training for the hospital staff/operation and maintenance manuals
- Inspections and regulatory approvals
- Moving people, equipment, and systems

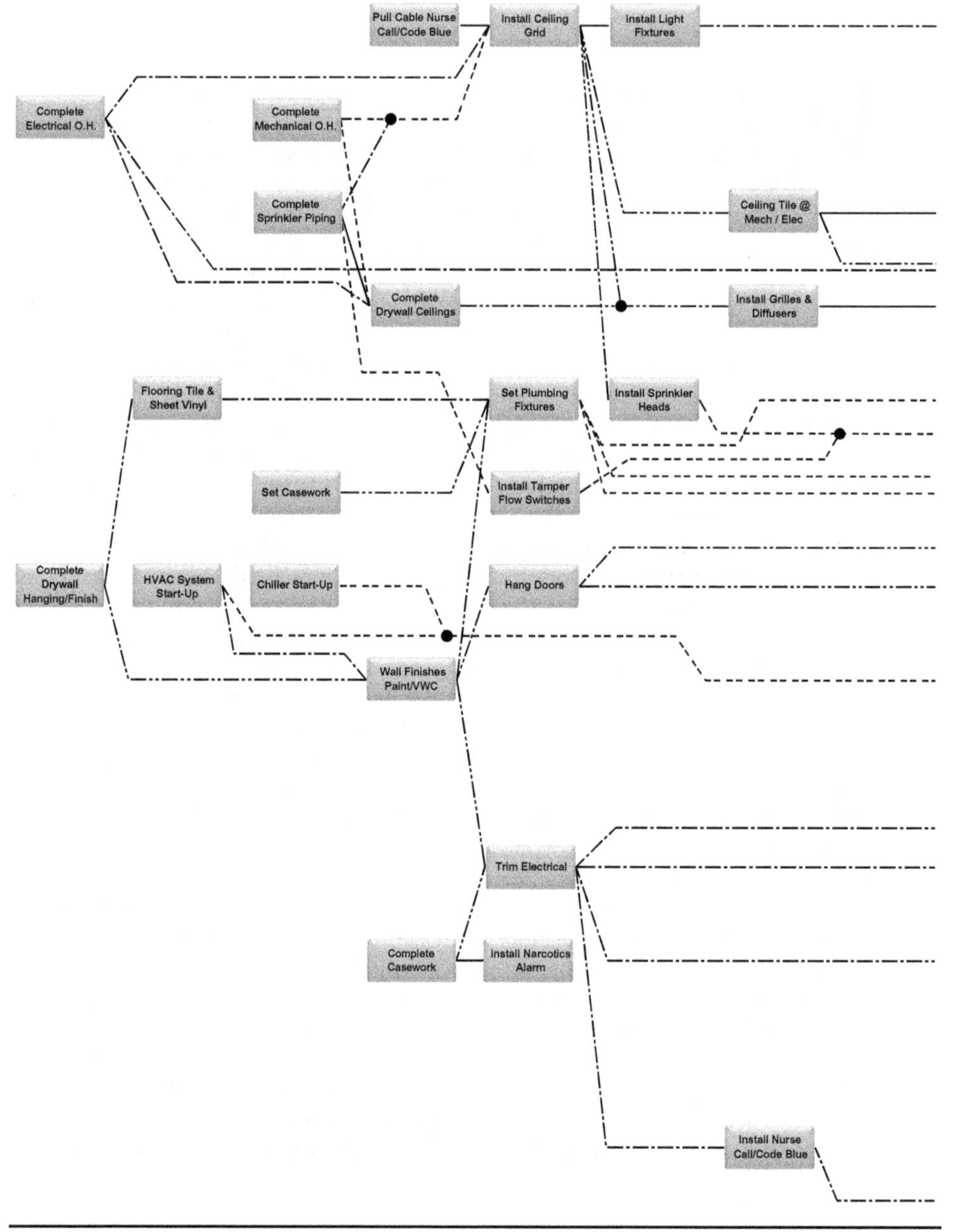

FIGURE 14.1 Graphic of last 90 days schedule.

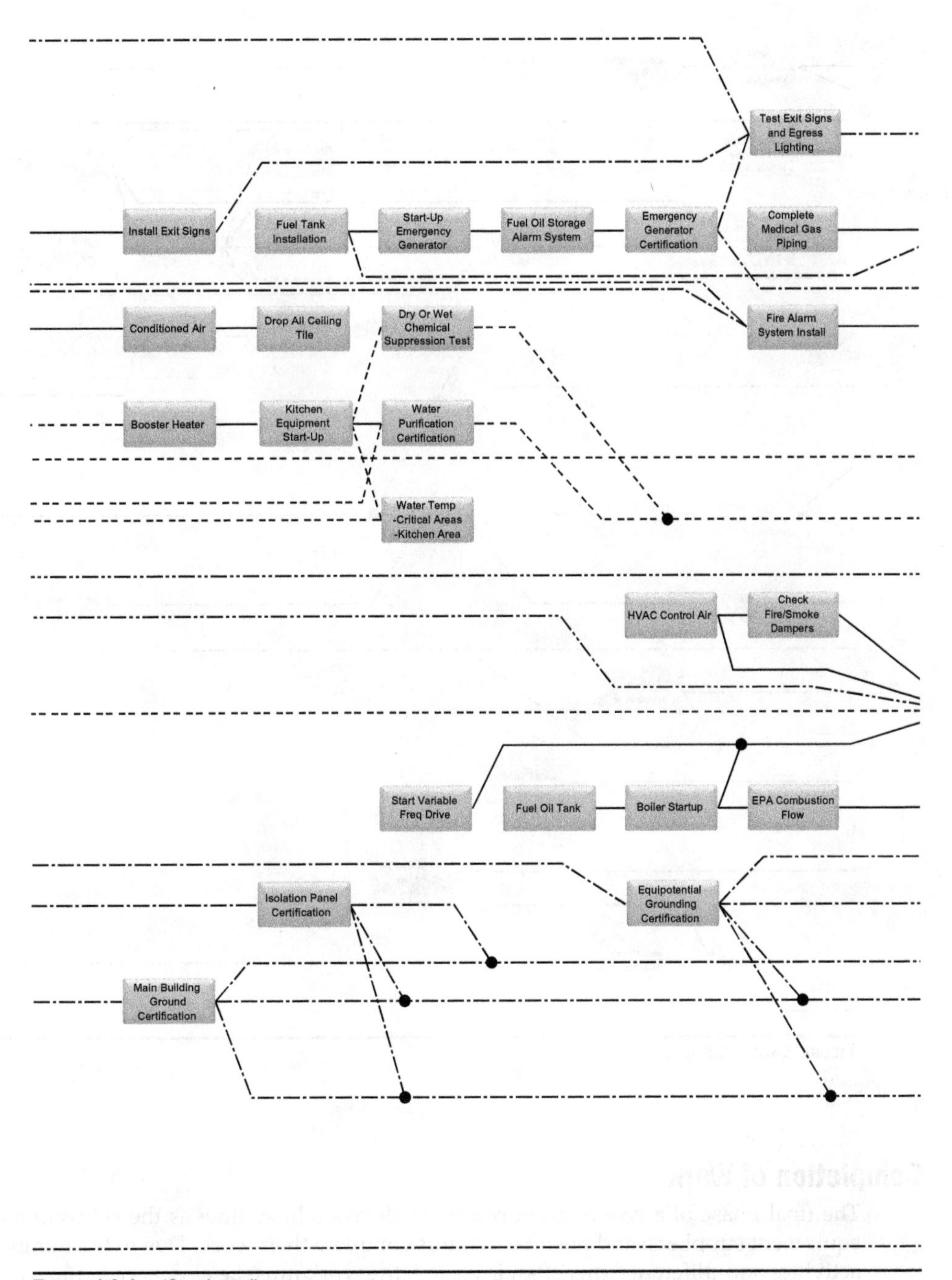

FIGURE 14.1 *(Continued)*

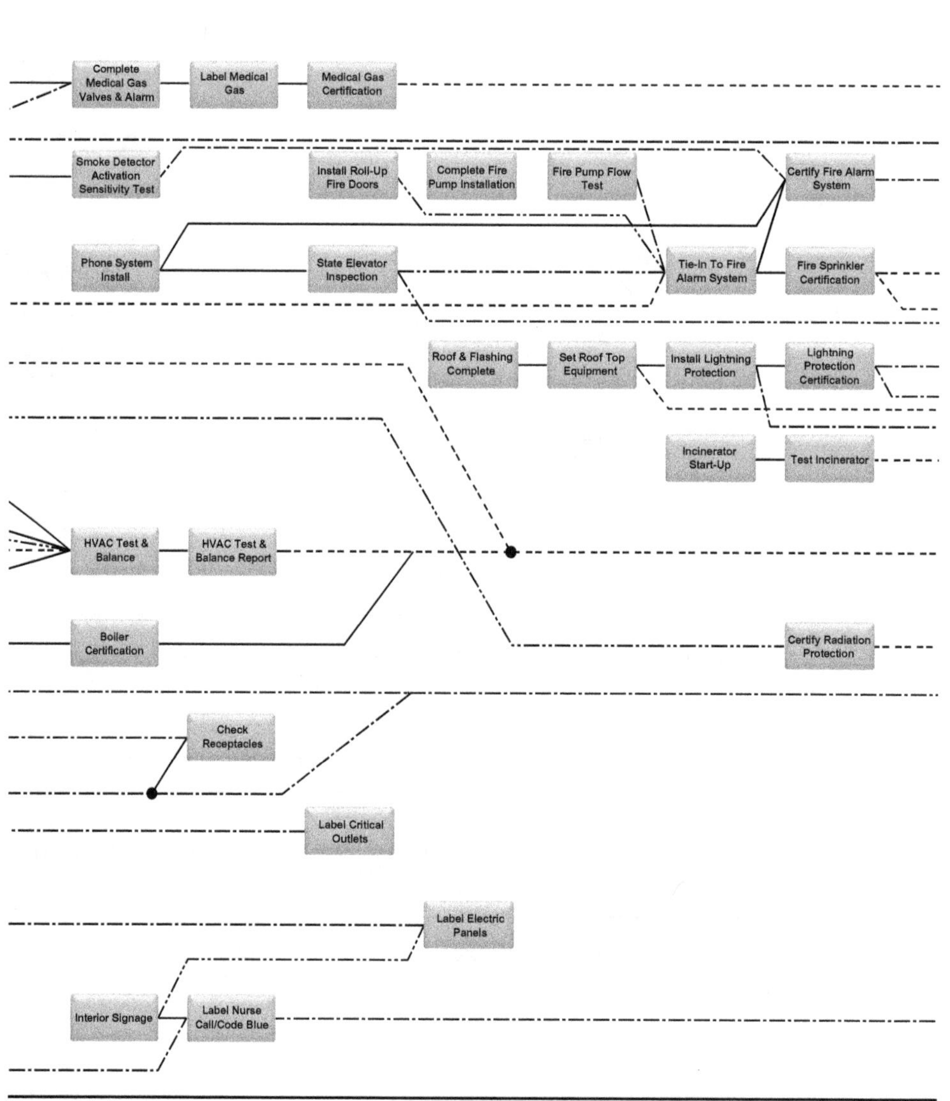

Figure 14.1 *(Continued)*

Completion of Work

The final phase of a construction project is always a busy time as the subcontractors, equipment suppliers, and hospital vendors complete their work. Due to the number of activities and different firms often involved, the work must be closely coordinated. For example, functional air-conditioning systems are often required prior to installation of sensitive imaging equipment, which typically starts weeks before actually seeing patients. Also, elevators must be inspected and ready for use by hospital vendors to move

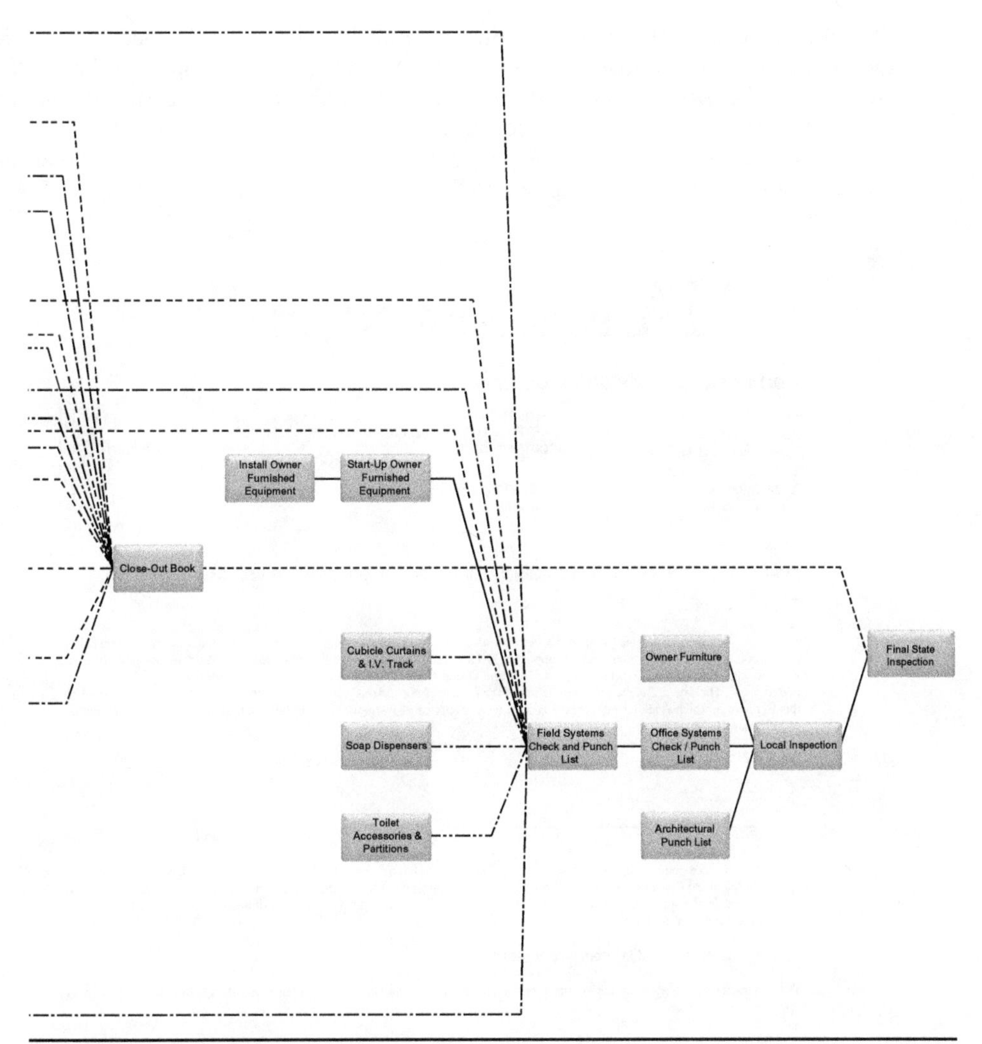

FIGURE 14.1 *(Continued)*

beds and equipment to the upper floors. These are just two of the many activities that are very interdependent between the builders and vendors and are typically under separate contracts directly with the hospital or hospital system. Thoughtful communication and collaboration are required for a successful move.

While the goal of all projects should be to have no construction deficiencies, there is always the need for quality control. The construction industry uses the term "punch list" to describe the process of closely inspecting the work, to verify compliance with the plans and specifications, and developing lists of deficiencies that are communicated to the building team to be corrected. The punch lists are done in different ways, depending upon contractual terms between the owner, designers, and builder. The most common approach is for the builder to have the subcontractors punch, or inspect and correct their own work, then the contractor punches the overall project. At that point the project is ready for inspection by the owner, or more commonly by the architect/engineer on the owner's behalf. If acceptable, the architect would certify the project as substantially complete, and ready for beneficial occupancy. The architect would issue a document to establish the date the project is legally or contractually turned over to the owner. A similar document is shown in Fig. 14.2.

AIA® Document G704™ – 2000

Certificate of Substantial Completion

PROJECT:
(Name and address)
Random Finalized AIA Contracts

PROJECT NUMBER: /
CONTRACT FOR: General Construction
CONTRACT DATE:

TO OWNER:
(Name and address)

TO CONTRACTOR:
(Name and address)

OWNER: ☐
ARCHITECT: ☐
CONTRACTOR: ☐
FIELD: ☐
OTHER: ☐

PROJECT OR PORTION OF THE PROJECT DESIGNATED FOR PARTIAL OCCUPANCY OR USE SHALL INCLUDE:

The Work performed under this Contract has been reviewed and found, to the Architect's best knowledge, information and belief, to be substantially complete. Substantial Completion is the stage in the progress of the Work when the Work or designated portion is sufficiently complete in accordance with the Contract Documents so that the Owner can occupy or utilize the Work for its intended use. The date of Substantial Completion of the Project or portion designated above is the date of issuance established by this Certificate, which is also the date of commencement of applicable warranties required by the Contract Documents, except as stated below:

Warranty **Date of Commencement**

ARCHITECT	BY	DATE OF ISSUANCE

A list of items to be completed or corrected is attached hereto. The failure to include any items on such list does not alter the responsibility of the Contractor to complete all Work in accordance with the Contract Documents. Unless otherwise agreed to in writing, the date of commencement of warranties for items on the attached list will be the date of issuance of the final Certificate of Payment or the date of final payment.

Cost estimate of Work that is incomplete or defective: $0.00

The Contractor will complete or correct the Work on the list of items attached hereto within Zero (0) days from the above date of Substantial Completion.

CONTRACTOR	BY	DATE

The Owner accepts the Work or designated portion as substantially complete and will assume full possession at (time) on (date).

OWNER	BY	DATE

The responsibilities of the Owner and Contractor for security, maintenance, heat, utilities, damage to the Work and insurance shall be as follows:
(Note: Owner's and Contractor's legal and insurance counsel should determine and review insurance requirements and coverage.)

FIGURE 14.2 Sample certificate of substantial completion. (*Courtesy of the AIA.*)

The project can be substantially complete with a punch list of items that are not significant enough to prevent occupancy. However, this point is sometimes open to dispute, because the substantial completion certificate has legal and financial implications. Through this process, the responsibility for insurance, utilities, maintenance, security, and full ownership risks are transferred to the owner. It is imperative for all to understand their roles and accountability. On a new hospital, the utilities for a month could be tens of thousands of dollars, so the transfer point can have significant financial impact. The same is true for insurance, as a loss due to fire or theft around the time of final completion could have significant costs, and even delay opening the facility. Good clear communication and documentation are needed in all of these areas.

Technology has drastically improved the punch list process. There are now electronic punch list systems that can be utilized to expedite the process, with many that include photograph and email capabilities. The effectiveness of the system to document and communicate the list is important, as well as who is responsible for inspecting the numerous building components and systems for the owner. It is important to inspect both inside and outside of the building. A firm and reasonable timeline for the contractor and subcontractor to correct the deficiencies must be established. Once they leave the project site, it is sometimes hard to receive timely response times for corrective work. In addition to the physical punch list, the commissioning of the mechanical, electrical, and plumbing systems is an important part of completion, to provide an efficiently operating physical plant. This is discussed in Chap. 13 of this book.

During completion of a project, and as move-in begins, an issue that should be addressed is protection of the new facility or the addition/renovation, especially finishes. The activities of vendors moving equipment, furniture, and supplies into the recently completed new area should be closely monitored. For example, one of the authors saw new sheet vinyl floor damaged with deep grooves apparently from moving in new imaging equipment without proper protection. The floor had to be replaced, and it was difficult to hold the vendors financially accountable for the repairs, as there were many vendors moving material in the same corridor. Sometimes the new work gets damaged, such as a drywall corner, and there is no way to really tell who did it. The contractor and subcontractors can incur significant additional expenses if this is not controlled. The need to protect the new work should be well communicated to the vendors, and physical protection should be installed, such as "hardboard" sheets over the new floors (Fig. 14.3), pads on elevator cab walls, and cardboard or corner guards on exposed corners.

Completing the construction work on schedule, and allowing time to final clean and correct punch lists, is always the best approach and the easiest. There is less cost for the builder, as they can more easily access areas, and not have to work around equipment, furniture, and other obstacles. There is less disruption for the hospital, and they can more effectively start treating patients.

In-Service Training and Operation and Maintenance Manuals

Typically, the general requirements for this process are found in the project specifications, although often not all in one place. The overall goal is to provide the owner with the information, training, and materials needed to start up and operate the new project. Warranties and guarantees for building components, such as roofs and doors, as well as copies of regulatory approvals, should also be included in the close-out package. The hospital facility management staff must be provided with demonstrations on the

Figure 14.3 New flooring protected (FLEXBOARD) (www.protectiveproducts.com/products/flexboard.html).

start-up and maintenance of equipment, such as boilers, chillers, electrical switchgear, and many others. Sometimes, these are videotaped for future use in training new staff, and electronic access to information regarding parts, maintenance schedules, filters, etc., are provided. Building information modeling (BIM) tools are being utilized to provide owners with electronic plans showing building elements or equipment that is electronically tagged to provide online operation and maintenance information.

In addition to information and training, the contractor and subcontractors are required to furnish the owner with a limited amount of start-up material and spare parts. For example, typically the builder must turn over extra ceiling tiles and floor tiles for the owner to utilize for patching, so colors will match. This is called *attic stock,* and these items are to be furnished in the amounts shown in each section of the specifications. This is often required for mechanical and electrical work, as well. Electronic keying is being used to some degree in hospitals, but many still also have some traditional metal keys. If an electronic security system is being used, then training and the turnover of blanks and keying equipment are required. With either system, the keys for the new building or addition/renovation must be turned over using a systematic and secure process that allows the owner to control access, and to distribute the right access to the right people. The keying approach is often complex, with master and sub master keys, to limit access to some users, such as for an individual office, but also give greater access to some, such as security or housekeeping personnel. The keys can control sensitive areas that may contain drugs, such as pharmacies or med rooms, or offices with billing information or patient medical records. For this reason, the hand-off of the keys from contractor to owner should be well documented, so the responsibility for security is clearly defined during this transition. All keys held by the construction team should be returned to the hospital. An example of operation and maintenance manual table of contents is shown in Fig. 14.4.

Example of Operation and Maintenance Manual Information from Mechanical Subcontractor for Hospital Project

Table of Contents

Domestic Water Pumps
Circulating Pump 1 & 2
Sump Pumps
Domestic Water Softeners
Domestic Water Heaters
Plumbing Fixtures
Med Gas Equip
Control Devices
Fuel Oil Storage
Hydronic Pumps
Exhaust Fans Centrifugal & Belted
Kitchen Make up Air Unit
VAV's
RTU AHU 1-10
Boiler B1 & 2
Packaged Air Cooled chiller RM 1 & @
RTU AC 1 & 2
Computer Room Units AC 4 & 5
Unit Heaters
Electric Heaters
Humidifiers H-4 5 7 8 & Cond Coolers

FIGURE 14.4 Operations and maintenance manual—table of contents.

Inspections and Regulatory Approvals

At the completion of construction, prior to occupancy by the hospital staff and patients, there are numerous approvals to be obtained. The agencies that regulate healthcare facilities are often local, state, and federal. Chapter 2 explores the many codes that impact healthcare. The manager responsible for capital projects should investigate the agencies and regulations that govern their particular location at the start of the job to avoid problems. At the project conclusion, the regulatory agencies typically are scheduled by the builders to inspect the completed facility or addition/renovation. The agencies can include the state hospital regulatory agency (e.g., AHCA in Florida and OSHPD in California), the local or state fire marshal, the local building department (which often has separate mechanical, electrical, and plumbing inspectors), the Center for Medicare services (CMS-federal agency for reimbursement of Medicare/Medicaid), the Joint

Commission, and maybe others. Some states have licensing agencies that are separate from the organization that inspects the facility. Other utilities and public works agencies may also be required to approve some part of the project. For example, the public works or transportation department must approve tie-ins to water/sewer lines and connectors, or turn-ins, from the hospital to public roads. One of the challenges faced by builders is the various agencies enforcing different versions or editions of the codes. This may involve the life safety codes, such as NFPA 101 or 99, the IBC (International Building Code), IFC (International Fire Code), IPC (International Plumbing Code), IMC (International Mechanical Code), IECC (International Energy Conservation Code), and the Guidelines for the Design and Construction of Health Care Facilities.

The inspection process varies with the location, size of project, and AHJ (authorities having jurisdiction). Typically, the inspections are done by architects and engineers employed by the regulatory agency, a fire marshal, the Joint Commission, or federal regulators. The primary focus of the inspections typically is as follows

- Life safety and emergency egress, such as firewalls, lighting, fire alarm, and sprinkler
- Systems operations, such as generators, medical gas, and elevators
- Accessibility through ADA compliance and, depending on state regulations, compliance with state accessibility standards or the International Building Code and ICC A117.1
- Patient and staff safety features of the building, such as nurse call, security, and monitoring
- Air balance and volume, to provide proper pressure relationships and air changes
- Mechanical, electrical, and plumbing systems
- Medical equipment and technology

In addition to the inspection process, the agencies may also want documentation to verify compliance with the codes, as well as with the plans and specifications. Written letters of certification are sometimes required from the architect and engineer of record. Certified reports for testing of the medical gas system and the HVAC test and balance reports are often required, as well as documentation of the fire ratings of materials. These can range from drywall assemblies in fire walls to flame spread ratings on cubicle curtains. One good approach is to prepare a binder (or electronic database) with fire ratings and material safety data sheets (MSDSs) for all the material on the project, to provide quick reference for the inspectors. It is best for the builder and designers to help the inspection run as smoothly as possible for the inspector, so the facility will be approved for occupancy. Significant problems can be created if the inspection fails the first time and must be reinspected at a later time. Often the hospital has already hired staff, scheduled vendors and supplies for deliveries, and assumed the start of revenues. In some states, it may require several weeks for the AHJ to reschedule. This leads to a delay in opening the new facility or area, causing issues for the hospital. The contractor must pay for additional overhead, utilities, insurance, and, potentially, more trips to the project for reinspections. In addition to a thorough pre-inspection by the builder and design team, prior to the AHJ's final inspection, another successful approach is to have work crews on standby during the regulatory agencies' final inspection. With people and materials ready, most deficiencies can be corrected as they are found by the AHJ, so there are not significant issues remaining at the completion of the inspection, which could lead the AHJ to refuse approval to occupy the newly constructed space. Ladders, radios or cell phones, flashlights, tools, and materials, such as drywall, fire caulking, ceiling tiles, and door hardware, are essential items for a final inspection. It is also important to consider the right people from the design and construction team, as well as from the hospital, to accompany the inspectors. Typically, it is best to have someone that is knowledgeable of the project and the codes, and has good political skills, as there can be a good deal of subjectivity when interpreting the requirements of the regulations.

Moving Equipment, Systems, and People

Depending on the type of project, the moving process may vary widely. For a small addition, such as adding a few treatment rooms to an emergency department, the hospital staff may move the furniture in. For a major addition or a new facility, professional moving companies will be required for the large number of items that must be moved in and set up in the new spaces. Generally, vendors include the labor and equipment needed to install the materials in their contracts. This may include hospital beds, patient or waiting room furniture, drapes or window treatments, kitchen equipment, televisions, shelving, office furniture, or various other items. Often, vendors do not have the in-house staff needed, so they outsource to third party moving companies, which can require additional supervision and coordination. Without this, the finishes in the new project, as well as the items being moved, can be damaged, and the items can be placed in the wrong location, or even stolen. Security of the building is of utmost importance at this time, as there are many people entering and exiting the facility, including hospital staff, vendors, contractors/subcontractors, inspectors, etc. New televisions and computers are a prime target during this transition phase.

The other issue that often emerges is the desire by all parties to access the loading dock and elevators at the same time, to meet deadlines and complete work prior to the grand opening. Trucks are often in line to make deliveries, which can lead to traffic jams, frustration, and delays. Simultaneously, there can be large quantities of cardboard boxes, wrapping materials, trash, and debris that need to be removed from the new building/renovated area. For these reasons, a well-defined and coordinated plan needs to be established in advance of the move-in date, to minimize disruption and to allow everyone to complete their portion of work in a safe, timely, and effective manner. The multidiscipline transition planning team must provide the direction. This is why it is important to have the representation from the right constituencies, as those involved in this phase of the project may be employees of the hospital, a vendor for the hospital, a contractor or subcontractor, or an inspector with one of the regulatory agencies.

Medical equipment is another critical element of the occupancy process. Equipment is discussed further in Chap. 11. This may range from complex imaging modalities, such as an MRI, sterilizers, and lights for operating rooms, physiological monitoring systems in the ICU, or even basic, portable x-ray machines that merely plug into the wall. The installation of complex equipment generally begins weeks or months before occupancy, as it must be installed, powered-up, tested, and certified prior to use with patients. An MRI or a CT scanner may literally take several months to install, while an

OR light can be completed in a day. In spaces using radiation, a physicist must test and certify the shielding is properly installed, to provide the protection needed for staff and patients. This written certification must be included in the close-out documents or O&M manual. This diverse variety of equipment requires a wide range of time periods, tools, and manpower to install and ensure everything is working properly.

The medical equipment is often sensitive to temperature or humidity changes. Also, it needs "clean" power sources during the installation, start-up, and testing phases. Close coordination is required by the contractor with the mechanical and electrical subcontractors, to make sure the appropriate services are provided for the medical equipment installation. This can be further complicated during the commissioning and testing of the mechanical, electrical, and plumbing (MEP) systems. For example, during the testing of the emergency generator, the normal power is shut down to verify the automatic transfer switch works properly, and to start the generator and transfer the load in the code-required 10 seconds. An unexpected loss of normal power could cause significant problems for the equipment installer working on the computer that controls the PET scanner. This coordination, communication, and scheduling would be the responsibility of the transition planning team.

The relocation, or initial installation, and start-up of computer and information technology systems can be as complicated as the medical equipment. These systems might include electronic medical records, the picture archiving and communication (PAC) system, accounting or billing systems, surgery or emergency department scheduling, and others. The first step is to have all hardware installed and operational. This involves installation of the cabling, the equipment in the distribution rooms, and the actual computers or systems that will be utilized by the hospital staff. This often also includes printers, display monitors, or other periphery devices. As with medical equipment, the power and air-conditioning systems must be fully operational for the systems' installations to be successfully implemented. Often these systems require links to the Internet for access to data, or to interface with the owner's organization, so this link must also be complete. These computer and information technology systems must be installed and tested prior to occupancy, before patient care can begin.

The final step for moving in or occupying the new facility or addition/renovation is moving the supplies and instruments, staff, and patients. Generally, the regulatory agencies do not want the supplies, such as linens, drugs, surgical instruments, food, and other operational necessities moved in until their inspections are complete. This is different for the Joint Commission, or other licensing agencies, because they are surveying not only the facility, but also the operational functions of the hospital.

Once the construction work is complete, the regulatory approvals are received, and the certificate of substantial completion is executed, thus moving responsibility to the owner, the hospital can officially occupy the new building or space and begin treating patients. Typically, the staff would move in first to start training on the new systems or equipment, stocking the needed supplies, and prepare to safely receive patients. Depending on the scope and type of project, the staff may need days or weeks to adequately prepare to start providing healthcare. In some cases, the patients may be slowly transferred into one new space at a time, or the transition can happen with several being moved simultaneously. For example, if a new ICU wing has been added to an existing facility, then the hospital may just transfer a few patients at a time to one wing, and slowly ramp up to full occupancy. On the other hand, if a new emergency department has been added, then the move into the new space may occur at one time, so the ambulance entrance and emergency services can be quickly relocated, and become functional during a slow period of the day, such as mornings. The most complicated is obviously a complete new hospital, where the patients must be transported from an existing facility to the replacement. A move of this scale can require numerous ambulances, and months of planning, to prepare. Typically, elective surgeries are delayed, and as many patients as practical are discharged, to minimize the number of transfers.

It is generally a good practice for the contractors and subcontractors to have manpower and materials available during the move. There will always be something that needs to be fixed or adjusted during this busy time. Some of the key subcontractors that should be included are the mechanical, electrical, and elevator. Supervisory staff from the contractor should also be available to help coordinate work, answer questions from the hospital, and provide support as needed. This is a prime opportunity for the builder to enhance the owner's perception of their value and the benefit of having them on the project.

The occupancy process can also be very complicated on multiphased additions and renovations. The move-ins may occur numerous times for small areas over the course of the project, or multiple times in just a few weeks. Many renovation projects involve a domino-like sequence, where the hospital moves out of a small area to allow the contractor to complete the work there, and then they move back in to open up another area for construction. An example of this may be a vertical expansion on a patient tower, where the hospital would move out of two to six rooms, to allow the contractor to extend the plumbing and duct risers up to the new floor. Once these are done, another few rooms would be taken out of service for construction, and this sequence would be replicated over the entire floor until all the risers are connected. It is a time-consuming and costly approach, but often necessary to keep adequate patient rooms, or other services, open for patient care. The contractor must take special precautions while working adjacent to patients, which are further explained in Chap. 12.

Conclusion

While all phases of the project are important, this hand-off point is often a busy and complicated time that requires good planning and communication. There may be many different stakeholders, including the hospital staff and physicians, facility management, equipment and furnishing vendors, regulatory agencies, and the builders. The best approach is to develop a multidisciplinary team, or to outsource to an expert, to develop and implement a detailed transition plan.

Figure 14.5 shows a sample table of contents from a transition planning team that was used on the move into a replacement hospital. This provides insight into some of the key issues to be addressed, and tools used, for occupying a new facility.

This is also an important time for the builder from the standpoint of the owner's perception. At this point, the hospital will start to use and evaluate the quality of the project's construction. Until now, they may have only seen it from a distance or during brief tours. Now they will actually experience and use the facility that will lead to perceptions, feelings, and maybe conclusions, about the builder's performance. It is

Typical Elements of a Hospital Transition Plan

Team charter
Transition vision/guiding principles
Move assumptions
Move structure
Job Descriptions
Patient flow
Transfer guideline
Acuity classification
Department move plans
Move sequence
Sequencing worksheets
Transfer order set
Move record
Mock scenarios
Mock move evaluation forms
Patient/physician/office setters
Training agendas
Patient evaluation

FIGURE 14.5 Transition planning team—table of contents.

worth the constructor or builder's investment to have staff on-site to correct any problems and provide support during the move-in. This is often the end of the hospital's interaction with the contractor and subcontractors, so their perception of a successful project can lead to excellent references and more work in the future, or to a contrary opinion. The goal is a smooth transition that allows the clinical staff to start safely, efficiently, and cost-effectively operating the new healthcare facility.

CHAPTER 15

The Future of Healthcare Construction

While the Egyptians stitched wounds and set bones, their facilities are thought to have been more like temples to healing gods, and the physicians made house calls. The earliest traces of a facility dedicated to housing sick people have been linked to Mesopotamia or the India/Sri Lanka region. Architectural evidence from ruins in Sri Lanka, from the 9th century AD, appears to suggest 13- by 13-foot patient rooms, which are surprisingly similar to today's room size. The early Roman military hospitals appeared to use a three-patient ward concept, while physicians made house calls for private citizens. The ward concept continued to be popular as the church took an increasing role in public healthcare. It is reported that the rich often made larger donations to the church to have better accommodations in the hospitals, many of which were attached to the cathedrals, as ordered by Charlemagne in the 6th century. The wards for the poor grew until they often became unsafe, such as at Hotel-Dieu in Paris in the mid-1700s, which had wards with over 100 beds, and multiple patients in each bed. The Nightingale era helped improve the unsanitary conditions as the "pavilion" plan was developed to provide fresh air and daylight to improve patient care and reduce infections. The perception of hospitals as a place to die was gradually being replaced as a place for treatment and recovery. Two early and notable facilities used the "pavilion" plan—St. Thomas Hospital in London and Johns Hopkins in Baltimore, which had 24 bed wards. One of the earliest noted cases of "pay" hospitals was at Hopkins, which had two "pay" floors with single rooms. One hospital built in London 1842 had eight single beds that were used on a "pay" basis. As medical and technological advances were made, there was more need for "in patient" services for the poor, as well as the "pay" patients. While house calls by physicians continued in the United States up until the 1940s, hospitals were expanding and providing more services. The size of the wards slowly declined as evidenced by the construction of Mount Sinai in New York City in 1920, with 26 bed wards, while Montefiore Hospital built in the city in 1950 with four bed wards and some private rooms.

The Hill-Burton Act, passed by Congress in 1946, led to significant hospital construction across the country. The goal was "to assist the several States in the carrying out of their programs for the construction and modernization of such public or other nonprofit community hospitals ... to furnish adequate hospital, clinic, or similar services to all their people." The effort was also "to stimulate the development of new or

This chapter was written by Sanjiv Gokhale.

improved types of physical facilities and to promote research, experiments, and demonstrations relating to the effective development and utilization of hospital, clinic, or similar services." The other key points that continue to have an impact on hospitals today were the requirements for "standards of construction and equipment" and "criteria for determining needs." The requirement for standards led to the development of written guidelines that eventually evolved into today's Guidelines for the Design and Construction of Health Care Facilities. The criteria for need eventually developed into the certificate of need laws that are present in many states in the eastern part of the United States. Through a better understanding of the hospital environment and evidence-based design concepts, the ward concept for the most part came to a close in the United States with the 2010 Guidelines that require private patient rooms in all medical/surgical hospitals.

Technological advances have greatly increased over the last 40 years with the advent of the computer. As medical procedures continue to improve, the technology and facilities continue to evolve and become more complex. Hybrid ORs for image-guided surgery, proton therapy for cancer treatment, and CT scanners for cardiac studies are just a few of the advancements that have increased the technical requirements for construction of a hospital. Given this background, the hospital industry and the public healthcare system continue to adapt to the incredible medical and technological advancements as well as an increasingly political environment. This leads to a complex picture for the future of hospital construction as there are many competing issues:

- The "baby boomers," in record numbers, are reaching the age that will result in the need for more healthcare on top of the increasing overall population
- The hospital infrastructure is aging, which will result in the need for significant capital improvements
- More healthcare is being provided on an outpatient basis that could result in the need for fewer in patient beds
- The increasing role of the government in healthcare, in response to the growing overall cost to our economy, which is now 17 percent of the gross domestic product (GDP)

While there are just a few major factors, there are many other issues affecting the more than 5000 hospitals in the United States. It is clearly beyond the scope of this book to project the future of healthcare in our country or others; however, these issues will directly impact the type of hospital facilities to be built in the future. According to Health Policy Center (Urban Institute—http://www.urban.org/health_policy/uninsured/), there are currently 46 million nonelderly Americans without health insurance, including both children and adults. Once fully implemented, the new healthcare reform law—the Patient Protection and Affordable Care Act (PPACA) of 2010, is expected to cut the uninsured rate by roughly 60 percent. It is possible that these newly insured may seek more routine and preventative care at the physician's office, thus reducing the demand for emergency room services. For the past 10 years, there have been many emergency department expansions, as this was considered the "front door" to the hospital, due to the large number of admissions. Improvements with stents and interventional cardiology have reduced the number of open heart surgeries, which may lead to construction of fewer operating rooms and more "cath labs." The healthcare trends will drive the construction needs and projects, so it is important for the hospital construction management professional to stay in tune with the industry through trade publications, conferences, and involvement in developing the codes and regulations. Understanding the direction will help one be better prepared from a technical and career perspective.

There appear to be some consistent trends that are having an impact on hospital construction:

Building Information Modeling (BIM)

While this technology is impacting all construction, it is gaining wide acceptance on hospital projects. The ability to better coordinate the complex systems in a hospital is improving the quality, cutting time, and reducing cost, as well as allowing the next trend—off-site prefabrication.

Prefabrication

Through the use of BIM modeling, the design details are much more accurate, which is allowing for prefabrication off-site using a manufacturing approach, rather than most of the work being built in the field. This is improving quality, reducing cost, and saving time as well as providing a safer environment for the workers. Sections of corridors of above-ceiling walls and mechanical/electrical systems are being built with incredible precision off-site, and shipped to the site to be hoisted in place. Entire bathrooms, with rough-ins complete, have been built off-site and lifted by crane into the new hospital. These approaches have the potential to significantly change the building techniques used on new construction and renovations.

Increasing Technology in the Hospital

There have been tremendous advances in medical imaging equipment and medical information technology. Electronic medical records are the standard today, and will force the need for more integration of diagnostic and therapeutic equipment and processes. Hospitals are becoming more advanced technologically. Electronic infrastructure will be required to support the medical systems and, as well, building systems, such as security, fire alarm, and energy management.

Sustainability/Energy Efficiency Measures

Numerous "green-" and LEED-certified hospitals have been completed in the past few years. This trend will continue as there is more pressure from a political, environmental, and financial standpoint. New ideas and techniques will be incorporated into hospital construction to reduce the "carbon footprint" and conserve our limited natural resources.

Importance of Cost in the New Economy

As the cost of healthcare has grown to 17 percent of the U.S. GDP, there is more pressure on many fronts to slow the steady increases that have been the norm for the past years. The Patient Protection and Affordable Care Act, while still hotly disputed, is intended to reduce healthcare costs while improving quality. This cost pressure has and will continue to have an impact on the types and number of hospital construction projects.

While there are many factors that will shape the direction of hospital construction, there are experts with insight into the how this industry will continue to grow and develop. This chapter provides a diversity of opinions for the reader through summaries of interviews with leaders in the healthcare capital development industry, giving their perspective on the future of hospital construction. These people include owners, designers, and builders with many years of experience working on hospitals. The authors reached out to these recognized leaders in the healthcare design and construction profession to solicit their opinion regarding the future of "healthcare construction" in order to provide the readers with the latest thinking on the direction of this dynamic field. Each person was asked the same series of six questions. They were encouraged to only answer those questions they felt they were best suited to answer.

Below is a compilation of the list of questions and *selected responses*.

Questions

1. Driven by the Affordable Care Act, the U.S. healthcare system is shifting to managed contracts and clinic-based care. The days of the big-box hospital may be numbered and, instead, the big box may be replaced by a smaller "mother ship" surrounded by community clinics, smaller tertiary clinics, and more retail-based clinics. Do you agree/disagree with the statement?
2. "Doing more, with less"—Owners are often faced with the challenge of providing the biggest "bang" for the lowest "buck." How can today's hospitals achieve the highest value from their capital improvement programs despite inevitable spending barriers?
3. What do you see as the "biggest trends" in design and construction of healthcare facilities?
4. What is/are the biggest construction material and/or construction process successes and failures that you have seen on hospital projects?
5. How can the hospital builder or construction manager have the greatest impact on improving the quality of hospital construction work and processes?
6. What delivery method for hospital projects do you think produces the best overall results, based on your experience?

Designers/Planners

James W. Bearden, AIA
CEO, Gresham, Smith and Partners
1400 Nashville City Center, 511 Union Street
Nashville, TN 37219
www.gspnet.com

Ray Brower, AIA, NCARB
Vice President, RTKL Associates Inc.
2101 L Street NW, Suite 200, Washington, DC 20037
www.rtkl.com

Sandra Shield Tkacz, AIA, ACHA, EDAC, NCARB
Director, Healthcare—URS
277 W, Nationwide Blvd., Columbus, OH 43215
www.urscorp.com/Healthcare

Richard L. (Dick) Miller, FAIA, EDAC
President/Principal

J. Todd Robinson, AIA, EDAC
Principal/Senior Designer
Earl Swensson Associates, Inc. (ESa)
2100 West End Avenue, Suite 1200
Nashville, TN 37203
www.esarch.com

Owners

Joey L. Abney
VP Design, Construction & Plant Operations
IASIS Healthcare, Inc., Franklin, TN 37067
www.iasishealthcare.com

Robert F. McCoole
Senior Vice President, Facilities Resource Group
Ascension Health, St. Louis, MO 63134
www.ascensionhealth.org/

Louis C. Saksen, AIA, FHFI
Senior Vice President, Parkland Health and Hospital System
2222 Medical District Drive, Suite 300, Dallas, TX 75235
www.parklandhospital.com/

Contractors

Robert J. Nartonis
Senior Vice President
Mortenson Construction, Minneapolis, MN 55422
www.mortenson.com

Andrew S. Quirk
Sr. Vice President and National Director
Skanska USA Building, Healthcare Center of Excellence, Franklin, TN 37067
www.skanska.com

1. **Driven by the Affordable Care Act, the U.S. healthcare system is shifting to managed contracts and clinic-based care. The days of the big-box hospital may be numbered and, instead, replaced by a smaller "mother ship" surrounded by community clinics, smaller tertiary clinics, and more retail-based clinics. Do you agree/disagree with the statement?**

   [Abney] Agree. IASIS Healthcare currently has a freestanding ED under construction in Roy, Utah. The FSED is 8 miles from our Davis Hospital & Medical Center in Layton, Utah. The New ED will act as a feeder to Davis's main campus, while providing a much needed service to the community in Roy, Utah.

[Bearden] Agree and disagree. The delivery system will change, but the need for hospital beds based on acuity will still be present. Granted, the new system of delivery will be more about managing the entire needs of the person. This means a variety of needs and a variety of types of environments to provide them in.

[Brower] Two strategic approaches are being implemented by our clients simultaneously, and I believe that the era of "big box hospitals" is alive and well, but a significant shift toward ambulatory care is occurring within the marketplace for many reasons.

1. Hospital consolidation into a few dominant healthcare systems within each market eliminates weaker, single, stand-alone, nonaligned providers, creating health system strength to negotiate profitable insurance contracts. These larger healthcare systems are dominating markets, maintaining efficient delivery of centralized services coordinated among a group of hospitals, and, therefore, the "big box hospitals" in certain markets remain alive and well. On the other hand, academic medical centers burdened with high costs are struggling to meet the value targets of insurance plans, and are looking for innovative approaches to delivering care at lower cost settings. Lower cost settings include merging with suburban hospitals and developing ambulatory facilities. So, in larger "regional" medical centers in suburban markets, I think the "big box model" will continue to thrive. It's not a one size fits all approach to strategy that drives hospital development.
2. Recent cuts by CMS to "disproportionate share" providers will hurt major academic medical centers financially, and this will probably continue to limit growth of core medical centers for the foreseeable future. This is mostly due to financial pressure, not strategic desire to improve/replace aging facilities and expand in place.
3. Expansion of ambulatory services is still a major part of strategy to provide community-based care at the lowest cost setting, but physician subspecialists (especially in academic environment) still want to maintain proximity to the major medical centers for convenience and efficiency.
4. The major medical centers continue to expand and dominate delivery of inpatient and subspecialty care. Subspecialists now being employed by hospitals want to be located on campus for convenience, as they are primarily "proceduralists" (ob/gyn, cardiac interventionist and surgeons), so the main campuses continue to expand with new medical office buildings, ASCs, "cardiac institutes," "orthopedic centers," etc., located on campus. Bed tower expansion is now on the back burner, but main campus expansion still continues in many institutions, with a focus on ambulatory services. I think we've built so many new bed towers over the past 10 years that demand for these is now minimal. Once again, I don't think it's so much a change of strategy, but rather a change of priorities, as most facilities solved their inpatient deficiencies over the past decade. Those that did not are not going to survive anyway.
5. As stated earlier, the health systems continue to expand their ambulatory care network to effectively respond to risk-based contracts that require a greater population base. Ambulatory care centers under development, located away from the medical center, are either suburban health systems defending their

market from inner-city academic providers who must expand their presence outside of the city, or inner-city providers expanding into major population centers. While this is the most visible impact of the Affordable Care Act (ACA) on facility development today, it's not the only strategic driver.

6. Insurance companies are gaining the upper hand and driving patients to the highest value providers (cost & quality), which most likely include the inner-ring suburban, nonteaching, healthcare systems that already have established primary care ambulatory networks, and are well positioned to meet the challenges of the ACA. In this case, they will probably limit expansion of inpatient facilities and push ambulatory care center development, both on campus and at community-based facilities.

[McCoole] I somewhat disagree, in terms of one aspect of the issue. I understand the delivery of care is changing, but how are we going to afford to demolish the big boxes? How will we also afford to build the mother ships? The big boxes may be too big for continued use, but they will need to be repurposed in some way to utilize them. In addition, the big boxes will continue to need maintenance and upgrades to delay obsolescence.

[Miller and Robinson] We agree, to some extent, with the statement. With the population still growing, more care is given in the ambulatory setting, but there is still a need for tertiary care. One trend is toward patients being treated in neighborhoods. There will still be specialty hospitals, but they may be smaller. Market share is being strengthened by mergers and acquisitions. We are starting to see fewer independent hospitals.

[Nartonis] We generally agree, given the increased number of insured parties who will presumably receive their care through scheduled appointments, as opposed to through visits to urgent care and emergency room venues. Also, the apparent drive toward patient satisfaction criteria seems to be leading healthcare facilities to be located closer to the patients, at less acute care facilities, with the feeder system supporting more specialized care at centralized locations.

[Quirk] This is exactly where the industry is headed. The industry, due to the ACA, is no longer a volume-based sector, but rather is quality based. Having inpatients in the hospital setting is a higher risk business than treating patients in an outpatient setting. Reimbursement will come from successful clinical outcomes and patient satisfaction. Both are better achieved in an outpatient setting. The hospital will eventually be reserved for more serious cases, while most other cases will be managed in the medical office setting, through retail (minute clinics), and even remotely from the home. This transformation within the industry has been underway for many years and has begun to show itself in the built environment. More projects are making their way to the AEC industry, and including freestanding EDs that are placed in the communities the hospital serves, ASCs, larger MOBs that include education centers, and multiple specialty services.

[Saksen] While I do agree that the "mother ship" hospital, particularly academic medical center teaching hospitals, will be part of a community-based series of outpatient clinics and surgery centers, stand-alone emergency, and urgent care centers, I disagree that the need for inpatient care will decrease in the near term. The health of the population will not improve simply because the Affordable Care Act is implemented.

Clearly outpatient care is less expensive than inpatient; however, the poor lifestyle choices and the aging of the population continue to increase undeterred by any legislation. While healthcare organizations follow the money and expand outpatient services into the community, it will simply provide better access for patients that will eventually be funneled to the "mother ship" for serious interventions and inpatient services. Chronic diseases, like diabetes and high blood pressure, are on the increase, in both young and old populations. This adds comorbidities and complications to other diagnoses, which result in avoidable but necessary inpatient care. The overall aging of the population, and particularly the extended lifespan due to improved healthcare, make it very hard for me to believe that the need for inpatient beds will decrease in the future. Inpatient and outpatient healthcare use and cost will continue to increase for the foreseeable future.

[Tkacz] The move to deployed community-based care is about access—it's more than buildings. It's about creating pathways to access for low-acuity care, and creating appropriate settings at a lower cost.

I agree that the inpatient hospital campus will continue to be redefined, as outpatient services migrate off acute care campuses. Reallocation and renovation of hospital space can support resizing the inpatient enterprise to optimize for efficiencies of high-acuity care. Hospitals may reduce in size, but increase flexibility and adaptability for future technology and bed requirements.

2. **"Doing more, with less"—Owners are often faced with the challenge of providing the biggest "bang" for the lowest "buck." How can today's hospitals achieve highest value from their capital improvement programs despite inevitable spending barriers?**

[Bearden] The physical environments have to reflect the changed process they house. The ROI of the physical plant has to enhance the process, and do it in a manner that substantiates the value proposition to the user.

[Brower] Insurance companies define highest value providers as those who control cost and deliver high-quality care. What that means in terms of hospital building design is that architects and medical planners must improve their understanding of hospital operations, to help clients improve operational efficiency and improve space utilization (less capital cost). This requires a robust team of medical planners and operational consultants to maximize space utilization and lower capital investments.

Meeting the demands of "high quality" also means having contemporary facilities to deliver care within, both from a functional and aesthetic point of view, to drive patient satisfaction. This means that, at some point, the "big box" hospital projects will return to address aging facilities in some markets. These projects may be more driven by emergency, cardiology, surgery, and outpatient services than bed towers, but the need will be there once the current financial concerns stabilize and recent regulatory impact is better understood.

[McCoole] This is easy to say, but hard to do—eliminate spending capital on anything that is not contributing closely enough to patient outcomes, patient safety, associate safety and satisfaction, and energy efficiency. It is the "closely enough" that is hard to pin down. Do radius soffits above the patient's head contribute or are they fluff? The list of questions is endless. Unnecessary space and features must be eliminated.

[Miller and Robinson] We are working through processes of Lean design and providing efficiencies. There is pressure to ratchet down operational costs. Convenience is a driver. Many groups are taking lesser risks by starting with environments that are not all comprehensive, but with intentions to grow to that level. There is the continual challenge to be as cost effective as possible, while meeting the evidence-based design criteria. We may see more modular construction, especially in areas of redundancy and consolidation of services. Fragmentation of services has created redundant staff and technology, while creating additional costs to the system. More creative scheduling and IT will achieve better utilization and less downtime of technology. There will be more remote readings for imaging and more centralized labs.

[Nartonis] This certainly presents a balance that weighs patient care outcomes against the patient care experience. Advancements in diagnostic and treatment equipment drive organizations to value their capital spend on bricks and mortar less, until it begins to impact their patient's and staff's experience and their ability to attract and retain the best people. This drives our organization to seek ways to reduce the cost of bricks and mortar, without sacrificing the quality of the facilities and the effectiveness of functional and healing design solutions. Healthcare is becoming quite unique in this three-dimensional squeeze of capital, especially given the reductions in philanthropy and reimbursements caused by the overall economic conditions and the recent legislation.

[Quirk] There are three key areas to be examined in today's environment;—efficiency, value, and ROI. Efficiency rules the day, and everything in a facility is questioned. Today's programs are being scrutinized at every level to make sure that there is no overbuild, or waste in the system. Departmental data is more thoroughly questioned, and demographics examined, for better information on trends. Business informatics are being used to a greater level by institutions, to assist in analyzing data to help make more informed decisions. The industry also needs to be more focused on asking what the return is (ROI) for every square foot employed in a facility. This holds true for both new and renovation projects.

Value creation is measured not only by the results in a facility, but by the measurable value partners (AEC industry) bring to the table to assist in solutions. Lean, while not new to the industry, is also being more widely implemented, to squeeze as much waste from operations and the built environment.

[Saksen] Unfortunately, many healthcare organizations mistake their capital program for their strategic plan. The next new facility will address overcrowding or outdated facilities, but fails to address the core needs of the community. Each institution should continually access the population data in its immediate area, and try to impact specific public health issues that are present. Simply put, the best way to reduce capital costs is to improve the community's health, therefore, avoiding the need for either inpatient or outpatient care.

Another strategy that deserves more attention is the use of technology to remotely diagnose and monitor patients. More than just telemedicine, there are evolving nanotechnologies that can provide vital data 24/7/365. These technologies can make providers, and all caregivers, more efficient and effective, thereby reducing the need for capital investment in facilities, and for unnecessary patient trips to outpatient or inpatient settings. These devices can provide the necessary information for family and other in-home caregivers to provide needed interventions, and prevent the need to seek in person medical care.

Finally, there will always be a need for facility improvements and new facilities as buildings age and medical technology continues to advance. The single biggest change that healthcare institutions should make, in order to maximize their capital dollars, is to become an active participant on the project team, and openly share the responsibility for the mistakes and successes of executing their capital development programs. Simply put, this requires hands-on participation, responsive decision making, and prompt payment. The healthcare owner should be an active participant in prefabrication, and design assist initiatives, building information management (BIM), and owner-controlled insurance programs that reduce costs and improve quality. This does increase the owner's risk, when they become active participants in the design and construction process, but the rewards of lower costs and higher quality make it more than worthwhile.

[Tkacz] Value, defined as the highest quality care at the lowest necessary cost, for me sums up the charge—high-performance, operationally driven, planning and design solutions must align optimal operational concepts with strategic initiatives and ROI realities. Planning well and leveraging the power of a multidisciplinary, integrated team are more critical than ever.

The "Quality Value Toolkit" must include every available and conceivable way to approach problem-solving to weigh quality initiatives to investment. Evidence-based design, smart building strategies, sustainability, new delivery methods, and prefabrication can add value. At the end of the day, it's about the people who are committed to the enterprise's success.

3. **What do you see as the "biggest trends" in design and construction of healthcare facilities?**

[Abney] Design/Planning: Freestanding EDs, demographic planning to help right size the facilities to better serve the community.

Construction: BIM modeling is a tool that has really helped with coordination of above-ceiling MEP components. Contractors are now able to fully build the project on the computer and, using clash detection, work out the above-ceiling coordination, prior to physically doing the work.

Technology/Equipment/IT: IASIS has a low-voltage consultant that manages all IT infrastructure design and implementation. I tell my architects and contractors all the time that the building is useless to us if our operational software is not up and running ahead of the opening of the project.

[Bearden] Design/Planning: The addition of new environments in a coordinated manner; acting in a revised fashion of healthcare delivery.

Construction: total team design efforts with the team being defined as design, construction, and processes; IPD is a shorter answer.

Technology/Equipment/IT: IT will continue to demand higher dollars being spent until we conquer the EMR. Equipment will be more generic, be used in a wider array of applications, and everyone will not have the same duplicated items.

[Miller and Robinson] Cost is the number one issue. Ambulatory care and freestanding emergency departments are major drivers, as is avoidance of duplication; and integration of physicians.

Design/Planning: More multifunctional spaces are being planned with adaptive design for future flexibility, and there is allowance for incremental growth versus major expansions.

Construction: LEED design will become the standard, not the exception, in how we deal with windows, systems, orientations, and material selections. There will be more modular construction that will take place in warehouses, that will be moved to healthcare sites and implemented.

Technology/Equipment: The BIM process is becoming more standard. For both technology and equipment, there is the intent to become more flexible and to not duplicate.

[Nartonis] Design/Planning: Designing for efficient (lean) healthcare operations and flexibility of purpose/use.

Construction: Builder's contributing to the incorporation of strategies that reduce life cycle costs (and operations and maintenance costs) in the facilities they help plan and build.

Technology/Equipment/IT: Ultimately, decisions to leave behind the physical infrastructure distribution networks and advance to virtual, wireless technologies with robust technology backbones.

[Quirk] Design/Planning: Smaller, more efficient and flexible spaces.

Construction: Implementing innovative solutions to increase quality, efficiency, safety, sustainability, and energy.

Technology/Equipment/IT: BIM utilized more fully to include 6D (facilities management), applications, manufacturing techniques, virtual design and construction (VDC).

[Saksen] Design/Planning: The most astounding and fast-paced trend in design and planning is the integration of three-dimensional BIM technology into the overall process. The separation and compartmentalization of design from construction is vanishing as the builder participates, both the CMAR and the major mechanical, electrical, plumbing, and other subs, in design assist from the onset of the project. These pre-construction services, not just cost estimating, are integral to intelligent project delivery, providing vital useable build ability information to the owner and designer throughout the process. This eliminates potential design issues, and eventually eliminates the need for "value engineering" when plans exceed budgets. Value engineering many times has resulted in more reduced value than cost.

Construction: The biggest trend in what had happened in the field, with the means, methods, and procedures of building, is the proliferation of handheld technology, iPads, etc., which has virtually eliminated paper drawings. Comprehensive databases that are updated daily, and provide field supervisory personnel, as well as inspectors and project managers, with a single source of truth regarding what has been planned. There are no out-of-date drawings that mislead the builders' workers, no need to remove unspecified work, and, with the incorporation of three-dimensional BIM technology, no need to correct uncoordinated conflicts among trades in the field. Just 5 years ago, it was very difficult to get acceptance of handheld technology. Field supervisors relied on paper plans and specs on the table, in the plan room, in the trailer, or on plan tables throughout the site. These same individuals today lean on the plan table and look at their computer screen or handheld devices. Likewise, the project managers, quality control, insurance, and AHJ inspectors, as well as the owner and architect, have complete access to the design. Issues in the field can be recorded, photographed, and questioned, and become documented requests for information (RFI) or clarifications, without returning to the construction trailer. The amount of wasted time and effort that has been eliminated is enormous.

Technology/Equipment/IT: Modern buildings, not just healthcare occupancies, are no longer comprised of electrically/mechanically driven machines and devices that satisfy the owner's needs. They are truly digital environments, which are totally computer controlled. This includes the HVAC, medical gases, communications, security, visual surveillance, medical equipment, pneumatic tubes, trash, linen, patient monitoring, imaging, and OR integration systems. The future of all these services is becoming increasingly wireless, and usable by handheld technologies. The biggest growing demand is for convenient charging options for handheld devices. While wireless technology does require more than substantial amounts of wire above the ceiling, the mechanical outlet to plug in equipment will be as curious to future generations as the phone booth is to our children.

[Tkacz] In consideration of the three biggest trends impacting healthcare today—I see three design and construction trends focused on performance improvement strategies.

Three Biggest Trends Impacting Health Systems	Three Biggest Trends in Design and Construction
1. Shift from volume to value	1. Imperative for value-based strategies
2. Era of collaboration	2. Collaborative A/E/C team dynamics
3. Population management	3. Unbundling services from inpatient setting

4. **What is/are the biggest construction material or construction process successes and failures that you have seen on hospital projects?**

[Bearden] Success: IPD—it reengineers the process for better outcomes; failure: CM at Risk, when there is really no risk to the CM.

[Quirk] Success: Prefabrication and modular construction. We have seen incredible results from implementing these techniques. Results are benefitting both the AEC industry and the clients we serve. Positive outcomes include schedule reduction, increased quality, increased safety, sustainability, and logistical improvements.

[Saksen] Success—while we have had great successes at Parkland with BIM, field handheld technology, co-location of project team, and prefabrication, I think the most significant process change that we have implemented is the integration of our project management system, Prolog, with our financial management system, Lawson. In every other project I have worked on in the past 39 years, there has always had to be a monthly manual reconciliation of the two systems, in order to issue payment, and to avoid having the chief financial officer report different numbers than facilities management reports at the board level. Further, the manual reconciliation dramatically slows this payment process, forcing contractors to "finance" the project for 45, 60, or more days after the work is complete, before a check is released. Owners earn reputations in the construction industry for their late payment history, prices are elevated accordingly, and competition is reduced when some contractors choose not to bid late paying owners. By integrating our project and finance systems, we have made good on our promise to pay our contractors in 25 days or less, we have averaged between 22 and 23 days, and they in turn have paid their subcontractors in 3 to 5 days after receiving our electronic transfer. On our $1.3 billion project, which had been a virtual economic stimulus package for the Dallas community, the entire construction community has been very appreciative, and vocal, about the success of our payment program. This has not only resulted in greater competition, it has also resulted in small and MWBE companies being able to afford to compete and gain work on the project, and it has more than achieved the 30 percent MWBE goal that our board has set for management. We have completed $150,000,000 of the project with more than 40 percent MWBE participation. Further, for the first time in my experience, we retained a 100 percent MWBE design team, architect, and all engineers and designers.

Mistakes—I believe the biggest mistake was our decision to try to break up a project, in order for the MWBE market to have a better opportunity to be the lead architect on a response to and RFP. Specifically, we asked for proposals on the core and shell of a medical clinic building separately from the interior fit out. It is true that an MWBE firm is leading on the core and shell, but the additional coordination that is going to be required between the two architects-of-record presents the owner with unnecessary non-value added costs. Further, dabbling with the market is not the proper role of a public institution, particularly when the MWBE design business community in Dallas is robust, and has developed many very capable firms that could easily handle the entire clinic. I believe this underestimation of the strength and vitality of our MWBE design professionals may have inadvertently resulted in a disadvantage, rather than the advantage that was intended. Certainly our recent contract with the 100 percent MWBE team to design the project's connecting bridges, a complex design that requires coordination with two existing buildings and two new under construction buildings, is proof of their professional expertise.

5. **How can the hospital builder or construction manager have the biggest impact on improving the quality of hospital construction work and processes?**

[Abney] By placing qualified, healthcare experienced teams on projects that are willing to "over communicate" with the owner and their consultants.

[Bearden] Understand the whole process, not just pieces or parts.

[Brower] CMs are hired early in the design process to improve constructability, assist the design team to make good decisions regarding cost management, and confirm project schedules. Many CMs have very good predesign services, many also do not, and offer very little value to the client and project team during predesign. Nothing is more detrimental to a project than starting construction over-budget because the predesign estimate was poorly done. Delivering quality is all about team relationships. When everyone has their eyes on accomplishing the mission, and dealing effectively with problems when they arise, everyone is happy at the end of the day.

[McCoole] Have a very significant commitment to healthcare, invest in research in healthcare construction, and develop a high level of expertise in healthcare in their staffs—from execs to field supervision.

[Miller and Robinson] The biggest impact can be made by having knowledgeable, qualified staff that can make timely decisions. Maintaining a strong schedule and being proactive about understanding the job are additional ways that the builder or construction manager can have an impact.

[Nartonis] Construction Work/Construction Processes: Embrace infection control and quality control protocols during the design and planning phases, so what gets built, and how it gets built, leaves as little room as possible for error.

[Quirk] I have seen the best outcomes result from early involvement of not only the CM, but of all key parties. Understanding and sharing the project goals and vision of the client for the project make it very clear what the expectations are. It also results in increased communication between the stakeholders. This process results in a clear understanding of the end product, assists designers in designing and detailing the project, allows the client and end users better insight into the process, and can aid them in making timely decisions.

[Saksen] The successful hospital builder of the future must harness the IT technologies currently available to move to a paperless seamless delivery system. They must join the architectural and engineering design team at the outset in a co-location site with the owner and other consultants to be involved in every decision that affects quality, schedule, and cost. Further, they must thoroughly promote a comprehensive safety program that incorporates prefabrication, through the intensive utilization of BIM, placing the workers in a climate-controlled ergonomically designed environment, as much as possible, and eliminating jobsite conflicts among the trades. The construction manager must be very comfortable with open book accounting of costs and a comprehensive independent auditing process throughout the job, which results in a totally transparent and compliant project close-out.

I firmly believe that the owner should embrace a project delivery model in which there are independent contracts with the design team and specialty consultants, e.g., information technology, medical equipment, move logistics, etc., and the Construction Manager at Risk. As I mentioned above, I believe the entire design/consultant/builder team should be co-located with the owner's program management team, through the completion of the design development (DD) documents. Depending of the need to expedite delivery of the project, the guaranteed maximum price can be established at DD, with the full buy-in and the understanding of the design team and owner regarding the risks of this approach. The development of full contract documents (CDs) is preferable, although there may be a need to develop phased packages, e.g., foundation and site work, core and shell, and interior build out, in order to minimize the inherently slower schedule that full CD development requires. There will always be risks relative to unforeseen changes in technology, and bases of design equipment decisions that will require modifying, even in thoroughly cross-checked CDs, but the risks are more manageable if the owner has complete support from finance, IT, procurement, biomedical, and, plant engineering, and, most significantly, the clinical user groups.

6. **What delivery method for hospital projects do you think produces the best overall results, based on your experience?**

[Abney] There are so many new terms that basically describe the way we have been doing business for the last 15 years. It is essential to get a qualified GC on board early in the design process. Have the GC involve teams of qualified subcontractors and begin collaborating with the architects and engineers early, to insure that not only cost, lead times, and availability are major considerations in the final design.

[Bearden] Any answer I give will have one common theme—spend the right amount of time on the front end of the project, with the right team members, and it will pay off. Currently, a properly scoped project and IPD delivery is my choice.

[Brower] CM at Risk is still the gold standard at this point. Integrated project delivery (IPD) is being tested, but it is rarely formalized as such, still CM at Risk, for the most part, is how it is actually being delivered. Once again, the most important parts of any successful project are teamwork, trust, and execution, which are all about people and relationships.

[McCoole] A lean, negotiated team approach to design and construction. Not design-bid-build, and not design-build. The level of expertise necessary in the design team is significant and, I believe, less likely to be found in the design-build firms. Contractors must be on the team day one and held accountable for being a contributing factor in not just construction, but every element of the process.

[Miller and Robinson] The team should be developed early—this includes the contractor, architect, engineers, and major subcontractors. Ideally, the contractor should be involved early from the pre-construction phase, which includes proper sequencing with qualified people.

[Nartonis] The contract type and/or financial incentive program are irrelevant—what is absolutely essential is the entire team's commitment and follow-through to behave and innovate in a collaborative, integrated, and co-dependent manner. Design, planning, and construction team members are facing increasing pressures on their performance from plenty of outside forces as it is—the project team must face these together in order to deliver the greatest value to its customers.

[Quirk] The best project delivery method is somewhat of a hybrid. I will refer to it as a design-build partnering agreement. Whereby, the client hires a DB team, but remains intrinsically involved in the process, or part of the team. You could also refer to this as a hybrid IPD, only with a different contracting vehicle. This vehicle should also require complete transparency from all the team members (including the client), to include a single total project budget, showing not only construction costs, but all owner budget line items, like FF&E, project soft costs, debt payments, legal, etc. Co-location is also a good idea, when appropriate. This promotes communication and transparency, and provides team members with an "inside look" at how their partners' businesses are run, and how decision making affects all parties involved.

Reference

The Hospital Survey and Construction Act (or the Hill–Burton Act): http://www.hrsa.gov/gethealthcare/affordable/hillburton/.

APPENDIX

Mechanical, Plumbing, Electrical, and Telecommunication Schematic Design Narrative Example

Mechanical

New central station air handling units will be provided as follows:

A/C-1	Cancer center	19,000 cfm
A/C-2	Emergency dept.	22,000 cfm
A/C-3	Pharmacy/imaging	25,000 cfm
A/C-4	OR/Cen Ster	32,000 cfm
A/C-5	Outpatient staging	18,000 cfm
A/C-6	Lab	15,000 cfm
A/C-7	Pulm/Stor/IT	14,000 cfm
A/C-8	Kitchen/Dining	16,000 cfm
A/C-9	Admin	20,000 cfm
A/C-10	Admit/Lobby	19,000 cfm
A/C-11	Patient wing (west)	18,000 cfm
A/C-12	Patient wing (east)	16,000 cfm
A/C-13	ICU	23,000 cfm

Each air handling unit will be a rooftop air handling unit, double-wall construction with no through-metal thermal conduction, consisting of a return fan, control section with variable frequency drives (VFDs) for return and supply fans, outside air economizer, 2-inch MERV-7 prefilter, access section, hot water preheat coil, steam humidifier, access section, chilled water cooling coil, supply fan, diffuser, 12-inch cartridge MERV-14 final filters, and discharge plenum. Preheat coil shall be max. 600 fpm face

This appendix was provided by Rick Wood.

velocity, 180°F entering water, 45° to 55°F air temperature rise. Cooling coil shall be max 500 fpm face velocity, 43° to 59°F chilled water temperature rise, 80/67 to 50/50 air temperature drop. Each air handling unit will have an attached, enclosed 6-foot-wide service vestibule. Air handling units 2, 3, 4, 5, 6, 11, 12, and 13 will have a standby VFD with manual selector switch on the supply fan.

Each air handling unit will supply conditioned air via medium pressure ductwork to double wall, pressure independent, variable volume terminal boxes with hot water reheat coils. Tempered air from the boxes shall be supplied via low pressure ductwork to ceiling diffusers. Return air and exhaust air shall be fully ducted.

Approximately 245 terminal boxes will be provided in the first floor and 144 boxes on the second floor.

The connecting corridor from the hospital to cancer center and the main corridor south of dietary and admin will be served by chilled beams and perimeter hydronic heating. Chilled beams will be recessed, 2 × 8 feet, sized for 5000 Btu/h with 75 cfm of primary air at 70 dB/50 wb, with MERV-8 filters on secondary air side, and shall have a 2:1 secondary:primary air ratio. Hydronic heating panels shall be 9-inch-high wall-mounted Runtal Thermotouch, 400 Btu/lf, custom color to be selected by the architect. Provide a total of 18 chilled beams and 260 feet of hydronic heating panels for all of the corridors.

Air from spaces, such as soiled utility, toilets, etc. shall be ducted to the roof and exhausted via roof-mounted exhaust fans, approximately 15 fans at 5000 cfm each.

All valve and damper actuators and thermostats shall be DDC and connected to the building automation system.

All fans (supply, return, and exhaust) will have variable frequency drives.

Ductwork passing through 1-hour and 2-hour rated walls shall have type C curtain dynamic rated fire dampers. Ductwork passing through smoke barriers shall have air-foil blade fire/smoke dampers.

The MRI, linear accelerator, and CT areas will have medical chillers sized for the medical equipment. Chillers will have remote air-cooled condensers and domestic water switchover panels.

A glycol snow melting system will serve embedded coils in the slabs at the two patient entrance canopies and the outpatient canopy. Each location will have 1-inch plastic piping embedded on 6-inch centers, covering an area of 150 square feet.

Building automation system shall be a web-based, enterprise level system equal to Johnson Metasys or Siemens Apogee. Provide one operator workstation, complete with PC and printer, in the CEP.

Central Energy Plant Mechanical Equipment

The central energy plant will have the following equipment:

1. Two 150 psig firetube steam boilers, each at 500 bhp.
2. One dual tank deaerator (separate surge tank and deaeration section, each sized for 10 minutes storage), sized for 45,000 pounds per hour, 0.005 cubic centimeter O_2 removal, three boiler feed pumps, two condensate transfer pumps.

3. Two centrifugal water-cooled chillers, 650 tons each, 52 ent chilled water, 42 lvg chilled water, 86 ent condenser water, 390 kilowatts power demand at full load, with VFDs.
4. Two double-suction chilled water pumps, arranged for variable flow primary operation, 1300 gpm, 120-foot head, 60 hp, VFDs.
5. Two double-suction condenser water pumps, 1950 gpm, 90-foot head, 100 hp.
6. Two induced draft, cross flow, cooling towers, to cool 1950 gpm from 95 to 85°F at 75 wb (wet bulb), with stainless steel hot and cold water basins, VFD on motors. Towers shall sit on 10-foot raised structure.
7. Plate/frame heat exchanger for winter water side economizer operation, 1300 gpm each side.
8. Plate/frame heat exchanger to preheat domestic hot water with chiller condenser water, 500 gpm each side.
9. 12 million Btu/h steam to hot water heat exchanger system with two double-suction pumps, each at 900 gpm, 120 feet, 60 hp with VFDs.
10. 20,000 gallon, double wall, underground fuel oil tank with leak detection system, fuel oil storage monitoring system, and duplex packaged pumping system.
11. 90,000 Btu/h steam to glycol heat exchanger system for snow melting, with two inline circulating pumps.

Alternate #1—Chilled Beams for Second Floor

The second floor west and east patient wings and the LDRP area (patient rooms and core) will be served by chilled beams and perimeter hydronic heating. Each patient room will have a 2 × 8-foot recessed chilled beam, sized for 6000 Btu/h total sensible heat removal when supplied with 100 cfm of primary air at 70 dB/50 wb. Each chilled beam shall have MERV-8 filters on secondary air side and shall have a 2:1 secondary:primary air flow ratio. Chilled beams shall be Dadanco, Halton, or Trox. Two 90 gpm inline pumps and a three-way control valve shall circulate 58°F chilled water, through 3-inch chilled water supply and return mains in the second floor, to the chilled beams. Distance of piping will be approximately equal to length of reheat hot water piping for base bid. Perimeter room heating units will be Runtal Thermotouch, 9-inch-high wall-mounted, with custom color selected by architect, with capacity of 400 Btu/lf. Interior spaces shall utilize the chilled beam for heating. Provide chilled water and hot water control valves for each beam. ICU rooms and ICU core areas will be served by standard terminal boxes.

A/C unit sizes will be changed to the following sizes

A/C 11	Patient wing west	8,000 cfm
A/C 12	Patient wing east	7,500 cfm
A/C 13	ICU and LDRP	16,000 cfm

There will be approximately 130 beams and total of 43 terminal boxes on second floor. (Deduct 101 boxes on second floor from base bid for this alternate.) Each chilled beam will have individual space temperature control.

Both A/C units will be 100 percent outside air with glycol run-around coils in exhaust air and outside air streams; each coil will be 5 row/14 fpi coil, 13,500 cfm.

Materials of Construction

1. Medium pressure ductwork: galvanized steel sheet metal, SMACNA pressure class 6 inches
2. Low pressure supply ductwork: galvanized steel sheet metal, SMACNA pressure class 1 inch
3. Return and exhaust ductwork: galvanized steel sheet metal, SMACNA pressure class 2 inch
4. All ductwork longitudinal and transverse seams and joints shall be sealed
5. Ductwork insulation
   a. Concealed supply: 2-inch, 3/4-pound density fiberglass batt, alum foil facing
   b. Exposed supply in mechanical rooms: 2-inch, 1½-pound density rigid fiberglass, alum foil facing
   c. Exposed on roof: 2-inch, 1½-pound density rigid fiberglass, venture clad covering
6. Ceiling diffusers (except as noted below): Titus TMS louvered face, four-way blow
7. Ceiling diffusers (ORs, trauma, isolation rooms): 2 × 4-foot perforated face, Titus TLF, 3-inch HEPA
8. Ceiling diffusers (public waiting, nurse stations): slot diffusers, Titus MLF
9. Ceiling return registers: 1/2-inch egg crate aluminum grid
10. Sidewall supply and return registers: Titus 350
11. HVAC water piping: black steel, schedule 40, threaded or welded joints
12. Steam piping: seamless black steel, schedule 40, threaded or welded joints
13. Steam condensate piping: seamless black steel, schedule 80, threaded or welded joints
14. Piping insulation
    a. Chilled water: rigid phenolic
       i. Concealed and in mechanical rooms: ASJ covering
       ii. On roof: aluminum jacket
    b. Reheat hot water: rigid fiberglass
       i. Concealed and in mechanical rooms: ASJ covering
    c. Steam and condensate: rigid fiberglass
       i. Concealed and in mechanical rooms: ASJ covering
       ii. On roof: aluminum jacket

Plumbing/Medical Gas/Fire Protection

Domestic Water

The facility will be served by a new 6-inch domestic water line, refer to civil narrative for connection to utilities outside 5 feet from the building. The line will have two backflow preventers inside the CEP. A 3-inch line with two backflow preventers will feed the mechanical equipment. A 2-inch line with two backflow preventers will feed central sterile. Two 2-inch backflow preventers will feed the lab. A 3-inch line with backflow preventer shall feed the landscape irrigation system.

The linear accelerator and CT scan will require a backup water supply to the chilled water system. This line will include a 1-inch backflow preventer.

Domestic Hot Water

The facility will have two new Aerco B+05/1.25 at 85 gpm semi-instantaneous steam water heaters to provide 100 percent redundancy. Each water heater will be set to deliver 120°F water to fixtures.

The central sterile hot water system will be served with dual, Aerco B+05/1.00 semi-instantaneous steam water heaters to provide 100 percent redundancy. The water heaters are sized at 30 gpm each at 100°F rise to deliver 140°F water to central sterile equipment.

Two New Aerco B+ semi-instantaneous steam water heaters will be installed in the food service area. The dishwasher booster heater will be provided by the food service vendor.

Each hot water system listed above will have its own hot water recirculation system to maintain a 5°F drop in the hot water system, and will have circuit setters installed to provide hot water to all fixtures in a timely manner.

Domestic hot water lines more than 20 feet from the hot water circulation main will have a HWATT heat maintenance system. The system will be complete at 4 watts per foot and control panels located in each area. There will be six control panels.

Each heater will connect to the mechanical preheat heat exchanger to preheat the water feeding the heater. Refer to mechanical narrative.

Filtered Water

Central sterile and the lab will each have a filtered water system, provided by others, with a polypropylene pure water piped system. Complete with supply, return piping, and pump.

Waste and Vent

Acid waste and vent piping will be installed below grade to serve the new lab suite with a 200-gallon acid pit.

Provide a grease waste system for the kitchen with a grease trap located outside the building. Provide a Parks Equipment Co. GT-6000 (6000-gallon capacity) grease trap, with two manholes, that meets local and state codes.

The emergency department will have a 500-gallon decontamination tank connected to the drain in the decontamination shower. The tank will be complete with high level alarms that connect to the building management system and a local alarm.

Storm Water

The roof drains (primary system) will be collected together and routed, down through the building to below slab, to storm water mains that will exit the building and connect to the storm water system on site.

Storm water will exit 5 feet outside the building and connect to a rainwater collection system for irrigation that is provided on the civil portion of the contract.

Overflow roof drains (secondary system) will be collected together, routed down through the building, and will terminate to daylight at the ground floor.

Medical Gas

A new vacuum pump and air compressor will serve the building.

A new Beacon Medaes variable speed claw VH15T-240V-QCV 15-hp quadruplex medical vacuum packaged system. The system will be set up as a quadruplex system, initially with four 15 hp pumps, with ability to convert to pentaplex operation in the future. Piping, valves, tank, and panel will be sized for pentaplex system. The controls will be set up for quadruplex system.

A new Beacon Medaes variable speed scroll SAS-15Q quadruplex medical air packaged system. The system will be set up as a quadruplex system initially, with four 15-hp compressors, with the ability to convert pentaplex operation in the future. Piping, valves, tank, and panel will be sized for pentaplex system. The controls will be set up for quadruplex system.

New 2-inch oxygen will connect to a new bulk oxygen park provided by others. Coordinate with bulk supplier for exact scope of work.

New 1-inch nitrogen (N), 1-inch nitrous (N_2O) lines will be routed from the medical gas park to the building.

The security exam rooms in the emergency department will have vandal proof outlets.

Critical care areas and each patient bed wing will be monitored by area alarm panels located at the nurses' stations as required per NFPA 99.

Two master alarms will be installed.

Isolation locking valves will be required at the base of all medical gas mains, and on each branch main off the riser on each floor as required per NFPA 99. Area zone valves will be required to serve each treatment suite.

A new Powerex STD030 duplex air compressor shall feed the equipment in the central sterile area.

A new Powerex STD030 duplex air compressor shall feed the equipment in the lab.

Natural Gas

The proposed new gas entrance to the building shall be located in the utility yard. Coordinate with local gas company for exact scope of work. A 3-inch medium pressure gas line shall be routed into the building and shall connect to the new boilers. A second 3-inch low pressure line will feed the kitchen and other services in the building.

Automatic Sprinklers

Install a new 8-inch fire line entrance into the central energy plant. Install post indicator valve, alarm check valve assembly, backflow preventer, and fire department (Siamese) connection.

The facility will have a manual wet standpipe system. The standpipe in stairwell will serve automatic sprinkler zones on each floor.

The facility is to be fully sprinkled throughout and will meet all NFPA 13 and insuring agent requirements. Sprinkler heads will be quick response-type throughout. All major components will be UL/FM rated. Sprinkler heads will be semi-recessed type in all areas, unless noted otherwise. Fully recessed-type heads, with white cover plates, will be provided in finished areas. Sprinkler heads are to be located in the center of ceiling tiles. Upright heads will be provided in MEP equipment rooms and other spaces without ceilings. Extended coverage heads will not be allowed.

The linear accelerator, data room, and main electrical room will have a pre-action sprinkler system.

The new addition will have recessed fire extinguisher cabinets throughout to comply with NFPA, with wall-mounted fire extinguishers in equipment and electrical rooms.

Piping

Piping materials shall be as follows:

- Drainage/waste/vent/storm water
  - Above grade–std weight cast iron, no-hub bands
  - Below grade–std weight cast iron, bell/spigot, gaskets
- Domestic hot and cold water: type L copper, soldered joints
- Medical gas: type K seamless copper, brazed joints, per NFPA 99
- Automatic sprinkler piping will be schedule 40, black steel pipe for feed and cross main piping.

Piping shall have rigid fiberglass insulation with ASJ jacket. Thicknesses shall be per energy codes, with following minimum thicknesses

- Domestic cold—1/2 inch
- Domestic hot—1 inch
- Horizontal runs of rain water leaders and roof drain bodies—1 inch

Plumbing Fixtures

Plumbing fixtures shall be hospital grade.

Water closets will be floor-mounted, dual flush flushometer valves of Kohler Wellcomme series. Urinals will be 1/8 gallon flush wall-hung fixtures

Bariatric toilet: Acorn—ADA Bariatric On-Floor Toilet Dura-Ware 2125A Series

Flushometers: Sloan

Backflow preventers: Watts

Shower faucets: Symmons #1-117VT-FS-X-BFSC

Shower heads: Patient and ADA showers will have handheld Kohler Forte head with 24-inch bar and a 2.0-gpm fixed Kohler Forte shower head and diverter.

Lavatories/sinks

- Patient rooms—china
- Public restrooms—china
- Staff support in patient areas (nurse stations)—stainless steel
- Staff support (utility rooms, etc.)—stainless steel
- Substerile—stainless steel clinical double compartment
- Scrub sinks at ORs—double compartment (owner provided by equipment vendor)

Hose bibbs will be installed every 100 feet around exterior of building on the first floor, and around the second floor for the first floor roof. Two hose bibbs will be installed on the roof.

Meters

Meters will be provided in the following areas and connect to the building management system

- Cooling tower makeup and blowdown
- Incoming water to the project
- Purified water system (reverse osmosis and/or deionized)
- Water use in dietary department
- Outdoor irrigation systems
- Steam boiler systems makeup water
- Water use in laboratory
- Water use in central sterile and processing department
- Water use in physiotherapy and hydrotherapy treatment areas
- Cold water makeup for domestic hot water systems

HVAC Condensate Return System

Each mechanical room with an HVAC unit will have an HVAC condensate return system, which will collect the condensate and pump it to the cooling tower makeup. The system will be complete with 2-inch copper piping with fiberglass insulation and Shipco Duplex PC 45 gpm, 1-hp cast iron pump.

(Alternate) Rainwater Harvesting Reclamation System for Plumbing Flush Valves

Provide a 3-inch domestic water line from the discharge of the reclamation system pump that will feed a separate line to the plumbing flush water closets and urinals.

All reuse piping shall be painted purple and labeled "Rainwater Reuse Water." Vertical risers shall be routed to flush fixtures. Each toilet room shall have a shut off valve.

The system shall have an approximate 40,000 gallon underground storage concrete tank installed on site, near the central energy plant.

The tank shall receive rainwater from three 10-inch rainwater leaders. The three rainwater leaders will connect to a common 12-inch rainwater horizontal header line, which will connect to a J.R. Smith RH9521-12AC Rainwater Vortex Filter with base plate. The vortex filter shall be mounted high on the base plate with structural supports. The 8-inch discharge shall be routed to the inlet of the storage tank. The 12-inch discharge at the bottom of the filter shall be connected to the storage tank 12-inch overflow line to the city storm water main.

The 8-inch rainwater line from the vortex filter will flow into the storage tank. At the end of the line, provide a J.R. Smith #RH9530SI Smoothing Inlet.

There shall be a 12-inch tank overflow line, that is mounted high on the tank, to maintain approximately 40,000 gallons of water in the storage tank. The 12-inch overflow line shall be routed by gravity and connected to the 12-inch vortex filter discharge to the city storm water line.

The 12-inch overflow line shall have a J.R. Smith #RH9530DOK overflow device installed on the storage tank. The 3-inch water supply to the plumbing fixtures shall be connected to J.R. Smith #RH9532C Storage Tank Floating Filter and Hose.

The 3-inch water supply line will connect to a duplex bag filter system similar to the Hayward Industrial Products Polypropylene filter systems. An automatic chlorinator and dye injection system shall be installed downstream of the filter system. All systems shall be located in an area with maintenance access.

The 3-inch line shall be connected to a packaged Grundfos variable speed duplex domestic water booster pump system that shall provide 210 gpm flow with an inlet pressure of 0 psi and an outlet pressure at 70 psi. System pressure shall not exceed 80 psi. The pump system shall be controlled by a J.R. Smith #RH9542FSO float switch installed in the storage tank. Float shall shut off pumps with 10 percent capacity left in tank.

The 3-inch discharge line from the pumps shall be routed to the building flush valves as described above.

The 3-inch city domestic makeup water line shall be connected to a 3-inch city approved, reduced pressure backflow preventer. Downstream of the backflow preventer, the 3-inch line shall connect to a pressure valve that will open when water pressure in rainwater system is below 60 psi. The intent for this part of the system is to provide city backup water during times when the rainwater system is out of water with this sequence of operation. When the tank level is at 10 percent of capacity the float will shut down the rainwater booster pumps. As a result, water pressure will go below 60 psi. When water pressure is below 60 psi, the domestic water valve will open, allowing city water to supply the fixtures until the rainwater tank has water.

System shall be complete, including electrical and low voltage power connections to equipment and controls. System shall be installed complete with full warranty for 2 years. The system representative shall be available to meet with local codes officials and make changes as required to meet local and state regulations.

Electrical

Normal Power—Service Entrance

Normal power is to be provided from two separate utility feeds with automatic switching capability, serving two utility-supplied, pad-mounted transformers with a secondary voltage of 480/277 volts. This equipment is to be located within the service yard. Secondary conductors (contractor furnished and installed) will be routed underground from the equipment to the main switchboard in the main electrical room of the central energy plant.

Normal Power—Main Distribution

A new 4000-amp, 480/277-volt main-tie-main configuration main switchboard is to be provided in the main electrical room of the central energy plant. The switchboard will have group-mounted feeder breakers, to feed the normal side of automatic transfer switches, and 480/277-volt distribution panels in electrical rooms on the 1st and 2nd levels of the hospital, as well as the Medical Office Building (MOB). 208/120-volt power will be provided via step down transformers and panelboards in the electrical rooms on each floor.

Emergency Power—Main Distribution and Paralleling

Emergency power will be provided from a 480/277-volt, 3000-amp, 3-phase, 4-wire paralleling switchgear located in the emergency power electrical room of the central energy plant. This switchgear is to include a master control section, generator breaker sections, and distribution breaker sections with capability for future expansion. Transient voltage surge suppression shall be provided. Switchgear is to be provided with generator and distribution low voltage power circuit breakers. Paralleling system is to be capable of providing complete control and monitoring of all transfer switches, to include status, source available, load shed, load add, and remote transfer. Communication modules will be provided load information from each switch back to the master control. A remote master control station is to be provided for monitoring and control.

This switchgear will serve two 600-amp equipment branch automatic transfer switches, a 400-amp elevator equipment branch transfer switch, two 400-amp critical branch transfer switches, one 400-amp UPS-backed critical branch transfer switch, and one 150-amp life safety branch transfer switch. Transfer switches are to be four pole, bypass isolation, open transition, with full metering, control, and communications.

Each transfer switch will feed a distribution panel, which will, in turn, feed subpanels on each floor and the MOB. Step down transformers will be utilized to provide 208/120-volt power on each floor. Critical branch distribution will serve two 10-kVA, 480-volt primary/120-volt secondary isolated power panels per operating room and C-section. The isolated power system will be provided with code-required remote alarming capability, integration of alarms to the building management system, and remote monitoring system by the manufacturer.

A separate 480/277-volt 2500-amp switchboard section, with a main circuit breaker and two 600/3 keyed interlocked circuit breakers (to be located in the emergency electrical room), is to be provided to serve a 2500-amp, 600-volt-rated NEMA 3R roll up generator termination box, to be located in the service yard for connection to a temporary generator. Underground feeders are to be routed between termination box and switchboard.

Equipment-specific materials and options for switchgear, switchboards, distribution and branch panelboards, and dry-type transformers

- Copper bus or windings
- TVSS
- Keyed interlock on main and tie breakers
- Main and tie circuit breakers to be low voltage power circuit breakers with power management-type trip units, with the following metering functions: current RMS, voltage RMS (line to line, line to neutral), energy kWh, demand kWh, peak demand kW, real power kW, total (apparent) power kWA
- Feeder circuit breakers, 1200 amp and below, to be molded case with feeder protection unit to include the following metering function: current RMS, each phase
- Feeder circuit breakers 1600 amp and larger to be insulated case with feeder protection unit, to include the following metering function: current RMS, each phase
- Lighting and power distribution panelboards will be separate to provide adequate load readings for metering purposes

Emergency Power—Diesel Generators

Two 1-megawatt, 480/277-volt, emergency standby diesel generators will be provided for emergency power generation. The generators will be located in the generator room of the central energy plant, and will be served from the site diesel fuel storage system. Generators will be integrated and controlled by the paralleling switchgear system.

Sound attenuation should be provided for generator system.

Uninterruptible Power Supply

A 300-kVA flywheel UPS will be used to provide "ride through capability" on one of the emergency power critical branches. Power distribution from this branch will be distributed only to telecommunication equipment, patient care medical equipment, and selected facility systems that require continuous power to ensure proper function.

The UPS will serve a 400-amp, 480/277-volt distribution panel that will serve the emergency load side of an automatic transfer switch. Power will be distributed through panelboards and transformers as part of the emergency power distribution system.

Sub-Metering

All electrical loads are to be sub-metered within the hospital and MOB. Segregation of power distribution, branch power, and lighting loads will be required for adequate metering. Metering equipment and communications equipment and cabling are to be provided for integration of the metering system to the building management system for monitoring and data collection.

Lighting

Interior building lighting will consist generally of 277-volt 2 × 4 recessed fluorescent fixtures and LED down lights with dimming capability.

In clinical spaces, back-of-house areas, and non-public corridors, fixtures will typically be 2 or 3 lamp 2 × 4 recessed fluorescent fixtures with prismatic acrylic lenses. Public space lighting will be as directed by the architect. Typical color temperature for fluorescent lighting will be 4100 K.

Patient room lighting will consist of multifunction medical-type troffers with low voltage controller interfaces to the nurse call system. Additional room lighting will be provided with LED recessed fixtures and fixtures specified by the architect.

All accent, sign, display, and exterior building lighting will be as directed by the architect.

Site lighting will be pole-mounted LED-type fixtures with 100 percent cutoff.

Lighting Control

A networked relay-based circuit control system will be utilized. This system will control and monitor all lighting in public spaces, common areas, support spaces, utility system spaces, and exterior building and site lighting. The system will control and monitor through the use of dual technology occupancy sensors, day lighting sensors, photocells, and digital time clocks. The system will be provided with a central station control with remote monitoring, in the central energy plant, and the ability to be integrated into the building management system. Override stations will be placed in areas for local control. Meeting/conference rooms will be provided with local stations with multi-scene capability.

Fire Alarm

A complete, non-coded, addressable, intelligent, microprocessor-based, reporting fire alarm system with Class A wiring will be provided, in accordance with the applicable version of the IBC and NFPA 72 requirements for the institutional and business (MOB) occupancies.

A main control panel is to be provided for monitoring and control, and will be provided with 25 percent future capacity. Photoelectric intelligent, addressable smoke detectors, intelligent, addressable heat detectors, combination audio/visual devices, manual pull stations, duct-mounted intelligent, addressable smoke detectors, magnetic door holders, addressable circuit interface modules, remote monitors and printers, remote indicators and test switches, and remote LCD annunciators will be provided as part of the overall system.

Ceiling-mounted smoke detectors will be provided in public and common spaces, corridors, storage and utility areas, operating rooms, elevator lobbies, and patient sleeping rooms. Duct-mounted smoke detectors will be installed in the air handling units in the supply and return ducts and at fire/smoke dampers at rated walls. Manual pull stations will be located at each exit, and will be no more than 200 feet apart. Audio/visual devices will be provided in public and common spaces, corridors, storage and utility areas, staff work areas, and public and handicap toilets. Addressable circuit interface modules will be provided for interface of dampers, automatic doors, card-accessed spaces, Division 23 control equipment, and elevator controllers.

Electrical Materials

PVC Schedule 40 conduit shall be used underground for service entrance conduits. Transitions into the building and/or above grade are to be with rigid galvanized conduit.

Galvanized EMT shall be used above accessible ceilings or where installation is exposed above 6 feet, for branch circuits and feeders. Fittings shall be steel set screw or compression type.

Galvanized rigid conduit shall be used underground, where installation is exposed below 6 feet or where exposed to severe mechanical damage.

Flexible metal raceway shall be used for final connections to generators, transformers, motors, mechanical equipment, and other equipment subject to mechanical or electrical vibration. Length shall not exceed 6 feet. Maximum length concealed in walls shall not exceed 3 feet.

Conductors shall be copper of not less than 98 percent conductivity, 600-volt insulation THHN, except for isolated power distribution to be provided with XHHW.

Receptacles in patient care areas to be hospital grade, 20-amp, 125-volt devices, color-coded per branch of system.

Telecommunication

Telecommunication Distribution Room

The telecommunication distribution room will house all communication systems head end equipment and should be sized no smaller than 10 × 10 feet, with the preferred size being 10 × 16 feet, to accommodate all systems hardware and cable terminations. The room shall be environmentally conditioned to maintain proper systems operating temperatures and adequate electrical power and grounding systems shall be provided to support all systems. A reflective ceiling is not required. The walls shall be lined with ¾-inch AC grade fire retardant plywood, painted with a gray fire retardant paint. Appropriate floor covering is required. The general contractor shall be responsible for the construction of the room, including floor and wall penetrations to allow for horizontal and vertical cable installation. Any required cable tray, equipment racks and hardware will be provided by the communications contractors.

Cable Tray

The electrical contractor shall install wire basket cable trays in the main corridors, adequately sized to accommodate cable for all communications systems. Sizes will range from 4 × 4 inches to 4 × 24 inches. The electrical contractor will provide properly sized horizontal and vertical sleeves, as required, to install all communications cabling.

Structured Cabling System

All tele-data rooms (TDRs) will be fed from the main server room on the first floor. Voice and data risers shall consist of single mode, armor jacketed, riser rated optical fiber cable and Category 3 riser rated copper cable distributed from the main server room to each TDR. Redundant fiber risers are planned for the TDRs.

Work area outlets (WAOs) will be served by Category 6 Unshielded Twisted Pair (UTP) cabling routed in cable tray to the TDRs on each floor. The number of cables per face plate will vary according to need, but the standard face plate will contain two cables terminated on Category 6 RJ45 jacks in a six-port face plate at the WAO, and on Category 6 patch panels in the TDR. Cabling for all wireless access points shall be installed as part of the structured cabling system (SCS).

Cabling for physiological monitoring shall be included as part of the SCS. This cable will be installed as required by the monitoring vendor.

Nurse Call

A nurse call system shall be provided as required by state and local code requirements. The system will consist of nurse master stations, patient stations, pillow speakers with light and television controls, bath stations, emergency call stations, duty stations, staff stations, staff registry stations, and room management stations. These devices will be installed as required by state and local codes and by the owner.

The nurse call system shall be an expandable network system to provide for current and future needs. The nurse call system shall interface with a real-time locating system (RTLS), wireless telephones, pocket pagers, and shall provide an HL7 interface to the hospital's admit, discharge, and transfer (ADT) system. The system shall be accessible from any workstation on the hospital LAN to allow access to administrative reports and to make staff assignments.

Audio-Visual Systems

The board room, conference rooms, and classrooms shall be equipped with projection screens, video monitors, video projectors, teleconferencing equipment, and sound reinforcement, as required, to provide video conferencing and audio/video capabilities. Staff lounges will be provided with A/V input to a television, to facilitate staff training and education.

Security

An access control system shall be provided to control both external and internal access, as required. The system shall consist of software and hardware required for proper system operation. Door control cabinets and panels, and associated power supplies, shall reside in the TDRs. Card readers/keypads, request to exit switches, and door position switches will be installed at all controlled doors. A system server and some number of system licenses shall be provided for system operation and administration. A camera, badge printer, and sufficient badges shall be provided for staff identification. The system must interface with the fire alarm system as required by state and local safety codes.

A video surveillance system consisting of indoor, outdoor, fixed and pan-tilt-zoom (PTZ) IP cameras, network video recorders with digital storage capabilities, and system monitors shall be provided. A system server with system software and some number of system licenses shall be provided. This system shall interface with the access control system to provide common system alarm notification and monitoring capabilities.

Panic or duress buttons shall be provided as required by the owner. These buttons shall ring to the security office or other areas designated by the owner.

Point-to-point intercom systems will be provided at designated access-controlled doors to provide proper identification of persons prior to allowing access. These systems will have audio, video, and remote door release capability.

An infant protection system shall be provided in the women's center. This system will consist of a system server, application software, infant and mother identification badges, and monitoring hardware. Access to and from the center will be controlled by this system, and will provide mother-baby matching capabilities. The system must interface with the fire alarm and access control systems as required by state and local safety codes.

Public Address/Music System

A public address/music system consisting of mixers, amplifiers, speakers, volume controls, microphones, and telephone interface devices shall be provided. The system shall provide all call and zone paging, and background music, in those areas designated by the owner.

Television

A CATV cable distribution system shall be installed that will support basic cable television service, as well as a patient education entertainment system. The system shall consist of coaxial riser and horizontal cable, Category 6 horizontal cable, system amplifiers, taps, splitters, and terminations. Each television outlet shall be wired with a Category 6 UTP cable and an RG-6 coaxial cable from the TDR to the outlet.

Master Clock

A Primex GPS Master Clock System shall be provided. The system will consist of receivers, transmitters, wireless analog clocks, wireless digital clocks, and digital elapsed timers with controls. The analog clocks will be battery powered. The digital clocks and timers will be AC powered. The global positioning system (GPS) receiver captures a time signal from the GPS and sends it to the transmitter. The transmitter then will broadcast the time to all GPS clocks, down to the second hand.

Time Clocks

Kronos clocks will be used for time and attendance, and will be located as designated by the owner.

Peripherals

All peripherals (PCs, printers, fax machines, scanners, copiers, wireless phones, etc.) shall be provided by the owner.

LAN/WLAN/WAN Hardware

All core switches, edge devices, and wireless switches will be owner provided.

Telephone System

A voice over Internet protocol (VoIP) telephone system will be owner provided.

Wireless Telephone Systems

Two wireless telephone applications will be provided. A Vocera system and wireless VoIP handsets, as part of the VoIP telephone system, will be deployed for wireless staff to patient and staff to staff communication.

(Alternate) Systems

Distributed Antenna System

A distributed antenna system (DAS) may be provided to support cellular, paging, and UHF handheld radios. This system would consist of head end equipment, switches, horizontal and riser coaxial cable, and antennae. The system is not intended to support 802.11 LAN traffic or physiological monitoring at this time.

Real-Time Location System

Real-time location system (RTLS) may be deployed for personnel, asset, and patient tracking. If installed this system minimally would be interfaced with the nurse call system, to allow automatic updates to patient calls with no action required from the nursing staff. It could be expanded at any time to add asset and/or patient tracking. The system could be IR, RFID, ultrasonic, or 802.11-based depending on the application chosen.

Patient Education and Entertainment System

If deployed, this system could provide basic cable television, movies, Internet, and gaming for patient entertainment. Educational videos could be viewed by the patient, with an interface to the electronic medical record (EMR) to post an entry as to the date and time the video was viewed by the patient. The patient could access the EMR to review such basic information as who has seen his/her record, and what members of the medical staff are assigned to their care. Other features available would include basic information about nonmedical hospital services, meal menus for in-room meal ordering, discharge document completion, and patient satisfaction surveys.

Index

A

C

D

E

J

K

O

P

Q

R

S

Creating an optimal healing environment begins with construction managers, contractors, and others involved in building facilities.

But those working to build the next generation of hospitals must be aware of the special requirements needed to work in the complicated health care setting. ASHE has created several programs to help those in the construction industry prepare for jobs in health care.

Employers often refer to ASHE's Health Care Construction (HCC) certificate as the "stamp of approval" that shows participants have been properly trained to work in the health care environment. Contractors and construction project managers who complete the HCC program will learn about construction risk assessments, *Life Safety Code®* compliance, medical gas systems, medical technology, and other information needed before working on a health care job.

ASHE's Health Care Construction Subcontractor Program provides subcontractors and specialty contractors with information on the intricacies of working in health care facilities through its e-learning platform. Participants often use their completion of this online program as a selling point to market their services.

The Certified Healthcare Constructor (CHC) certification offered through the American Hospital Association is another way for construction professionals to earn recognition in a competitive marketplace. CHC recipients are among the elite in the health care field. ASHE offers a CHC exam review course that helps poeple prepare for the CHC exam in addition to offering studying materials for the exam. The course highlights key topics, provides test taking tips, and shows sample questions to help participants prepare for the CHC exam.

For more information on ASHE programs, visit **ashe.org/learn**.

American Society for Healthcare Engineering
of the American Hospital Association

The American Society for Healthcare Engineering (ASHE) is a personal membership group of the American Hospital Association. More than 11,000 ASHE members—including facility management professionals, health care construction managers, contractors, architects, engineers, and others—are committed to creating safe and healing health care spaces. ASHE is one of the largest associations dedicated to the health care physical environment and is a trusted resource for those who design, build, operate, and maintain health care facilities.

ASHE provides its members with professional development, education opportunities, resources for codes and standards compliance, and representation in regulatory arenas. Learn more about the benefits of an ASHE membership at **ashe.org/membership**.

For more information about ASHE, visit ashe.org.

CPSIA information can be obtained
at www.ICGtesting.com
Printed in the USA
LVHW020709151221
706118LV00002B/59